Poorly Soluble Drugs

Poorly Soluble Drugs

Editor

Lino Messana

scitus
academics

Poorly Soluble Drugs
Edited by **Lino Messana**

ISBN: 978-1-68117-701-4
Library of Congress Control Number: 2016958865

www.scitusacademics.com

Notice

Printed in United States of America on Acid Free Paper

Preface

Solubility is the property of a solid, liquid, or gaseous chemical substance called solute to dissolve in a solid, liquid, or gaseous solvent to form a homogeneous solution of the solute in the solvent. The solubility of a substance fundamentally depends on the solvent used as well as on temperature and pressure. The extent of solubility of a substance in a specific solvent is measured as the saturation concentration where adding more solute does not increase its concentration in the solution. Solubility also plays a major role for other dosage forms like parenteral formulations as well. Many newly proposed drugs suffer from poor water solubility, thus presenting major hurdles in the design of suitable formulations for administration to patients. Consequently, the development of techniques and materials to overcome these hurdles is a major area of research in pharmaceutical companies.

Poorly Soluble Drugs provides a comprehensive overview of currently used formulation strategies for hydrophobic drugs discusses the main instrumentation, operation principles and theoretical background, with a focus on critical formulation features and clinical studies. It provides a comprehensive assessment of the application of fundamental principles of dissolution and drug release testing to poorly soluble compounds and formulations.

More than 40% new chemical entities developed in pharmaceutical industry are practically insoluble in water. These poorly water soluble drugs having slow drug absorption leads to inadequate and variable bioavailability and gastrointestinal mucosal toxicity. For orally administered drugs solubility is the most important one rate limiting parameter to achieve their desired concentration in systemic circulation for pharmacological response. Problem of solubility is a major challenge for formulation scientist. The improvement of drug solubility thereby its oral bio-availability remains one of the most challenging aspects of drug development process especially for oral-drug delivery system.

Table of Contents

Chapter 1

Drug Solubility: Importance and Enhancement Techniques

Ketan T. Savjani, Anuradha K. Gajjar, and Jignasa K. Savjani

Institute of Pharmacy, Nirma University, Sarkhej Gandhinagar Highway, Gujarat, Ahmedabad 382481, India

ABSTRACT

Solubility, the phenomenon of dissolution of solute in solvent to give a homogenous system, is one of the important parameters to achieve desired concentration of drug in systemic circulation for desired (anticipated) pharmacological response. Low aqueous solubility is the major problem encountered with formulation development of new chemical entities as well as for the generic development. More than 40% NCEs (new chemical entities) developed in pharmaceutical industry are practically insoluble in water. Solubility is a major challenge for formulation scientist. Any drug to be absorbed must be present in the form of solution at the site of absorption. Various techniques are used for the enhancement of the solubility of poorly soluble drugs which include physical and chemical modifications of drug and other methods like particle size reduction, crystal engineering, salt formation, solid dispersion, use of surfactant, complexation, and so forth. Selection of solubility improving method depends on drug property, site of absorption, and required dosage form characteristics.

1. INTRODUCTION

Solubility is the property of a solid, liquid, or gaseous chemical substance called solute to dissolve in a solid, liquid, or gaseous solvent to form a homogeneous solution of the solute in the solvent. The solubility of a substance fundamentally depends on the solvent used as well as on temperature and pressure. The extent of solubility of a substance in a specific solvent is measured as the saturation concentration where adding more solute does not increase its concentration in the solution [1].

The solvent is generally a liquid, which can be a pure substance or a mixture of two liquids. One may also speak of solid solution, but rarely of solution in a gas.

The extent of solubility ranges widely, from infinitely soluble (fully miscible) such as ethanol in water, to poorly soluble, such as silver chloride in water. The term insoluble is often applied to poorly or very poorly soluble compounds [2].

Solubility occurs under dynamic equilibrium, which means that solubility results from the simultaneous and opposing processes of dissolution and phase joining (e.g., precipitation of solids). Solubility equilibrium occurs when the two processes proceed at a constant rate. Under certain conditions equilibrium solubility may be exceeded to give a so-called supersaturated solution, which is metastable [3].

Solubility is not to be confused with the ability to dissolve or liquefy a substance, since these processes may occur not only because of dissolution but also because of a chemical reaction. For example, zinc is insoluble in hydrochloric acid, but does dissolve in it by chemically reacting into zinc chloride and hydrogen, where zinc chloride is soluble in hydrochloric acid. Solubility does not also depend on particle size or otherkinetic factors; given enough time, even large particles will eventually dissolve [4].

IUPAC defines solubility as the analytical composition of a saturated solution expressed as a proportion of a designated solute in a designated solvent. Solubility may be stated in units of concentration, molality, mole fraction, mole ratio, and other units [5].

Extensive use of solubility from different perspective has led to solubility being expressed in various manners. It is commonly expressed as a concentration, either by mass (g of solute per kg of solvent, g per dL (100 mL) of solvent), molarity, molality, mole fraction, or other similar descriptions of concentration. The maximum equilibrium amount of solute that can dissolve per amount of solvent is the solubility of that solute in that solvent under the specified conditions [6]. The advantage of expressing solubility in this manner is its simplicity, while the disadvantage is that it can strongly depend on the presence of other species in the solvent (e.g., the common ion effect).

Saturated solutions of ionic compounds of relatively low solubility are sometimes described by solubility constants. It is a case of equilibrium process. It describes the balance between dissolved ions from the salt and undissolved salt. Similar to other equilibrium constants, temperature would affect the numerical value of solubility constant. The value of this constant is generally independent of the presence of other species in the solvent.

The Flory-Huggins solution theory is a theoretical model describing the solubility of polymers. The Hansen Solubility Parameters and the Hildebrand solubility parameters are empirical methods for the prediction of solubility. It is also possible to predict solubility from other physical constants such as the enthalpy of fusion.

The partition coefficient (Log P) is a measure of differential solubility of a compound in a hydrophobic solvent (octanol) and a hydrophilic solvent (water). The logarithm of these two values enables compounds to be ranked in terms of hydrophilicity (or hydrophobicity).

USP and BP classify the solubility regardless of the solvent used, just only in terms of quantification and have defined the criteria as given in Table 1 [7, 8].

Table 1: USP and BP solubility criteria.

Descriptive term	Part of solvent required per part of solute
Very soluble	Less than 1
Freely soluble	From 1 to 10
Soluble	From 10 to 30
Sparingly soluble	From 30 to 100
Slightly soluble	From 100 to 1000
Very slightly soluble	From 1000 to 10,000
Practically insoluble	10,000 and over

The Biopharmaceutics Classification System (BCS) is a guide for predicting the intestinal drug absorption provided by the U.S. Food and Drug Administration. This system restricts the prediction using the parameters solubility and intestinal permeability.

Solubility is based on the highest-dose strength of an immediate release product. A drug is considered highly soluble when the highest dose strength is soluble in 250 mL or less of aqueous media over the pH range of 1 to 7.5. The volume estimate of 250 mL is derived from typical bioequivalence study protocols that prescribe administration of a drug product to fasting human volunteers with a glass of water [9].

The intestinal permeability classification is based on a comparison to the intravenous injection. All those factors are highly important, since 85% of the most sold drugs in the USA and Europe are orally administered.

All drugs have been divided into four classes: class I—high soluble and high permeable, class II—low soluble and high permeable, class III—low soluble and high permeable and class IV—low soluble and low permeable.

2. IMPORTANCE OF SOLUBILITY

Oral ingestion is the most convenient and commonly employed route of drug delivery due to its ease of administration, high patient compliance, costeffectiveness, least sterility constraints, and flexibility in the design of dosage form. As a result, many of the generic drug companies are inclined more to produce bioequivalent oral drug products [10].

However, the major challenge with the design of oral dosage forms lies with their poor bioavailability. The oral bioavailability depends on several factors including aqueous solubility, drug permeability, dissolution rate, first-pass metabolism, presystemic metabolism, and susceptibility to efflux mechanisms. The most frequent causes of low oral bioavailability are attributed to poor solubility and low permeability.

Solubility also plays a major role for other dosage forms like parenteral formulations as well [11]. Solubility is one of the important parameters to achieve desired concentration of drug in systemic circulation for achieving required pharmacological response [12]. Poorly water soluble drugs often require high doses in order to reach therapeutic plasma concentrations after oral administration. Low aqueous solubility is the major problem encountered with formulation development of new chemical entities as well as generic development. Any drug to be absorbed must be present in the form of an aqueous solution at the site of absorption. Water is the solvent of choice for liquid pharmaceutical formulations. Most of the drugs are either weakly acidic or weakly basic having poor aqueous solubility.

More than 40% NCEs (new chemical entities) developed in pharmaceutical industry are practically insoluble in water. These poorly water soluble drugs having slow drug absorption leads to inadequate and variable bioavailability and gastrointestinal mucosal toxicity. For orally administered drugs solubility is the most important one rate limiting parameter to achieve their desired concentration in systemic circulation for pharmacological response. Problem of solubility is a major challenge for formulation scientist [13].

The improvement of drug solubility thereby its oral bio-availability remains one of the most challenging aspects of drug development process especially for oral-drug delivery system. There are numerous approaches available and reported in literature to enhance the solubility of poorly water-soluble drugs. The techniques are chosen on the basis of certain aspects such as properties of drug under consideration, nature of excipients to be selected, and nature of intended dosage form.

The poor solubility and low dissolution rate of poorly water soluble drugs in the aqueous gastrointestinal fluids often cause insufficient bioavailability. Especially for class II (low solubility and high permeability) substances according to the BCS, the bioavailability may be enhanced by increasing the solubility and dissolution rate of the drug in the gastro-intestinal fluids. As for BCS class II drugs rate limiting step is drug release from the dosage form and solubility in the gastric fluid and not the absorption, so increasing the solubility in turn increases the bioavailability for BCS class II drugs [10, 13, 14].

The negative effect of compounds with low solubility include poor absorption and bioavailability, insufficient solubility for IV dosing, development challenges leading to increasing the development cost and time, burden shifted to patient (frequent high-dose administration) [11].

3. TECHNIQUES FOR SOLUBILITY ENHANCEMENT

Solubility improvement techniques can be categorized in to physical modification, chemical modifications of the drug substance, and other techniques.

Physical Modifications

Particle size reduction like micronization and nanosuspension, modification of the crystal habit like polymorphs, amorphous form and cocrystallization, drug dispersion in carriers like eutectic mixtures, solid dispersions, solid solutions and cryogenic techniques.

Chemical Modifications

Change of ph, use of buffer, derivatization, complexation, and salt formation.

Miscellaneous Methods

Supercritical fluid process, use of adjuvant like surfactant, solubilizers, cosolvency, hydrotrophy, and novel excipients.

4. PARTICLE SIZE REDUCTION

The solubility of drug is often intrinsically related to drug particle size; as a particle becomes smaller, the surface area to volume ratio increases. The larger surface area allows greater interaction with the solvent which causes an increase in solubility.

Conventional methods of particle size reduction, such as comminution and spray drying, rely upon mechanical stress to disaggregate the active compound. Particle size reduction is thus permitting an efficient, reproducible, and economic means of solubility enhancement. However, the mechanical forces inherent to comminution, such as milling and grinding, often impart significant amounts of physical stress upon the drug product which may induce degradation. The thermal stress which may occur during comminution and spray drying is also a concern when processing thermosensitive or unstable active compounds. Useing traditional approaches for nearly insoluble drugs may not be able to enhance the solubility up to desired level.

Micronization is another conventional technique for the particle size reduction. Micronization increases the dissolution rate of drugs through increased surface area, it does not increase equilibrium solubility. Decreasing the particle size of these drugs, which cause increase in surface area, improve their rate of dissolution. Micronization of drugs is done by milling techniques using jet mill, rotor stator colloid mills and so forth micronization is not suitable for drugs having a high dose number because it does not change the saturation solubility of the drug [15].

These processes were applied to griseofulvin, progesterone, spironolactone diosmin, and fenofibrate. For each drug, micronization improved their digestive absorption, and consequently their bioavailability and clinical efficacy. Micronized fenofibrate exhibited more than 10-fold (1.3% to 20%) increase in dissolution in at 30 minutes biorelevant media [16, 17].

5. SOLID DISPERSION

The concept of solid dispersions was originally proposed by Sekiguchi and Obi, who investigated the generation and dissolution performance of eutectic melts of a sulfonamide drug and a water-soluble carrier in the early 1960s [18]. Solid dispersions represent a useful pharmaceutical technique for increasing the dissolution, absorption, and therapeutic efficacy of drugs in dosage forms. The term solid dispersion refers to a group of solid products consisting of at least two different components, generally a hydrophilic matrix and a hydrophobic drug. The most commonly used hydrophilic carriers for solid dispersions include polyvinylpyrrolidone (Povidone, PVP), polyethylene glycols (PEGs), Plasdone-S630. Surfactants like Tween-80, docusate sodium, Myrj-52, Pluronic-F68, and sodium lauryl sulphate (SLS) also find a place in the formulation of solid dispersion.

The solubility of celecoxib, halofantrine, and ritonavir can be improved by solid dispersion using suitable hydrophilic carriers like celecoxib with povidone (PVP) and ritonavir with gelucire. Various techniques to prepare the solid dispersion of hydrophobic drugs with an aim to improve their aqueous solubility are listed here [19–21].

5.1. Hot-Melt Method (Fusion Method)

The main advantages of this direct melting method is its simplicity and economy.The melting or fusion method was first proposed by Sekiguchi and Obi to prepare fast release solid dispersion dosage forms. In this method, the physical mixture of a drug and a water-soluble carrier are heated directly until the two melts. The melted mixture is then cooled and solidified rapidly in an ice bath with rigorous stirring. The final solid mass is then crushed, pulverized, and sieved, which can be compressed into tablets with the help of tableting agents. The melting point of a binary system is dependent upon its composition, that is, the selection of the carrier and the weight fraction of the drug in the system [22].

An important requisite for the formation of solid dispersion by the hot-melt method is the miscibility of the drug and the carrier in the molten form. Another important requisite is the thermostability of both the drug and the carrier.

5.2. Solvent Evaporation Method

Tachibana and Nakamura [23] were the first to dissolve both the drug and the carrier in a common solvent and then evaporate the solvent under vacuum to produce a solid solution. This enabled them to produce a solid solution of the highly lipophilic β-carotene in the highly water soluble carrier povidone. Many investigators studied solid dispersion of meloxicam, naproxen, and nimesulide using solvent evaporation technique. These findings suggest that the above-mentioned technique can be employed successfully for improvement and stability of solid dispersions of poorly water soluble drugs [15, 17].

The main advantage of the solvent evaporation method is that thermal decomposition of drugs or carriers can be prevented because of the low temperature required for the evaporation of organic solvents. However, the disadvantages associated with this method are the higher cost of preparation, the difficulty in completely removing the organic solvent (a regulatory perspective), the possible adverse effect of the supposedly negligible amount of the solvent on the chemical stability of the drug, the selection of a common volatile solvent, and the difficulty in reproducing crystal forms [24].

5.3. Hot-Melt Extrusion

Hot-melt extrusion is essentially the same as the fusion method except that intense mixing of the components is induced by the extruder. Just like in the traditional fusion process, miscibility of the drug and the matrix could be a problem. High-shear forces resulting in high local temperature in the extruder is a problem for heat sensitive materials. However, compared to the traditional fusion method, this technique offers the possibility of continuous production, which makes it suitable for large-scale production. Furthermore, the product is easier to handle because at the outlet of the extruder the shape can be adapted to the next processing step without grinding [20].

6. NANOSUSPENSION

Nanosuspension technology has been developed as a promising candidate for efficient delivery of hydrophobic drugs. This technology is applied to poorly soluble drugs that are insoluble in both water and oils. A pharmaceutical nanosuspension is a biphasic system consisting of nano sized drug particles stabilized by surfactants for either oral and topical use or parenteral and pulmonary administration. The particle size distribution of the solid particles in nanosuspensions is usually less than one micron with an average particle size ranging between 200 and 600 nm [25, 26].

Various methods utilized for preparation of nanosuspensions include precipitation technique, media milling, high-pressure homogenization in water, high pressure homogenization in nonaqueous media, and combination of Precipitation and high-Pressure homogenization [27, 28].

6.1. Precipitation Technique

In precipitation technique the drug is dissolved in a solvent, which is then added to antisolvent to precipitate the crystals. The basic advantage of precipitation technique is the use of simple and low cost equipments; but the challenge is the addition of the growing drug crystals to avoid formation of microparticles. The limitation of this precipitation technique is that the drug needs to be soluble in at least one solvent and this solvent needs to be miscible with antisolvent. Moreover, precipitation technique is not applicable to drugs, which are simultaneously poorly soluble in aqueous and nonaqueous media [29]. Nanosuspension of Danazol and Naproxen have been prepared by precipitation technique to improve their dissolution rate and oral bioavailability. The size reduction of naproxen was also associated with an apparent increase in the rate of absorption by approximately 4-fold [30, 31].

6.2. Media Milling

The nanosuspensions are prepared by using high-shear media mills. The milling chamber charged with milling media, water, drug, and stabilizer is rotated at a very high-shear rate under controlled temperatures for several days (at least 2–7 days). The milling medium is composed of glass, Zirconium oxide, or highly cross-linked polystyrene resin. High energy shear forces are generated as a result of the impaction of the milling media with the drug resulting into breaking of microparticulate drug to nanosized particles [28].

6.3. High Pressure Homogenization

High-pressure homogenization has been used to prepare nanosuspension of many poorly water soluble drugs. In this method, the suspension of a drug and surfactant is forced under pressure through a nanosized aperture valve of a high pressure homogenizer. The principle of this method is based on cavitation in the aqueous phase. The cavitations forces within the particles are sufficiently high to convert the drug microparticles into nanoparticles. The concern with this method is the need for small sample particles before loading and the fact that many cycles of homogenization are required [32].

Dissolution rate and bioavailability of poorly soluble drugs such as spironolactone, budesonide, and omeprazole have been improved by reducing their particle size by high pressure homogenization [33–35].

6.4. Combined Precipitation and Homogenization

The precipitated drug nanoparticles have a tendency to continue crystal growth to the size of microcrystals. They need to be processed with high-energy forces (homogenisation). They are in completely amorphous, partially amorphous or completely crystalline forms which create problems in long term stability as well as in bioavailability, so the precipitated particle suspension is subsequently homogenized which preserve the particle size obtained after the precipitation step.

7. SUPERCRITICAL FLUID (SCF) PROCESS

Another novel nanosizing and solubilisation technology whose application has increased in recent years is particle size reduction via supercritical fluid (SCF) processes. Supercritical fluids are fluids whose temperature and pressure are greater than its critical temperature (Tc) and critical pressure (Tp), allowing it to assume the properties of both a liquid and a gas. At near-critical temperatures, SCFs, are highly compressible allowing moderate changes in pressure to greatly alter the density and mass transport characteristics of the fluid that largely determine its solvent power. Once the drug particles are solubilised within the SCF (usually carbon dioxide), they may be recrystallised at greatly reduced particle sizes. The flexibility and precision offered by SCF processes allows micronisation of drug particles within narrow ranges of particle size, often to submicron levels. Current SCF processes have demonstrated the ability to create nanoparticulate suspensions of particles 5–2,000 nm in diameter. Several pharmaceutical companies, such as Nektar Therapeutics and Lavipharm, are specializing in particle engineering via SCF technologies for particle size reduction and solubility enhancement. Several methods of SCF processing have been developed to address individual aspects of these shortcomings, such as precipitation with compressed antisolvent process (PCA), solution enhanced dispersion by SCF (SEDS), supercritical antisolvent processes (SAS), rapid expansion of supercritical solutions (RESS), gas anti solvent recrystallization (GAS), and aerosol supercritical extraction system (ASES) [36, 37].

8. CRYOGENIC TECHNIQUES

Cryogenic techniques have been developed to enhance the dissolution rate of drugs by creating nanostructured amorphous drug particles with high degree of porosity at very low-temperature conditions. Cryogenic inventions can be defined by the type of injection device (capillary, rotary, pneumatic, and ultrasonic nozzle), location of nozzle (above or under the liquid level), and the composition of cryogenic liquid (hydrofluoroalkanes, N_2, Ar, O2, and organic solvents). After cryogenic processing, dry powder can be obtained by various drying processes like spray freeze drying, atmospheric freeze drying, vacuum freeze drying, and lyophilisation [38–40].

8.1. Spray Freezing onto Cryogenic Fluids

Briggs and Maxvell invented the process of spray freezing onto cryogenic fluids. In this technique, the drug and the carrier (mannitol, maltose, lactose, inositol, or dextran) were dissolved in water and atomized above the surface of a boiling agitated fluorocarbon refrigerant. Sonication probe can be placed in the stirred refrigerant to enhance the dispersion of the aqueous solution [41].

8.2. Spray Freezing into Cryogenic Liquids (SFL)

The SFL particle engineering technology has been used to produce amorphous nanostructured aggregates of drug powder with high surface area and good wettability. It incorporates direct liquid-liquid impingement between the automatized feed solution and cryogenic liquid to provide intense atomization into microdroplets and consequently significantly faster freezing rates. The frozen particles are then lyophilized to obtain dry and free-flowing micronized powders [42].

8.3. Spray Freezing into Vapor over Liquid (SFV/L)

Freezing of drug solutions in cryogenic fluid vapours and subsequent removal of frozen solvent produces fine drug particles with high wettability. During SFV/L the atomized droplets typically start to freeze in the vapor phase before they contact the cryogenic liquid. As the solvent freezes, the drug becomes supersaturated in the unfrozen regions of the atomized droplet, so fine drug particles may nucleate and grow [43].

8.4. Ultra-Rapid Freezing (URF)

Ultra-rapid freezing is a novel cryogenic technology that creates nanostructured drug particles with greatly enhanced surface area and desired surface morphology by using solid cryogenic substances. Application of drugs solution to the solid surface of cryogenic substrate leads to instantaneous freezing and subsequent lyophilization (for removal of solvent) forms micronized drug powder with improved solubility. Ultra rapid freezing hinders the phase separation and the crystallization of the pharmaceutical ingredients leading to intimately mixed, amorphous drug-carrier solid dispersions, and solid solutions [45].

9. INCLUSION COMPLEX FORMATION-BASED TECHNIQUES

Among all the solubility enhancement techniques, inclusion complex formation technique has been employed more precisely to improve the aqueous solubility, dissolution rate, and bioavailability of poorly water soluble drugs.

Inclusion complexes are formed by the insertion of the nonpolar molecule or the nonpolar region of one molecule (known as guest) into the cavity of another molecule or group of molecules (known as host). The most commonly used host molecules are cyclodextrins. The enzymatic degradation of starch by cyclodextrin-glycosyltransferase (CGT) produces cyclic oligomers, Cyclodextrins (CDs). These are nonreducing, crystalline, water soluble, and cyclic oligosaccharides consisting of glucose monomers arranged in a donut shaped ring having hydrophobic cavity and hydrophilic outer surface as illustrated in Figure 1. Three naturally occurring CDs are α-Cyclodextrin, β-Cyclodextrin, and γ-Cyclodextrin [46].

The surface of the cyclodextrin molecules makes them water soluble, but the hydrophobic cavity provides a microenvironment for appropriately sized non-polar molecules. Based on the structure and properties of drug molecule it can form 1 : 1 or 1 : 2 drug cyclodextrin complex as illustrated in Figure 2.

Various technologies adapted to prepare the inclusion complexes of poorly water soluble drugs with cyclodextrins are briefly described below.

9.1. Kneading Method

This method is based on impregnating the CDs with little amount of water or hydroalcoholic solutions to convert into a paste. The drug is then added to the above paste and kneaded for a specified time. The kneaded mixture is then dried and passed through a sieve if required. In laboratory scale, kneading can be achieved by using a mortar and pestle. In large scale, kneading can be done by utilizing the extruders and other machines. This is the most common and simple method used to prepare the inclusion complexes and it presents very low cost of production [48].

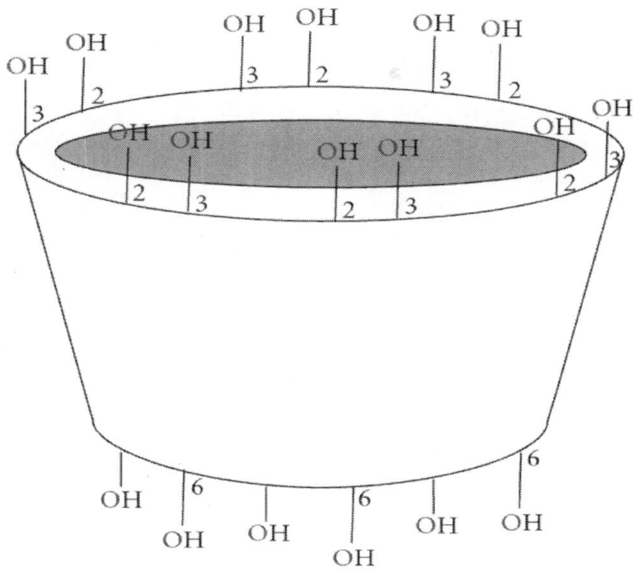

Figure 1: Representations of hydrophobic cavity and hydrophilic outer surface of cyclodextrin [44].

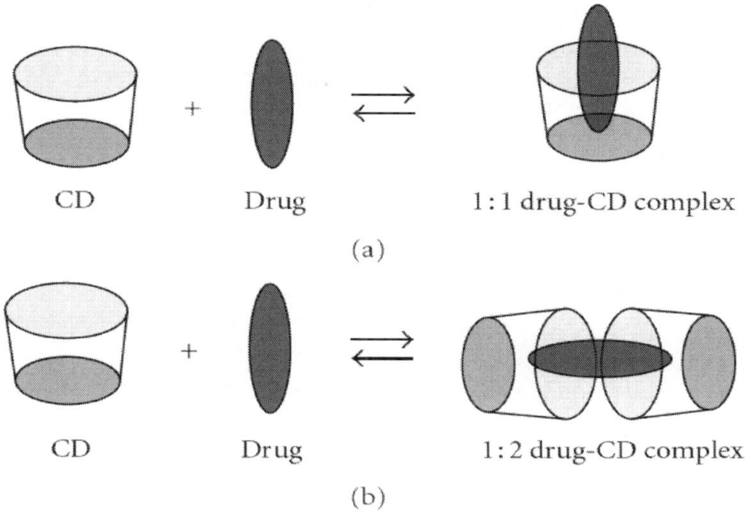

Figure 2: 1 : 1 and 1 : 2 drug cyclodextrin complexes [47].

9.2. Lyophilization/Freeze-Drying Technique

In order to get a porous, amorphous powder with high degree of interaction between drug and CD, lyophilization/freeze drying technique is considered suitable. In this technique, the solvent system from the solution is eliminated through a primary freezing and subsequent drying of the solution containing both drug and CD at reduced pressure. Thermolabile substances can be successfully made into complex form by this method. The limitations of this technique is the use of specialized equipment, time consuming process, and yield poor flowing powdered product. Lyophilization/freeze drying technique is considered as an alternative to solvent evaporation and involve molecular mixing of drug and carrier in a common solvent [49].

9.3. Microwave Irradiation Method

This technique involves the microwave irradiation reaction between drug and complexing agent using a microwave oven. The drug and CD in definite molar ratio are dissolved in a mixture of water and organic solvent in a specified proportion into a round-bottom flask. The mixture is reacted for short time of about one to two minutes at 60°C in the microwave oven. After the reaction completes, adequate amount of solvent mixture is added to the above reaction mixture to remove the residual uncomplexed free drug and CD. The precipitate so obtained is separated using whatman filter paper, and dried in vacuum oven at 40°C. Microwave irradiation method is a novel method for industrial scale preparation due to its major advantage of shorter reaction times and higher yield of the product [50].

10. MICELLAR SOLUBILIZATION

The use of surfactants to improve the dissolution performance of poorly soluble drug products is probably the basic, primary, and the oldest method. Surfactants reduce surface tension and improve the dissolution of lipophilic drugs in aqueous medium. They are also used to stabilise drug suspensions. When the concentration of surfactants exceeds their critical micelle concentration (CMC, which is in the range of 0.05–0.10% for most surfactants), micelle formation occurs which entrap the drugs within the micelles. This is known as micellization and generally results in enhanced solubility of poorly soluble drugs. Surfactant also improves wetting of solids and increases the rate of disintegration of solid into finer particles [11]. Commonly used nonionic surfactants include polysorbates, polyoxyethylated castor oil, polyoxyethylated glycerides, lauroyl macroglycerides, and mono- and di-fatty acid esters of low molecular weight polyethylene glycols. Surfactants are also often used to stabilize microemulsions and suspensions into which drugs are dissolved [51, 52].

Examples of poorly soluble compounds that use Micellar solubilization are antidiabetic drugs, gliclazide, glyburide, glimepiride, glipizide, repaglinide, pioglitazone, and rosiglitazone [53].

11. HYDROTROPHY

Hydrotrophy is a solubilisation process, whereby addition of a large amount of second solute, the hydrotropic agent results in an increase in the aqueous solubility of first solute. Hydrotropic agents are ionic organic salts, consists of alkali metal salts of various organic acids. Additives or salts that increase solubility in given solvent are said to "salt in" the solute and those salts that decrease solubility "salt out" the solute. Several salts with large anions or cations that are themselves very soluble in water result in "salting in" of non electrolytes called "hydrotropic salts"; a phenomenon known as "hydrotropism." Hydrotrophy designate the increase in solubility in water due to the presence of large amount of additives. The mechanism by which it improves solubility is more closely related to complexation involving a weak interaction between the hydrotrophic agents like sodium benzoate, sodium acetate, sodium alginate, urea, and the poorly soluble drugs [54, 55].

The hydrotropes are known to self-assemble in solution. The classification of hydrotropes on the basis of molecular structure is difficult, since a wide variety of compounds have been reported to exhibit hydrotropic behaviour. Specific examples may include ethanol, aromatic alcohols like resorcinol, pyrogallol, catechol, α and β-naphthols and salicylates, alkaloids like caffeine and nicotine, ionic surfactants like diacids, SDS (sodium dodecyl sulphate), and dodecylated oxidibenzene. The aromatic hydrotropes with anionic head groups are mostly studied compounds. They are large in number because of isomerism and their effective hydrotrope action may be due to the availability of interactive pi (π) orbital [56].

Hydrotropes with cationic hydrophilic group are rare, for example salts of aromatic amines, such as procaine hydrochloride. Besides enhancing the solubilization of compounds in water, they are known to exhibit influences on surfactant aggregation leading to micelle formation, phase manifestation of multicomponent systems with reference to nanodispersions and conductance percolation, clouding of surfactants and polymers, and so forth [57].

12. CRYSTAL ENGINEERING

The surface area of drug available for dissolution is dependent on its particle size and ability to be wetted by luminal fluids. This particle size, which is critical to drug dissolution rate, is dependent on the conditions of crystallization or on methods of comminution such as impact milling and fluid energy milling.

The comminution techniques can produce particles which are highly heterogeneous, charged, and cohesive, with the potential to cause problems in downstream processing and product performance. Hence, crystal engineering techniques are developed for the controlled crystallization of drugs to produce high purity powders with well-defined particle size distribution, crystal habit, crystal form (crystalline or amorphous), surface nature, and surface energy [58]. By manipulating the crystallization conditions (use of different solvents or change in the stirring or adding other components to crystallizing drug solution), it is possible to prepare crystals with different packing arrangement; such crystals are called polymorphs.

As a result, polymorphs for the same drug may differ in their physicochemical properties such as solubility, dissolution rate, melting point, and stability. Most drugs exhibit structural polymorphism and it is preferable to develop the most thermodynamically stable polymorph of the drug to assure reproducible bioavailability of the product over its shelf-life under a variety of real-world storage conditions. A classic example of the importance of polymorphism on bioavailability is that of chloramphenicol palmitate suspensions. It was shown that the stable polymorph of chloramphenicol palmitate produced low serum levels, whereas the metastable polymorph yielded much higher serum levels when the same dose was administered [59]. In another study, it was found that tablets prepared from the form A polymorph of oxytetracycline dissolved significantly more slowly than the tablets with form B polymorph [60]. The tablets with form A polymorph exhibited about 55% dissolution at 30 min, while the tablets with form B polymorph exhibited almost complete (95%) dissolution at the same time.

Crystal engineering approach also involves the preparation of hydrates and solvates for enhancing the dissolution rate. During the crystallization process, it is possible to trap molecules of the solvent within the lattice. If the solvent used is water, the resultant crystal is a hydrate; if any other solvent is used, it is referred to as solvate. The dissolution rate and solubility of a drug can differ significantly for different solvates. For example, glibenclamide has been isolated as pentanol and toluene solvates, and these solvates exhibited higher solubility and dissolution rate than two nonsolvated polymorphs [61]. It is possible for the hydrates to have either a faster or slower dissolution rate than the anhydrous form. The most usual situation is for the anhydrous form to have a faster dissolution rate than the hydrate. For example, the dissolution rate of theophylline anhydrate was faster than its hydrate form [62]. In certain cases, hydrate form of the drug may show rapid dissolution rate than its anhydrous form. Erythromycin dihydrate was found to exhibit significant differences in the dissolution rate when compared to monohydrate and anhydrate forms [63]. In general, it is undesirable to use solvates for drugs and pharmaceuticals as the presence of organic solvent residues may be toxic. Also all the organic solvents have specific limits for daily exposure to human as residual solvent in the formulated preparation.

Crystal engineering offers a number of routes to improved solubility and dissolution rate, which can be adopted through an indepth knowledge of crystallization processes and the molecular properties of active pharmaceutical ingredients. The process involves dissolving the drug in a solvent and precipitating it in a controlled manner to produce nanoparticles through addition of an antisolvent (usually, water) [58].

Pharmaceutical cocrystals open a new avenue to address the problems of poorly soluble drugs. They contain two or more distinct molecules arranged to create a new crystal form whose properties are often superior to those of each of the separate entities. The pharmaceutical cocrystals are formed between a molecular or ionic drug and a cocrystal former that is a solid under ambient conditions [64]. These are prepared by slow evaporation from a drug solution containing stoichiometric amounts of the components (cocrystal formers); however, sublimation, growth from the melt, or grinding of two or more solid cocrystal formers in a ball mill are also suitable methodologies [65].

Carbamazepine: saccharin cocrystal was shown to be superior to crystal forms of carbamazepine alone in terms of stability, dissolution, suspension stability, and oral absorption profile in dogs [66]. In another study by Childs et al., fluoxetine HCl succinic acid cocrystal was found to exhibit an approximately twofold increase in aqueous solubility after only 5 min [67]. It was observed that the itraconazole L-malic acid cocrystal exhibited a similar dissolution profile to that of the marketed formulation [68].

Traditional crystallization methods include sublimation, crystallization from solutions, evaporation, thermal treatment, desolvation, or grinding/milling. These are being replaced with novel methods of crystal engineering such as SCF technologies [69, 70] to produce pharmaceutical solids with desired dissolution rate and stability.Melt sonocrystallization is yet another emerging technology that uses ultrasonic energy to produce porous fast dissolving particles for hydrophobic drug molecules [71]. Based on these exciting reports, it appears that crystal engineering techniques need to be exploited more for enhancing the dissolution rate of poorly soluble drugs.

Other techniques that enhance the solubility of poorly water soluble drugs include salt formation, change in dielectric constant of solvent, Chemical modification of the drug, use of hydrates or solvates, use of Soluble prodrug, application of ultrasonic waves, and spherical crystallization.

13. CONCLUSION

Dissolution of drug is the rate determining step for oral absorption of the poorly water soluble drugs and solubility is the basic requirement for the absorption of the drug from GIT. The various techniques described above alone or in combination can be used to enhance the solubility of the drugs. Proper selection of solubility enhancement method is the key to ensure the goals of a good formulation like good oral bioavailability, reduce frequency of dosing and better patient compliance combined with a low cost of production. Selection of method for solubility enhancement depends upon drug characteristics like solubility, chemical nature, melting point, absorption site, physical nature, pharmacokinetic behavior and so forth, dosage form requirement like tablet or capsule formulation, strength, immediate, or modified release and so forth, and regulatory requirements like maximum daily dose of any excipients and/or drug, approved excipients, analytical accuracy and so forth.

REFERENCES

1. L. Lachman, H. Lieberman, and J. L. Kanig, The Theory And Practise of Industrial Pharmacy, Lea & Febiger, 3rd edition, 1986.
2. M. Clugston and R. Fleming, Advanced Chemistry, Oxford Publishing, Oxford, UK, 1st edition, 2000.
3. P. B. Myrdal and S. H. Yalkowsky, "Solubilization of drugs in aqueous media," in Encyclopedia of Pharmaceutical Technology, J. Swarbrick, Ed., p. 3311, Informa Health Care, New York, NY, USA,, 3rd edition, 2007.
4. A. Martin, Solubility and Distribution Phenomena, Physical Pharmacy and Pharmaceutical Sciences, Lippincott Williams and Wilkins, 6th edition, 2011.
5. "IUPAC gold book," http://goldbook.iupac.org/S05740.html.
6. M. Aulton, "Dissolution and solubility," in Pharmaceutics: The Science of Dosage form Design, M. E. Aulton, Ed., p. 15, Churchill Livingstone, 2nd edition, 2002.

7. The United States Pharmacopeia, USP 30-NF 25, 2007.

8. British Pharmacopoeia, 2009.

9. G. L. Amidon, H. Lennernäs, V. P. Shah, and J. R. Crison, "A theoretical basis for a biopharmaceutic drug classification: the correlation of in vitro drug product dissolution and in vivo bioavailability,"Pharmaceutical Research, vol. 12, no. 3, pp. 413–420, 1995.

10. S. R. K. Yellela, "Pharmaceutical technologies for enhancing oral bioavailability of poorly soluble drugs,"Journal of Bioequivalence & Bioavailability, vol. 2, no. 2, pp. 28–36, 2010.

11. K. H. Edward and D. Li, "Solubility," in Drug Like Properties: Concept, Structure, Design and Methods, from ADME to Toxicity Optimization, p. 56, Elsevier, 2008.

12. V. R. Vemula, V. Lagishetty, and S. Lingala, "Solubility enhancement techniques," International Journal of Pharmaceutical Sciences Review and Research, vol. 5, no. 1, pp. 41–51, 2010. •View at Scopus

13. D. Sharma, M. Soni, S. Kumar, and G. D. Gupta, "Solubility enhancement—eminent role in poorly soluble drugs," Research Journal of Pharmacy and Technology, vol. 2, no. 2, pp. 220–224, 2009.

14. A. Kumar, S. K. Sahoo, K. Padhee, P. S. Kochar, A. Sathapathy, and N. Pathak, "Review on solubility enhancement techniques for hydrophobic drugs," Pharmacie Globale, vol. 3, no. 3, pp. 001–007, 2011.

15. N. Blagden, M. de Matas, P. T. Gavan, and P. York, "Crystal engineering of active pharmaceutical ingredients to improve solubility and dissolution rates," Advanced Drug Delivery Reviews, vol. 59, no. 7, pp. 617–630, 2007.

16. M. Vogt, K. Kunath, and J. B. Dressman, "Dissolution enhancement of fenofibrate by micronization, cogrinding and spray-drying: comparison with commercial preparations," European Journal of Pharmaceutics and Biopharmaceutics, vol. 68, no. 2, pp. 283–288, 2008.

17. J. C. Chaumeil, "Micronization: a method of improving the bioavailability of poorly soluble drugs,"Methods and Findings in Experimental and Clinical Pharmacology, vol. 20, no. 3, pp. 211–215, 1998.

18. K. Sekiguchi and N. Obi, "Studies on absorption of eutectic mixtures. I.A. comparison of the behaviour of eutectic mixtures of sulphathiazole and that of ordinary sulphathiazole in man," Chemical and Pharmaceutical Bulletin, vol. 9, pp. 866–872, 1961.

19. P. Gupta, V. K. Kakumanu, and A. K. Bansal, "Stability and solubility of celecoxib-PVP amorphous dispersions: a molecular perspective," Pharmaceutical Research, vol. 21, no. 10, pp. 1762–1769, 2004.

20. A. M. Abdul-Fattah and H. N. Bhargava, "Preparation and in vitro evaluation of solid dispersions of halofantrine," International Journal of Pharmaceutics, vol. 235, no. 1-2, pp. 17–33, 2002.

21. S. Sinha, M. Ali, S. Baboota, A. Ahuja, A. Kumar, and J. Ali, "Solid dispersion as an approach for bioavailability enhancement of poorly water-soluble drug ritonavir," AAPS PharmSciTech, vol. 11, no. 2, pp. 518–527, 2010.

22. W. L. Chiou and S. Riegelman, "Pharmaceutical applications of solid dispersion systems," Journal of Pharmaceutical Sciences, vol. 60, no. 9, pp. 1281–1302, 1971.

23. T. Tachibana and A. Nakamura, "A methode for preparing an aqueous colloidal dispersion of organic materials by using water-soluble polymers: dispersion of β-carotene by polyvinylpyrrolidone," Colloid and Polymer Science, vol. 203, no. 2, pp. 130–133, 1965.

24. "Nanosuspension drug delivery technology and application—nanotech—express pharma pulse.htm," http://www.expresspharmapulse.com/.

25. R. H. Muller, C. Jacobs, and O. Kayer, "Nanosuspensions for the formulation of poorly soluble drugs," inPharmaceutical Emulsion and Suspension, F. Nielloud and G Marti-Mestres , Eds., pp. 383–407, Marcel Dekker, New York, NY, USA, 2000.

26. R. A. Nash, "Suspensions," in Encyclopedia of Pharmaceutical Technology, J. Swarbrick and J. C. Boylan , Eds., vol. 3, pp. 2045–3032, Marcel Dekker, New York, NY, USA, 2nd edition, 2002.

27. K. P. R. Chowdary and B. L. R. Madhavi, "Novel drug delivery technologies for insoluble drugs," Indian Drugs, vol. 42, no. 9, pp. 557–564, 2005.

28. V. B. Patravale, A. A. Date, and R. M. Kulkarni, "Nanosuspensions: a promising drug delivery strategy,"Journal of Pharmacy and Pharmacology, vol. 56, no. 7, pp. 827–840, 2004.

29. R. H. Muller, B. H. L. Bohm, and J. Grau, "Nanosuspensions: a formulation approach for poorly soluble and poorly bioavailable drugs," in Handbook of Pharmaceutical Controlled Release Technology, D. Wise, Ed., pp. 345–357, 2000.

30. E. Merisko-Liversidge, G. G. Liversidge, and E. R. Cooper, "Nanosizing: a formulation approach for poorly-water-soluble compounds," European Journal of Pharmaceutical Sciences, vol. 18, no. 2, pp. 113–120, 2003.

31. G. G. Liversidge and P. Conzentino, "Drug particle size reduction for decreasing gastric irritancy and enhancing absorption of naproxen in rats," International Journal of Pharmaceutics, vol. 125, no. 2, pp. 309–313, 1995.

32. C. M. Keck and R. H. Müller, "Drug nanocrystals of poorly soluble drugs produced by high pressure homogenisation," European Journal of Pharmaceutics and Biopharmaceutics, vol. 62, no. 1, pp. 3–16, 2006.

33. P. Langguth, A. Hanafy, D. Frenzel et al., "Nanosuspension formulations for low-soluble drugs: pharmacokinetic evaluation using spironolactone as model compound," Drug Development and Industrial Pharmacy, vol. 31, no. 3, pp. 319–329, 2005.

34. C. Jacobs and R. H. Müller, "Production and characterization of a budesonide nanosuspension for pulmonary administration," Pharmaceutical Research, vol. 19, no. 2, pp. 189–194, 2002.

35. J. Möschwitzer, G. Achleitner, H. Pomper, and R. H. Müller, "Development of an intravenously injectable chemically stable aqueous omeprazole formulation using nanosuspension technology," European Journal of Pharmaceutics and Biopharmaceutics, vol. 58, no. 3, pp. 615–619, 2004.

36. G. Sunkara and U. B. Kompella, "Drug delivery applications of supercritical fluid technology," Drug Delivery Technology, vol. 2, pp. 44–50, 2002.

37. L. Manna, M. Banchero, D. Sola, A. Ferri, S. Ronchetti, and S. Sicardi, "Impregnation of PVP microparticles with ketoprofen in the presence of supercritical CO2," Journal of Supercritical Fluids, vol. 42, no. 3, pp. 378–384, 2007.

38. H. Leuenberger, "Spray freeze-drying—the process of choice for low water soluble drugs?" Journal of Nanoparticle Research, vol. 4, no. 1-2, pp. 111–119, 2002.

39. M. Mumenthaler and H. Leuenberger, "Atmospheric spray-freeze drying: a suitable alternative in freeze-drying technology," International Journal of Pharmaceutics, vol. 72, no. 2, pp. 97–110, 1991.

40. R. Q. Williams, "Process for production of nanoparticles and microparticles by spray freezing into liquid," US Patent no. 20030041602, 2003.

41. A. R. Briggs and T. J. Maxvell, "Process for preparing powder blends," US Patent no. 3721725, 1973.

42. T. L. Rogers, J. Hu, Z. Yu, K. P. Johnston, and R. O. Williams, "A novel particle engineering technology: spray-freezing into liquid," International Journal of Pharmaceutics, vol. 242, no. 1-2, pp. 93–100, 2002.

43. I. R. Buxton and J. M. Peach, "Process and apparatus for freezing a liquid medium," US Patent no. 4470202, 1984.

44. "Cyclodextrins in pharmaceuticals: an overview," http://www.pharmainfo.net/pharma-student-magazine/cyclodextrins-pharmaceuticals-overview-0.

45. T. Purvis, M. E. Mattucci, M. T. Crisp, K. P. Johnston, and R. O. Williams, "Rapidly dissolving repaglinide powders producers by the ultra-rapid freezing process," AAPS PharmSciTech, vol. 8, no. 3, article 58, 2007.

46. K. Uekama, F. Hirayama, and T. Irie, "Cyclodextrin drug carrier systems," Chemical Reviews, vol. 98, no. 5, pp. 2045–2076, 1998.

47. http://www.nature.com/nrd/journal/v3/n12/fig_tab/nrd1576_F3.html.

48. R. K. Parikh, N. S. Mansuri, M. C. Gohel, and M. M. Soniwala, "Dissolution enhancement of nimesulide using complexation and salt formation techniques," Indian Drugs, vol. 42, no. 3, pp. 149–154, 2005.

49. F. Cao, J. Guo, and Q. Ping, "The physicochemical characteristics of freeze-dried scutellarin- cyclodextrin tetracomponent complexes," Drug Development and Industrial Pharmacy, vol. 31, no. 8, pp. 747–756, 2005.

50. X. Wen, F. Tan, Z. Jing, and Z. Liu, "Preparation and study the 1:2 inclusion complex of carvedilol with β-cyclodextrin," Journal of

Pharmaceutical and Biomedical Analysis, vol. 34, no. 3, pp. 517–523, 2004.

51. A. Martin, Physical Pharmacy, Willaims and Wilkins, Baltimore, Md, USA, 4th edition, 1993.

52. C. D. Rangel-Yagui, A. Pessoa, and L. C. Tavares, "Micellar solubilization of drugs," Journal of Pharmacy and Pharmaceutical Sciences, vol. 8, no. 2, pp. 147–163, 2005.

53. C. H. Hsu, Z. Cui, R. J. Mumper, and M. Jay, "Micellar solubilization of some poorly soluble antidiabetic drugs," AAPS PharmSciTech, vol. 9, no. 2, pp. 939–943, 2008.

54. A. A. Rasool, A. A. Hussain, and L. W. Dittert, "Solubility enhancement of some water-insoluble drugs in the presence of nicotinamide and related compounds," Journal of Pharmaceutical Sciences, vol. 80, no. 4, pp. 387–393, 1991.

55. A. A. Badwan, L. K. El Khordagui, A. M. Saleh, and S. A. Khalil, "The solubility of benzodiazepines in sodium salicylate solution and a proposed mechanism for hydrotropic solubilization," International Journal of Pharmaceutics, vol. 13, no. 1, pp. 67–74, 1983.

56. B. K. Roy and S. P. Moulik, "Functions of hydrotropes (sodium salicylate, proline, pyrogallol, resorcinol and urea) in solution with special reference to amphiphile behaviors," Colloids and Surfaces A, vol. 203, no. 1–3, pp. 155–166, 2002.

57. S. V. Patil and S. K. Sahoo, "Pharmaceutical overview of spherical crystallization," Der Pharmacia Lettre, vol. 2, no. 1, pp. 421–426, 2010.

58. N. Blagden, M. de Matas, P. T. Gavan, and P. York, "Crystal engineering of active pharmaceutical ingredients to improve solubility and dissolution rates," Advanced Drug Delivery Reviews, vol. 59, no. 7, pp. 617–630, 2007.

59. A. J. Aguiar, J. Krc, A. W. Kinkel, and J. C. Samyn, "Effect of polymorphism on the absorption of chloramphenicol from chloramphenicol palmitate," Journal of Pharmaceutical Sciences, vol. 56, no. 7, pp. 847–853, 1967.

60. W. Liebenberg, M. M. De Villiers, D. E. Wurster, E. Swanepoel, T. G. Dekker, and A. P. Lötter, "The effect of polymorphism on powder compaction and dissolution properties of chemically equivalent oxytetracycline hydrochloride powders," Drug Development and Industrial Pharmacy, vol. 25, no. 9, pp. 1027–1033, 1999.

61. M. S. Suleiman and N. M. Najib, "Isolation and physicochemical characterization of solid forms of glibenclamide," International Journal of Pharmaceutics, vol. 50, no. 2, pp. 103–109, 1989.

62. E. Shefter and T. Higuchi, "Dissolution behavior of crystalline solvated and nonsolvated forms of some pharmaceuticals," Journal of Pharmaceutical Sciences, vol. 52, pp. 781–791, 1963.

63. P. V. Allen, P. D. Rahn, A. C. Sarapu, and A. J. Vanderwielen, "Physical characterization of erythromycin: anhydrate, monohydrate, and dihydrate crystalline solids," Journal of Pharmaceutical Sciences, vol. 67, no. 8, pp. 1087–1093, 1978.

64. P. Vishweshwar, J. A. McMahon, J. A. Bis, and M. J. Zaworotko, "Pharmaceutical co-crystals," Journal of Pharmaceutical Sciences, vol. 95, no. 3, pp. 499–516, 2006.

65. N. Shan and M. J. Zaworotko, "The role of cocrystals in pharmaceutical science," Drug Discovery Today, vol. 13, no. 9-10, pp. 440–446, 2008.

66. M. B. Hickey, M. L. Peterson, L. A. Scoppettuolo et al., "Performance comparison of a co-crystal of carbamazepine with marketed product," European Journal of Pharmaceutics and Biopharmaceutics, vol. 67, no. 1, pp. 112–119, 2007.

67. S. L. Childs, L. J. Chyall, J. T. Dunlap, V. N. Smolenskaya, B. C. Stahly, and G. P. Stahly, "Crystal engineering approach to forming cocrystals of amine hydrochlorides with organic acids. Molecular complexes of fluoxetine hydrochloride with benzoic, succinic, and fumaric acids," Journal of the American Chemical Society, vol. 126, no. 41, pp. 13335–13342, 2004.

68. J. F. Remenar, S. L. Morissette, M. L. Peterson et al., "Crystal engineering of novel cocrystals of a triazole drug with 1,4-dicarboxylic acids," Journal of the American Chemical Society, vol. 125, no. 28, pp. 8456–8457, 2003.

69. K. Moribe, Y. Tozuka, and K. Yamamoto, "Supercritical carbon dioxide processing of active pharmaceutical ingredients for polymorphic control and for complex formation," Advanced Drug Delivery Reviews, vol. 60, no. 3, pp. 328–338, 2008.

70. I. Pasquali, R. Bettini, and F. Giordano, "Supercritical fluid technologies: an innovative approach for manipulating the solid-state of pharmaceuticals," Advanced Drug Delivery Reviews, vol. 60, no. 3, pp. 399–410, 2008.

71. A. Paradkar, M. Maheshwari, R. Kamble, I. Grimsey, and P. York, "Design and evaluation of celecoxib porous particles using melt sonocrystallization," Pharmaceutical Research, vol. 23, no. 6, pp. 1395–1400, 2006.

Chapter 2

The Prodrug Approach: A Successful Tool for Improving Drug Solubility

Daniela Hartmann Jornada, Guilherme Felipe dos Santos Fernandes, Diego Eidy Chiba, Thais Regina Ferreira de Melo, Jean Leandro dos

Faculdade de Ciências Farmacêuticas, UNESP—University Estadual Paulista, Rodovia Araraquara Jaú Km 01, 14801-902 Araraquara, São Paulo, Brasil

ABSTRACT

Prodrug design is a widely known molecular modification strategy that aims to optimize the physicochemical and pharmacological properties of drugs to improve their solubility and pharmacokinetic features and decrease their toxicity. A lack of solubility is one of the main obstacles to drug development. This review aims to describe recent advances in the improvement of solubility via the prodrug approach. The main chemical carriers and examples of successful strategies will be discussed, highlighting the advances of this field in the last ten years.

Keywords: prodrug; solubility; water-soluble prodrugs; solubility of prodrugs; molecular modification

1. INTRODUCTION

Poor solubility is one of the main problems faced by researchers during drug development. Commonly, even with the use of current computational "filters" to minimize this problem, compounds that are active *in vitro* may lack adequate pharmacokinetic properties and/or may be difficult to formulate [1]. A study conducted with the top 200 oral drug products in Japan, Great Britain, United States and Spain revealed that approximately 37% of drugs had solubilities of less than 0.1 mg/mL. One explanation may include the need for drugs that are highly potent in low doses, however, this issue represents a challenge in drug discovery [2].

Although the prodrug approach is often considered only when the prototype presents unexpected problems, this strategy offers a versatile approach to drug development and should not be considered a last resort. The prodrug approach is a promising molecular modification by which drug developers and designers can modulate drug pharmacokinetics, pharmacodynamics and toxicology [3].

A prodrug is a poorly active or inactive compound containing the parental drug that undergoes some *in vivo*biotransformation through chemical or enzymatic cleavage, enabling the delivery of the active molecule at efficacious levels [1,3,4]. Prodrugs are conventionally classified in two major classes: carrier-linked prodrugs and bioprecursors. Carrier-linked prodrugs can be classified as bipartite prodrugs, in which the carrier is linked directly to the parent drug, and tripartite prodrugs, in which a spacer links the carrier to the parent drug [5]. Carriers are commonly attached by chemical groups such as ester, amide, carbamate, carbonate, ether, imine, phosphate, among others [1,6,7] (Figure 1). Mutual prodrugs are a type of carrier-linked prodrug in which two active compounds are linked each acting as the carrier to the other. These prodrugs have increased effectiveness through synergistic action [5,8]. Another type of carrier-linked prodrug is the macromolecular prodrug; these prodrugs use polymeric backbones as carriers. Macromolecular prodrugs are commonly used to design prodrugs that will be cleaved inside of a cell and in targeted drug-delivery systems. This approach offers improved drug solubility, stability, drug release and pharmacokinetics. Additionally, it can facilitate the accumulation of a drug at the site of action and improve safety [9,10]. Bioprecursors are inactive compounds that do not have a carrier and are rapidly converted to an active drug after metabolic reactions (normally redox reactions) [5,8].

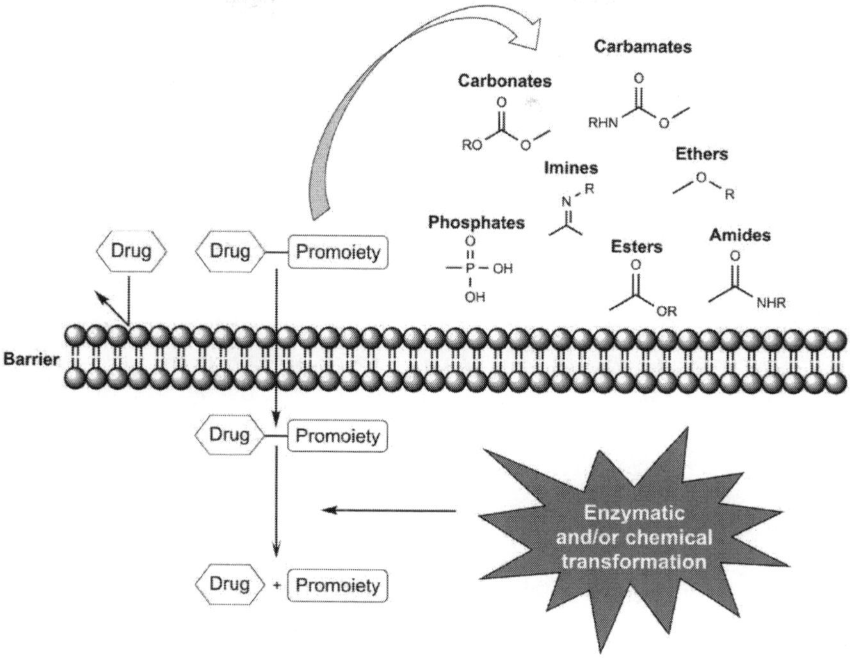

Figure 1. *In vivo* bioactivation of prodrugs by enzymatic and/or chemical transformations.

Undesirable properties, including poor aqueous solubility, chemical instability, insufficient oral or local absorption, fast pre-systemic metabolism, low half-life, toxicity and local irritation are commonly resolved using the prodrug approach. In addition, problems related to drug formulation and delivery can also be overcome by using this strategy [1,11,12,13]. Occasionally, strategies such as particle-size reduction, solubilizing excipients, complexation agents and use of surfactants fail to improve the solubility in water profile and reduce toxicity to desirable levels [13,14]. For example, some surfactants used in parenteral formulations frequently have toxic effects such as anaphylactic reactions [14].

However, in some cases, low-solubility compounds yield false-positive results on *in vitro* assays due to non-specific binding [15]. Moreover, in clinical trials, low-solubility drugs can lead to precipitation and crystalluria, raising additional safety concerns. Poorly soluble drugs have recently been discontinued in clinical assays for this reason [15,16].

For these reasons, the prodrug approach presents a safe and effective strategy by which to improve the solubility of drugs. This review aims to present the latest strategies in prodrug design used to obtain water-soluble compounds for oral and parenteral uses. We selected research from the last ten years (2005 through August of 2015) showing increased solubility through the prodrug approach. In the literature search, we used the following terms: "water-soluble prodrugs", "increased solubility prodrugs" and "enhanced solubility prodrugs" in published databases including PubMed, LILACS, Scielo, Cochrane, Web of Science and Scopus.

2. ESTER PRODRUGS

The features of an ideal prodrug include the following: (a) hydrolysis resistance during absorption; (b) weak or no activity; (c) aqueous solubility; (d) good permeability through the cells; (e) chemical stability at different pHs; (f) kinetics that allow release of the parental drug [17]. Among the chemical bonds used to link the parental drug and carrier, esters have proven to be promising due to their amenability to hydrolysis both *in vivo* and *in vitro*. Some examples of the use of esters in prodrug design are discussed below.

The enzyme thioredoxin–thioredoxin reductase plays an important role in thioredoxin system by catalyzing the reduction of thioredoxin. Specifically, the thioredoxin system participates in protecting DNA against oxidative damage and has been implicated in several diseases, including cancer and rheumatoid arthritis [18]. Among the described inhibitors, the naphthoquinone spiroketal compound palmarumycin (**1**) has shown *in vitro* anticancer activity; however, it failed in *in vivo*assays. The authors hypothesized that the lack of activity could be due to its high lipophilicity; therefore, they designed prodrugs containing amino-esters and two morpholine analogues. The glycyl ester derivative **2** (Figure 2) was more than seven times more soluble in water than its parent drug. Despite the higher solubility compared to the parent drug, the compound did not exhibit superior activity compared to palmarumycin CP_1 [19].

The adenosine A_{2A} receptor is a G-protein-coupled receptor expressed at high levels in blood platelets, GABAergic neurons, and cells of the thymus, the olfactory bulb and the spleen. This receptor has been widely known as a target since the approval of the drug regadenoson by the U.S. Food and Drug Administration (FDA). Among the activities described for adenosine antagonists are neuroprotective, antidepressant, and analgesic effects. In addition, this receptor has been explored as a target of new therapeutic

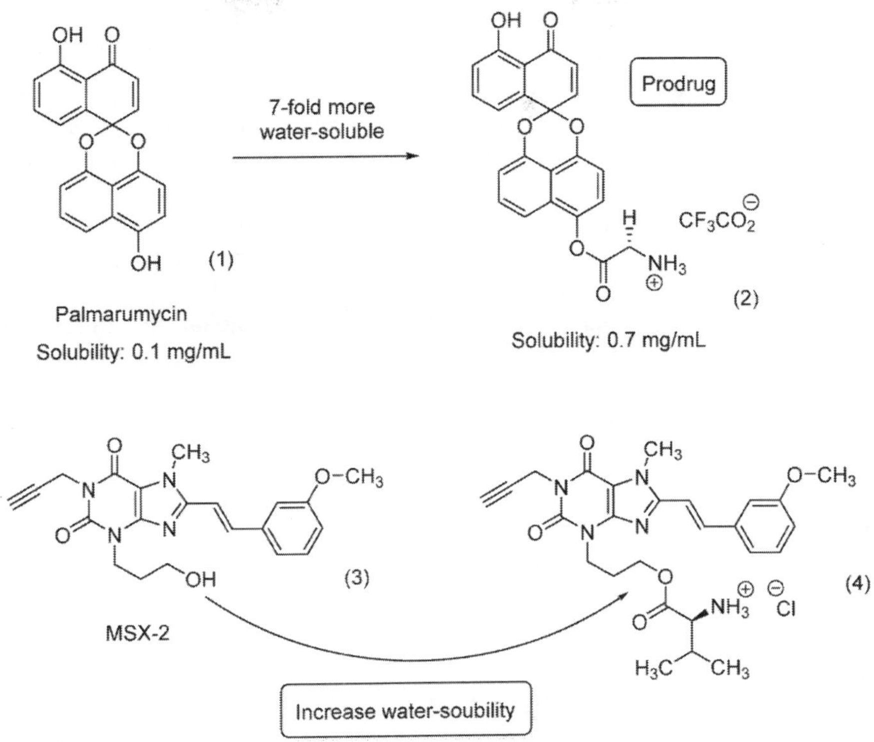

Figure 2. Glycyl ester and amino-acid-ester prodrugs [19,20].

agents to treat Parkinson's disease [21]. Compound **3**, known as MSX-2 (Figure 2), is a potent and selective antagonist of the adenosine A_{2A} receptor [22,23]. Despite its *in vitro* activity, this molecule shows low solubility in water, hindering *in vivo* studies. Two different approaches were undertaken to resolve this issue. The first involved the introduction of a phosphate subunit to increase solubility. Phosphate prodrugs have shown high solubility in water; however, it is not readily absorbed by the oral route. The second approach involved the introduction of amino acids, specifically L-valine. The solubility in water of compound **4** (Figure 2), as evaluated by a spectrophotometric method is 7.3 mg/mL, less than that of the phosphate prodrug (9.0 mg/mL) but superior to that of the parent drug. According to the authors, the valine prodrug was stable in aqueous solution at room temperature for up to four days; however, in the presence of pig liver esterase, the compound undergoes bioconversion, releasing the parental drug, with a half-life of 6.9 min [20].

New water-soluble antiviral compounds also use ester amino-acid prodrugs to increase solubility with a bicyclic nucleoside subunit. This scaffold was initially described as a by-product of the condensation of 5-iodo nucleosides within terminal alkynes catalyzed by palladium; however, *in vitro* studies revealed activity against several viruses, including herpes simplex (type 1 and type 2), varicella zoster and cytomegalovirus. One of the most active compounds was the *p*-penthylphenylbicyclic nucleoside analogue Cf1743; however, its low solubility in water limits its use [24]. The prodrug approach was used to resolve this limitation, with dipeptides as carriers of Cf1743 (Val, Asn, Lys, Asp) (Figure 3). The amino-acid pro-residue is recognized by the enzyme dipeptidyl-peptidase IV (DPPIV/CD26), which is expressed in leukocytes as well as epithelial, endothelial and fibroblast cells. It was hypothesized that leukocyte enzymes would activate the prodrug. Interestingly, the authors found an analog **5** with 4000-fold greater solubility in water and 7–15-fold greater bioavailability compared to the parent drug [25].

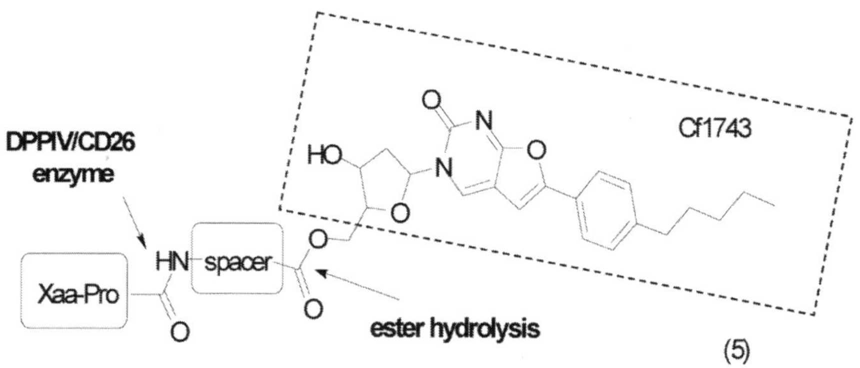

(5)

Xaa- Val, Asn, Lys, Asp
Spacer- Val, Ala, Leu, Phe

Figure 3. Bicyclic furanopyrimidine nucleoside analogue [25].

Amino acids were also used to improve the physicochemical properties of oleanolic acid (**6**). This triterpenoid molecule is a natural product found in Asian herbs whose pharmacological effects include anti-inflammatory, analgesic and antitumoral activities. Despite these effects, oleanolic acid has low bioavailability in rats (7%), probably due to its low solubility in water (0.0012 µg/mL). Attempts to improve the solubility using different formulations failed, probably due to its low permeability. To explore the transporters involved in the absorption of oleanolic acid, the researchers designed a series of prodrugs intended to be recognized by the PepT1 transporter. This transporter is important in drug absorption because it improves bioavailability. One commercially successful prodrug that exploits this transporter is valaciclovir, which has improved bioavailability (55%) compared to its parental drug, acyclovir (15%–20%) [26,27,28]. This transporter could also improve the absorption of oleanolic-acid derivatives. A series of novel ethylene-glycol-linked amino-acid-diester prodrugs of oleanolic acid was synthesized and tested for solubility and pharmacokinetic profile. Six analogs showed improved solubility in water. The compound **7** containing an L-valine subunit (Figure 4) was the most promising analog, exhibiting solubility greater than 25 µg/mL and increased intestinal perfusion and oral bioavailability in rats [29]. In a different paper, the same research group described a series of seven propylene glycol-linked amino-acid/dipeptide-diester prodrugs of oleanolic acid. In this study, propylene glycol (PEG) was chosen as a spacer due to its very low toxicity *in vivo*. The authors synthesized seven diester prodrugs and evaluated the influence of the linker on pharmacokinetic properties. Compound **8** was the most water soluble (up to 1.0 mg/mL; Figure 4). Compared to ethylene-glycol derivatives, the propylene-glycol series was the most soluble, stable and bioavailable and had the greatest affinity for the PepT1 enzyme [30].

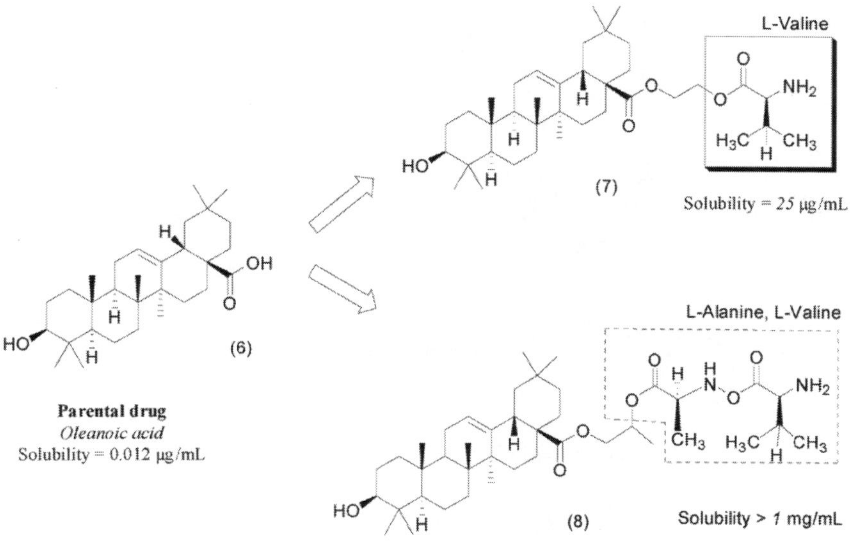

Figure 4. Ethylene glycol and propylene glycol prodrugs of oleanoic acid [29,30].

Another terpenoid with low solubility in water whose properties were enhanced by the use of the prodrug approach is oridonin (**9**). This natural product is found in *Rabdosia rubescens*, a Chinese medicinal plant, and it exhibited antitumoral activity against several types of cancer, including leukemia [31,32]. Exploring polyethylene glycol as carrier, the authors linked oridonin to PEG, using succinic acid as a spacer. Four different molecular weights of PEGs were tested to increase solubility in water. The greatest increase in solubility was observed for a low PEG-molecular-weight (5 kDa) conjugate **10**(Figure 5). This prodrug had 99.2 times the solubility of oridonin. A chemical hydrolysis study showed a sustained-release effect for the prodrugs. *In vivo* studies using the two intermediate conjugates (10 and 20 kDa) demonstrated a successful use of this prodrug approach, as these derivatives have better pharmacokinetic profiles than oridonin [33].

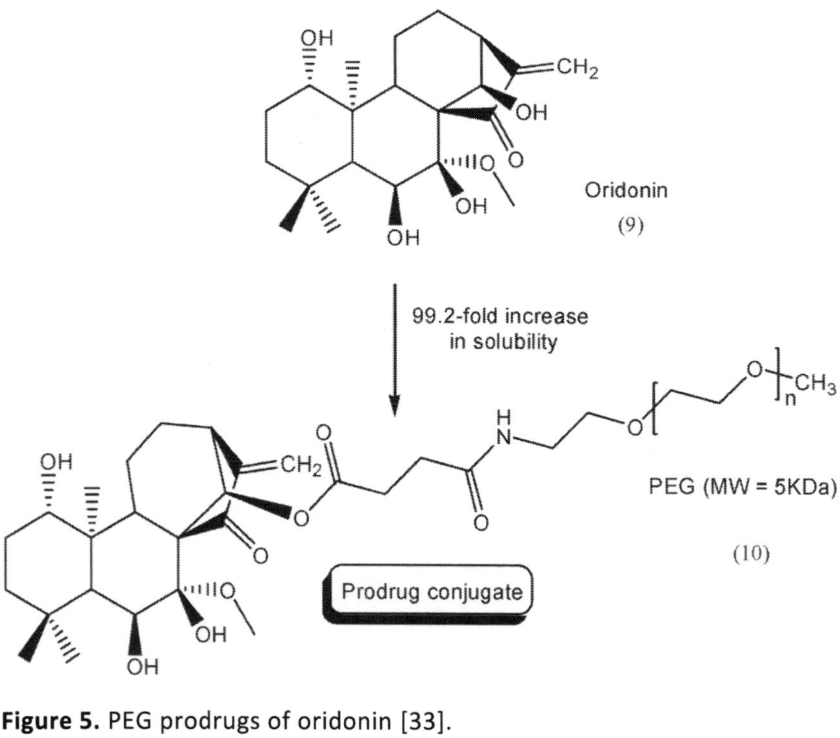

Figure 5. PEG prodrugs of oridonin [33].

A similar approach exploring the use of PEG to increase solubility was carried out using another anticancer compound, gambogic acid (**11**). This natural product inhibits tumor growth and induces apoptosis in various cancer cells [34]. The authors synthesized a series of PEGylate prodrugs **12** with molecular weights of 2 kDa, 4 kDa, 10 kDa and 20 kDa using L-leucine as a spacer (Figure 6). The solubility in water ranged from 645 mg/mL to 1750 mg/mL. The most soluble prodrug contained PEG-2 kDa-L-leucine-GA. For all prodrugs, the half-life measured in plasma ranged from 1.26 to 6.12 h [35].

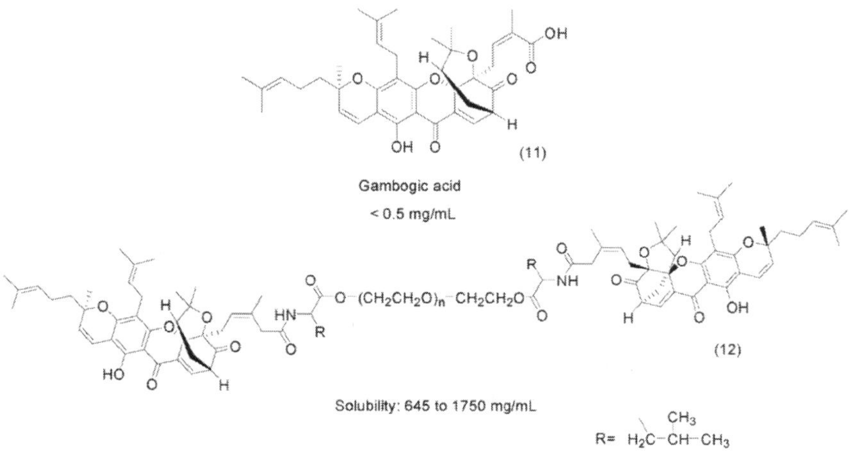

Figure 6. General structure of gambogic-acid PEG prodrugs [35].

The poor solubility in water of taxoids is well established in the literature [36]. Using the prodrug approach, Skwarczynski *et al.*, synthesized prodrugs whose solubilities ranged from 0.8 to 1.1 mg/mL. Interestingly, the authors identified a novel mechanism of parental-drug release: a pH-dependent O–N intramolecular acyl migration reaction (Figure 7). This strategy increased the stability of taxoids, released the parental drug at pH 7.4 and improved drug solubility. Some of these prodrugs containing the 2'-*O*-isoform of taxoids, such as compound **14**, exhibited 4000-fold greater solubility than the parent drug **13** [37].

One approach to improve the solubility of taxoids used docetaxel-glycopyranoside ester-linked prodrugs. The compounds were prepared by chemo-enzymatic reactions using β-xylosidase, lactase and β-galactosidase. All of them demonstrated improved solubility in water compared to the parental drug, docetaxel, I and the best compound **16** showed a 52-fold increase (Figure 7). In addition, the compounds were able to release the parental drug *in vitro* and exhibited cytotoxic effects against KB and MCF-7 human cancer cell lines [38].

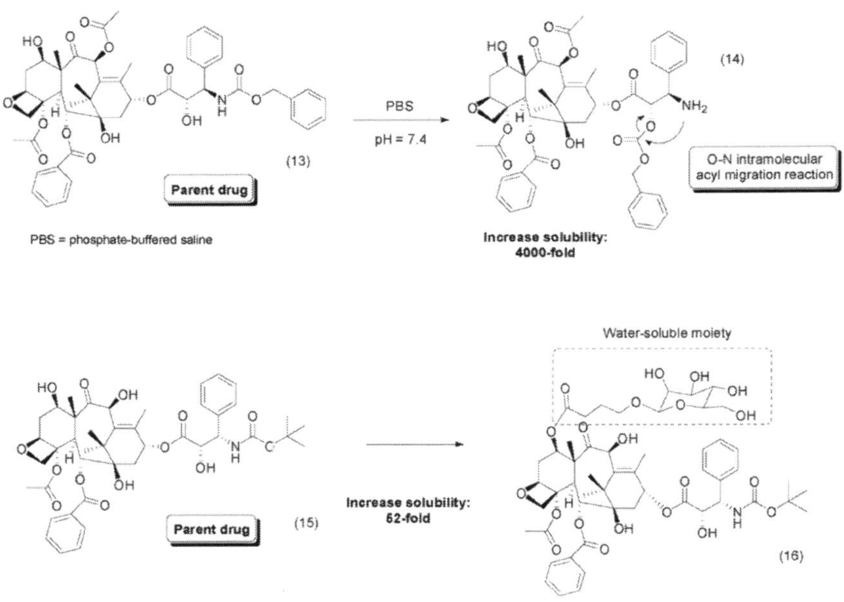

Figure 7. Taxoid prodrugs [37,38].

A paclitaxel prodrug containing a disulfide moiety was described by Gund *et al.*, as an anticancer compound. The rationale behind the design of these molecules is based on the recognition of disulfide subunit by the enzyme glutathione, which releases the paclitaxel (**17**). High levels of this enzyme have been described in cancer cells, and one hypothesis speculates that this enzyme is involved in resistance to many anticancer drugs [39,40]. The prodrug compounds were 6–100 times more soluble than paclitaxel. Moreover, the most active compound **18** (Figure 8) was 65 times more water soluble than paclitaxel. For this molecule, superior antitumoral activity was identified in all types of cells evaluated, except for MCF10A. After oral administration, the bioavailability in mice was five-fold greater than that of the parental drug [41].

Etoposide (**19**) is a topoisomerase inhibitor with variable pharmacokinetics and low solubility in water. Using the prodrug approach, a series of water-soluble etoposide ester prodrugs were produced. These prodrugs with an attached malic acid demonstrated solubility in water 23 to 120-fold greater than that of the parental drug. In addition, the most active compound **20** (Figure 9), demonstrated *in vitro* cytotoxic effects comparable to those of etoposide [42].

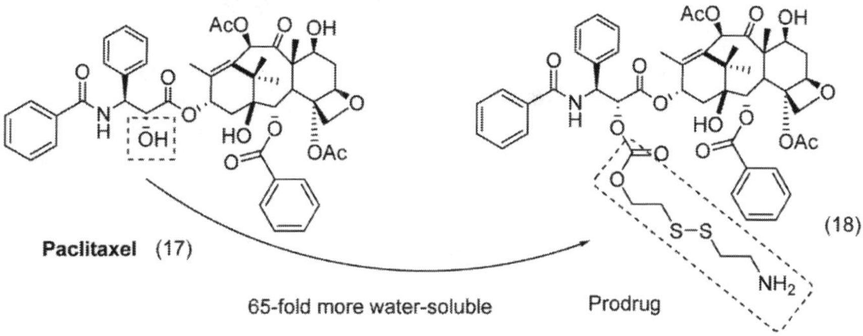

Figure 8. Paclitaxel-disulfide prodrug with anticancer activity [41].

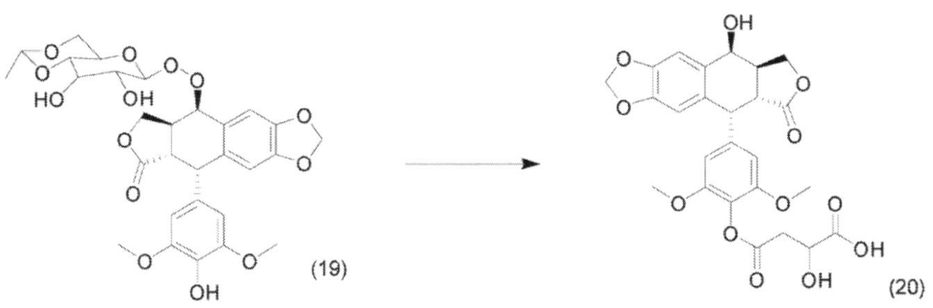

Parental drug - Etoposide
Water-solubility 0.1 mg/mL

Prodrug
Water-solubility~ 9.0 mg/mL

Figure 9. Etoposide ester prodrugs containing malic acid [42].

Non-steroidal anti-inflammatory drugs (NSAIDs) are widely prescribed worldwide. The use of NSAID-ester prodrugs is extensively reported in the literature. Controversially, some authors report that the esterification of NSAIDs can reduce the adverse gastric effects by masking the carboxylic acid function; however, the changes in the pharmacokinetic profile that provide sustained release can explain in part the reduced adverse effects seen in some cases [43,44]. In many cases, the lack of solubility limits the use and worsens the adverse effects of NSAIDs. Therefore, to improve the solubility in water of the NSAID 6-methoxy-2-naphthylacetic acid (6-MNA, **21**), a series of ester prodrugs intended for percutaneous drug delivery was designed and synthesized. The authors designed piperazine derivatives attached by an ester linkage to enhance the solubility of these compounds. They identified a compound **22** that proved to be water soluble and 11.2-fold more permeable through the skin than 6-MNA at pH 7.4 (Figure 10) [45].

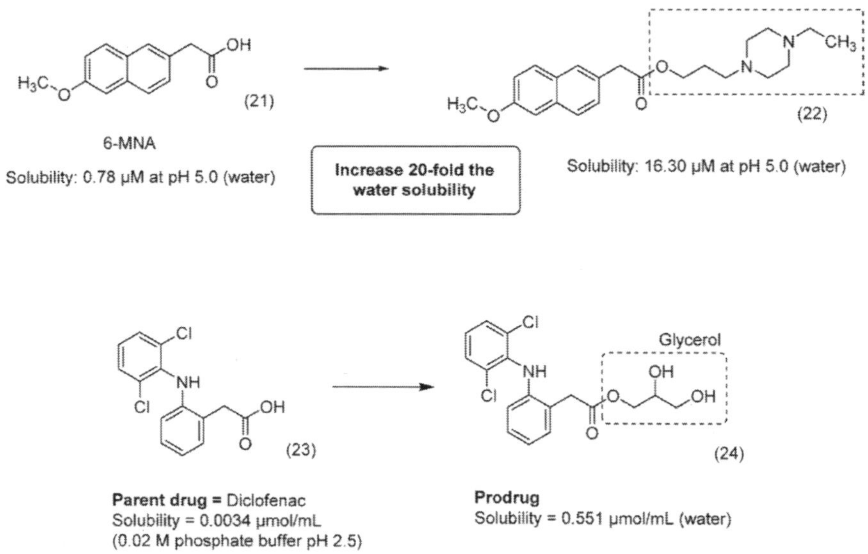

Figure 10. NSAID ester prodrugs [45,46].

Exploring transdermal delivery, Lobo *et al.*, have synthesized and evaluated series of diclofenac (**23**)-ester prodrugs. The most promising compound, glycerol diclofenac ester (**24**), demonstrated better solubility, a lower partition coefficient and higher flux across the skin than its parental drug (Figure 10). *In vitro* studies using rat plasma showed rapid conversion of prodrugs to diclofenac due to the activity of esterases [46].

The natural product quercetin (**25**) exhibited promising anti-inflammatory properties; however, its low skin permeation limits its use [47]. Therefore, to improve the passage of quercetin through the skin, a series of prodrugs was synthesized and evaluated. The researchers observed that prodrugs with long acyl chains were less soluble than the parent drug. However, compounds with shorter acyl chains were more water soluble, with an increase of 64-fold for compound **26**. Thus, the prodrug exhibited greater solubility and skin permeation than quercetin (Figure 11) [48].

2*R*-γ-Tocotrienol (γ-T3), which is present in vitamin E, and has antioxidant activity against brain microsomes and nitric oxide and inhibits cholesterol increases and cancer proliferation [49,50,51,52,53,54,55,56,57,58]; however, γ-T3 is poorly soluble in aqueous solutions, impairing its pharmacological use. Three aminoalkylcarboxylic acid esters generated as water-soluble prodrugs of γ-T3 showed better solubility in water compared to the parental drug. Compound **27** (Figure 11) was the most promising for parenteral use, with high solubility in water, stability and rapid hydrolysis in rat and human plasma, making it a good potential alternative for cancer therapy [59].

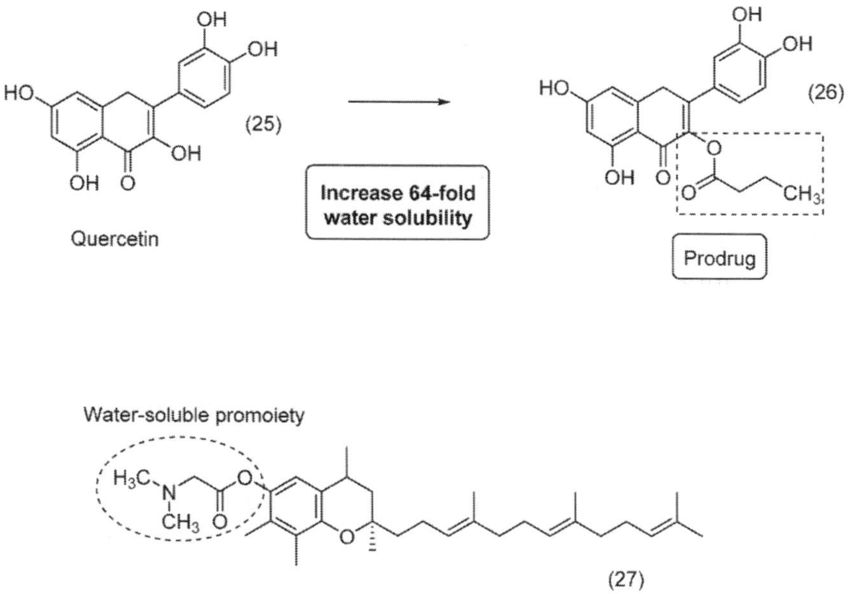

Figure 11. Prodrugs of quercetin and γ-T3 [48,59].

Screening of an in-house library was used to identify compound **28** as a promising candidate targeting Human Immunodeficiency Virus (HIV) replication; however, its poor solubility in water limits additional preclinical pharmacokinetic studies. Therefore, researchers have designed novel indazole derivatives. The most active molecule **29**(Figure 12) contains an *N*-acyloxymethyl group as a pro-moiety and was 300-fold more water soluble than its parent compound. Additionally, this prodrug was more active against HIV, exhibiting an EC_{50} of 2 μM, while the parental drug showed an EC_{50} of 3 μM [60].

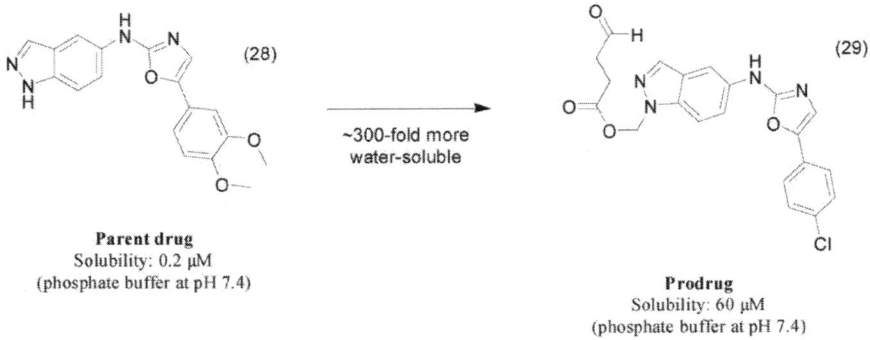

Figure 12. *N*-acyloxymethyl ester prodrug [60].

3. AMIDE PRODRUGS

Acyclovir (**30**) is an antiviral drug that has poor solubility in water and also has low bioavailability (between 10% and 20%) [61]. To improve these characteristics, two peptide prodrugs, an amide **31** and an ester prodrug **32** were developed (Figure 13). The first was designed to release acyclovir directly through the action of enzyme dipeptidyl peptidase IV (DPPIV or CD26), and the second requires two steps to release the parental drug. Both showed a significant increase in solubility (17-fold for amide and 9-fold for ester derivatives) compared to the parent drug. Furthermore, the compounds were stable in phosphate-buffered saline (PBS) and were rapidly cleaved by plasmatic enzymes. Although the antiviral activity was not superior to that of free acyclovir, both presented good antiviral activity against herpes simplex virus type 1 and type 2 [62].

SB-3CT (**33**) is a highly selective inhibitor of the endopeptidases matrix metalloproteinases 2 and 9 (MMP-2 and MMP-9), also known as gelatinases, which have attracted interest in research due to their participation in a significant number of human pathologies such as cancer, neuronal death, atherosclerosis and aneurysms [63]. Due to the limited solubility in water of SB-3CT (2.3 μg/mL), two series of prodrugs were developed using amino acids (Gly, L-Lys, L-Glu and L-Arg) and the dipeptide L-Arg-L-Arg as pro-moieties, linked by an ester or amide group. All prodrugs showed improved solubility in water (a more than 5000-fold increase). The ester compounds were unstable in aqueous solution, human plasma and blood, while amides demonstrated resistance to fast hydrolysis. The amide prodrugs released the active drug into the blood in 30 min. The arginyl-amide prodrug **34** (Figure 13) was metabolically stable in mouse, rat and human liver microsomes. Moreover, the prodrug **34** was non-mutagenic in an Ames assay [64].

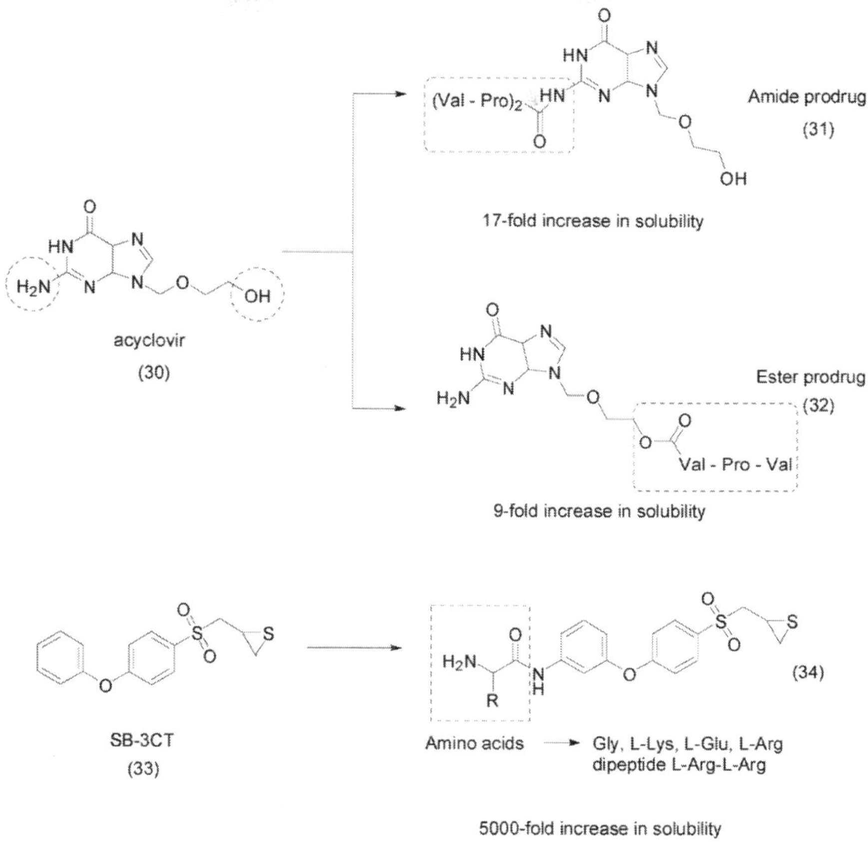

Figure 13. Peptide prodrugs of acyclovir and SB-3CT [62,64].

DW2282 (**35**) is a potent antiproliferative compound with a broad spectrum of action against various cancer cell lines, but in preclinical assays, this analog showed gastrointestinal toxicity [65]. A series of novel amino-acid conjugates were designed to improve the aqueous solubility of these analogs and prevent toxic effects. All synthesized compounds were more soluble and three were more active than the parental drug. The most promising analog **36** (Figure 14) had a good reconversion rate in human plasma and good bioavailability by the oral route in mice. Moreover, this molecule was more active and approximately 36-fold more soluble than DW2282 [66].

Bone tissues are difficult to treat because of limited drug delivery and retention and low sanguineous flow [67]. Hydroxyapatite (HA), a constituent of bones, can be a target for these drugs through the interaction between negative groups and the calcium ions in HA

[68,69,70,71]. Researchers designed and evaluated the solubility and pharmacokinetic profile of novel dendritic naproxen-peptide prodrugs, specifically L-aspartate and L-glutamate.

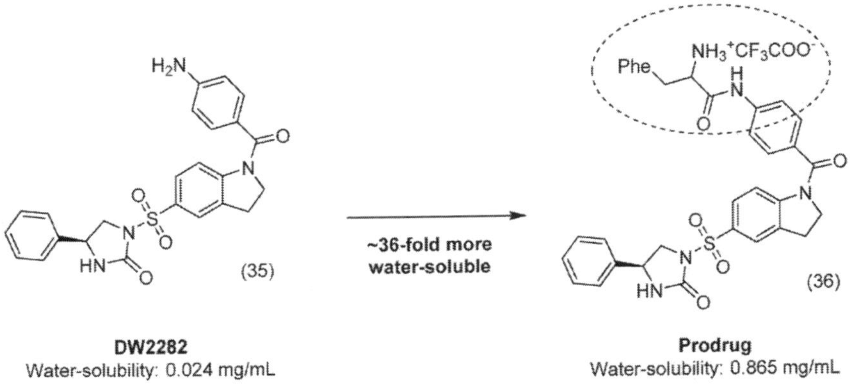

Figure 14. Prodrug derivative of DW2282 [66].

Their solubility in water was evaluated at three different pH values, and the prodrugs were more soluble at increased pH. The authors proposed that the solubility of the analogs can be related to surface area and carboxyl groups in dendritic peptides; in this way, the second-generation molecules have more interactions with water and thus are more water soluble than the first-generation drugs **39** (Figure 15). The most promising analog NAP-G2-Asp (**38**) attached to L-aspartate subunits showed an approximately 90-fold increased solubility in water at pH 7. In the HA-binding assay, NAP-G2-Asp showed 80% binding activity in 25 mg/mL in PBS. In 50% human plasma it acted as a prodrug, rapidly and effectively releasing the parent drug [72].

A benzamide derivative, PC190723 (**40**, Figure 16), was proposed as a potent antimicrobial agent targeting the FtsZ protein. This enzyme plays a key role in bacterial cell division, and studies suggest that it is a promising new target for antibiotic development. To overcome its poor formulation, two new prodrugs were developed. The first, the compound TXY436 (**41**, Figure 16) is a *N*-Mannich base derivative that was 2.8-fold more soluble than the parent drug in PBS (pH = 7.4) and 100-fold more soluble in 10 mM citrate solution (pH = 2.6). The latter is suitable for *in vivo* drug administration. Moreover, this compound is stable in citrate solution and is rapidly converted to the parent drug under physiological conditions (pH =

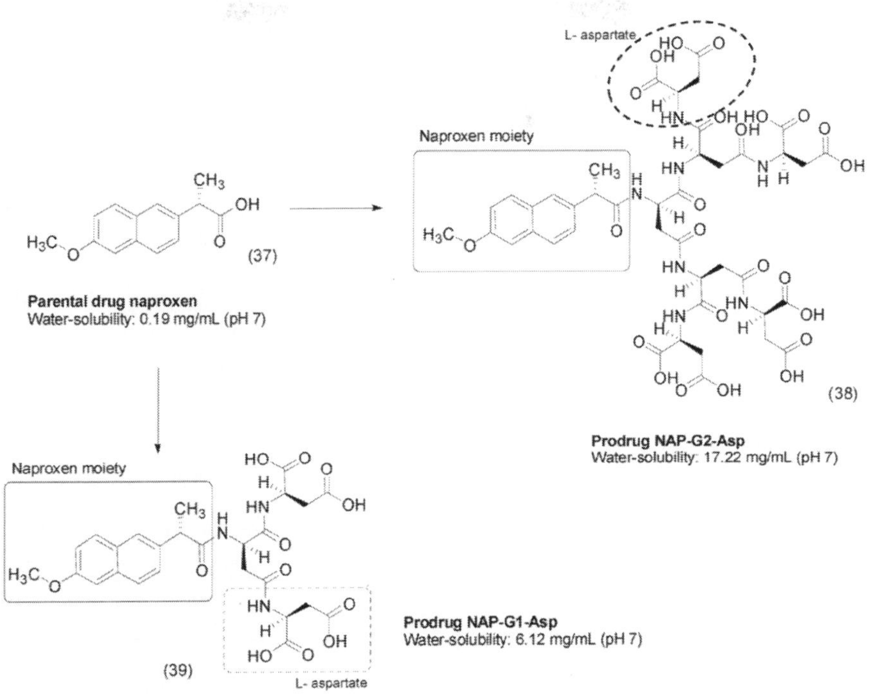

Figure 15. Dendritic naproxen-peptides prodrugs [72].

7.4). In addition, it maintained anti-staphylococcal activity due to *in vitro* FtsZ inhibition and showed 73% oral bioavailability. *In vivo* studies demonstrated that the prodrug was effective in a murine model of peritonitis with systematic infection with methicillin-susceptible *Staphylococcus aureus* (MSSA, survival of 100%) and methicillin-resistant*Staphylococcus aureus* (MRSA, survival of 67%), while the parent drug was not effective in this model [73].

The second prodrug TXY541 (**42**, Figure 16) is a carboxamide derivative with increased solubility in water (143-fold in citrate solution). Interestingly, the solubility in PBS decreased, to almost half the solubility of the parental compound. The stability study revealed that TXY541 is stable in citrate solution and is converted under physiological conditions to PC190723, but at slower rates than the parental compound is **41**. In mouse serum, the conversion occurred faster, which suggest an enzymatic conversion. The oral bioavailability of the prodrug **42** was 29.6%, and *in vivo* evaluation showed that this analog was effective against systematic infection with MSSA (83% survival) and MRSA (100% survival) [74].

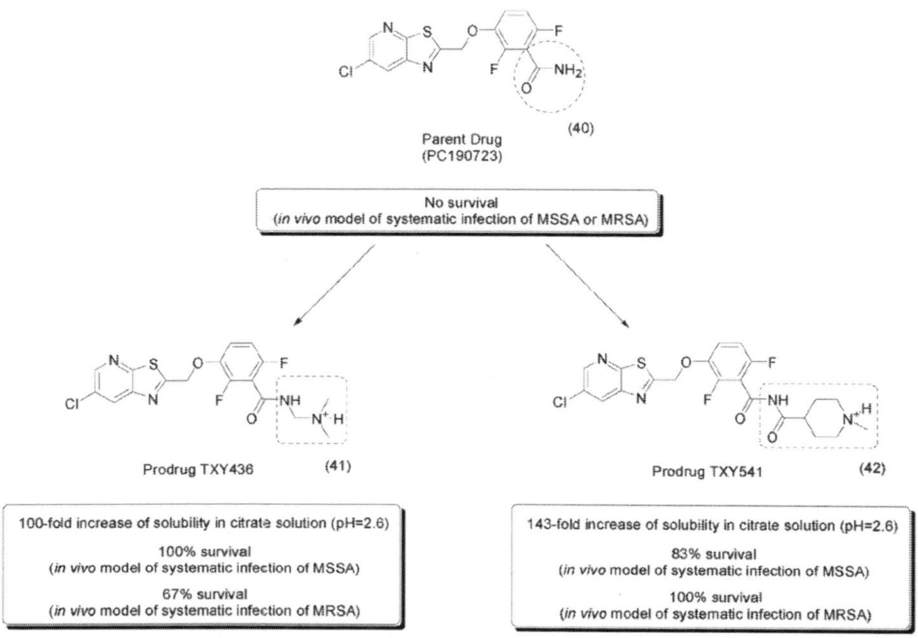

Figure 16. Benzamide and carboxamide prodrugs of PC190723 [73,74].

4. CARBAMATE PRODRUGS

The benzamide compound CI-994 (**43**) is a potent inhibitor of histone deacetylases (HDACs). HDACs have an essential role in the regulation of gene expression, and several types of inhibitors exhibit antitumor activity [75,76], although adverse effects and poor solubility in water limit its use. Two glucuronide prodrugs, one (compound **44**) linked to the glucuronide moiety by a spacer and the other linked directly by a carbamate group (compound **45**), were used to resolve these limitations (Figure 17). While the solubility of CI-994 is 0.08 mg/mL, those of the prodrugs were greater than 1 mg/mL. Both compounds were stable at different pH values (2.1 and 7.0). In addition, the prodrugs were evaluated *in vitro* using an antiproliferative assay with NCI-H661 non-small cell lung cancer cells. The results demonstrated that this compound, when incubated with β-glucuronidase, showed the same antiproliferative activity (IC_{50} = 20 μM) shown by CI-994. Without the enzyme, the prodrugs had decreased cytotoxicity, which may decrease the toxic effects of the parent drug [77].

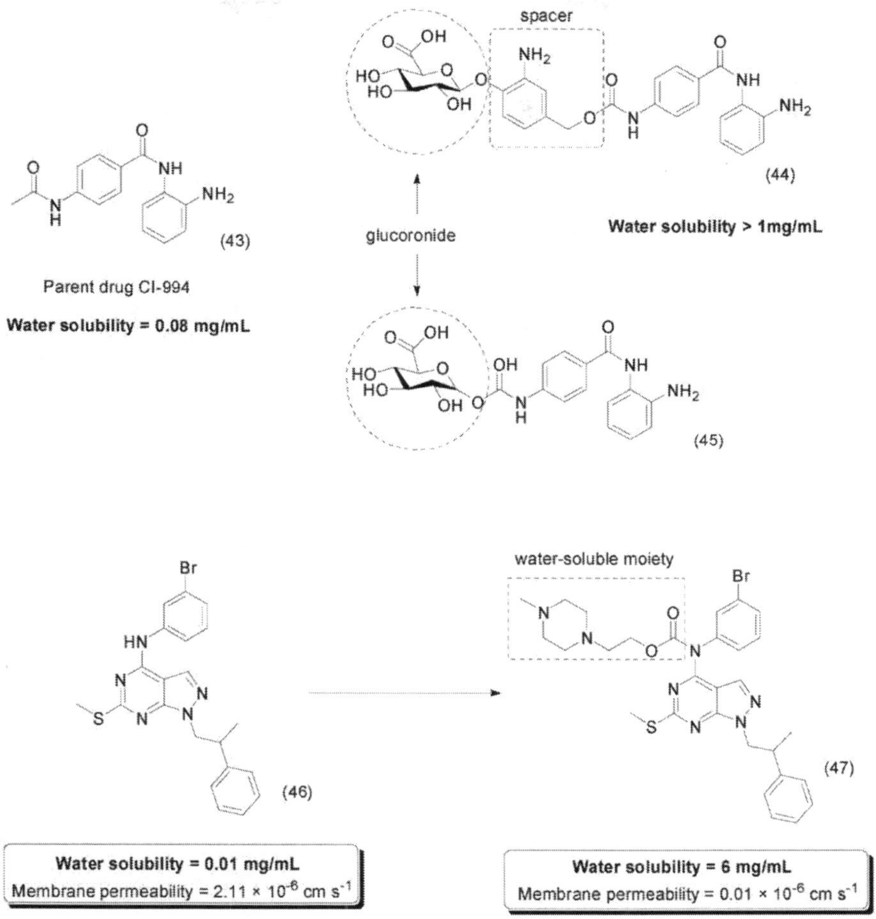

Figure 17. Glucuronide and pyrazolo[3,4-*d*]pyrimides prodrugs [77,78].

Pyrazolo[3,4-*d*]pyrimides compounds are able to inhibit oncogenic tyrosine kinases and may be used to treat cancer. In general, pyrazolo[3,4-*d*]pyrimides compounds have low solubility in water [79]. Prodrugs of pyrazolo[3,4-*d*]pyrimidine compounds were designed to enhance its pharmacokinetic properties using a water-soluble *N*-methylpiperazino promoiety linked by a *O*-alkyl carbamate linker. The prodrug **47** showed 600-fold improvement in solubility compared to the parent drug **46** (Figure 17). *In vitro* assay showed good increase in passive membrane permeability, from 0.01×10^{-6} cm·s^{-1} to 2.11×10^{-6} cm·s^{-1}. In addition, this compound was more cytotoxic compared to the parental drug; however, the mechanism of action is still not well characterized [78].

Photodynamic therapy is a non-invasive treatment for several types of cancer and infectious diseases [80]. This strategy consists in administration of a sensitizer and non-mutagenic substance that will be irradiated only in the affected area, where it will be activated by a specific wavelength of light [81]. This tool can be considered a prodrug approach, in that it uses the light to convert the photosensitizer (prodrug) to an active compound and in that both the selectivity and the solubility of the photosensitizer are important in design [80].

Three photopaclitaxel prodrugs were generated by linking paclitaxel (**17**) with a coumarinic derivative. The most soluble compound **48** showed at least 400,000-fold higher solubility in water compared to paclitaxel (Figure 18). This same compound was stable under physiological and storage conditions. In a photoconversion assay, the carbamate prodrug **48**rapidly generated paclitaxel with minimal tissue damage under 365 nm UV-A light irradiation at low power [82].

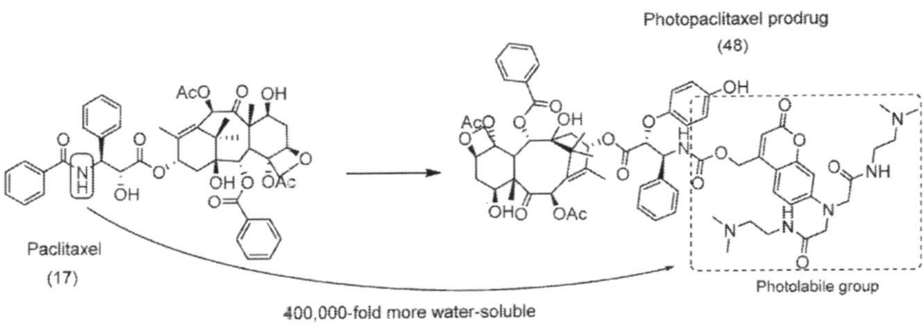

Figure 18. Photopaclitaxel prodrug [82].

5. CARBONATE PRODRUGS

In an attempt to improve the solubility of an anticancer compound, CHS8281 (**49**), a series of prodrugs was synthesized using a carbonate group to link a tetraethylene-glycol moiety to the parental drug. The prodrug EB1627 (**50**) was 600-fold more soluble than the parental drug at pH 5.5 (Figure 19). The selection of the carbonate group provided the required hydrolytic labiality and rapid release of the active drug *in vivo* [83].

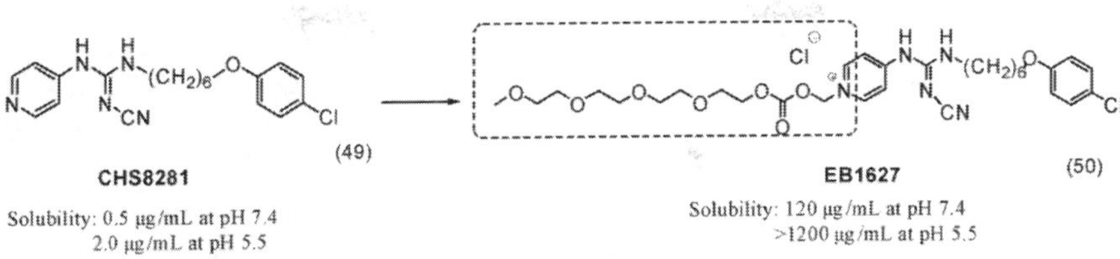

Figure 19. CHS8281 prodrug [83].

To overcome the poor solubility of paclitaxel (**17**) and release the drug selectively into malignant tissue, an *N*-(2-hydroxypropyl)-methacrylamide (HPMA) copolymer-drug conjugate **51** with an AB_3 self-immolative dendritic linker was designed. The HPMA was used as a water-soluble group, and the dendritic peptide linked the HPMA to the parent drug (Figure 20). This enzyme-cleavable linker served as a platform with a triple payload of paclitaxel attached by a carbonate group [84]. Interestingly, HPMA not only showed improved solubility but also selectively accumulated in tumors due to its enhanced permeability and retention [85].

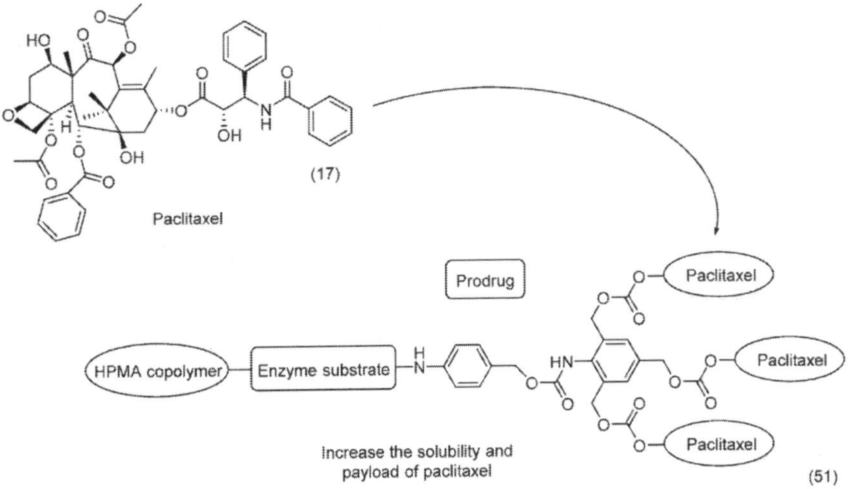

Figure 20. Copolymer-paclitaxel conjugate [84].

6. ETHER PRODRUGS

A glucuronide prodrug of the anticancer compound 10-hydroxycamptothecin (**52**) was reported by Leu *et al.* The authors used the prodrug approach to improve the solubility of the parental drug, which showed anticancer activity against various cancer cell lines [86,87]. To improve the solubility of 10-hydroxycamptothecin, a prodrug was designed using glucuronic acid as a water-soluble moiety linked by a 3-nitrobenzyl spacer. The compound **53** was 80-fold more soluble than the parental drug. The prodrug was activated by enzymatic cleavage followed by a 1,6-elimination reaction to release the parental drug (Figure 21) [88].

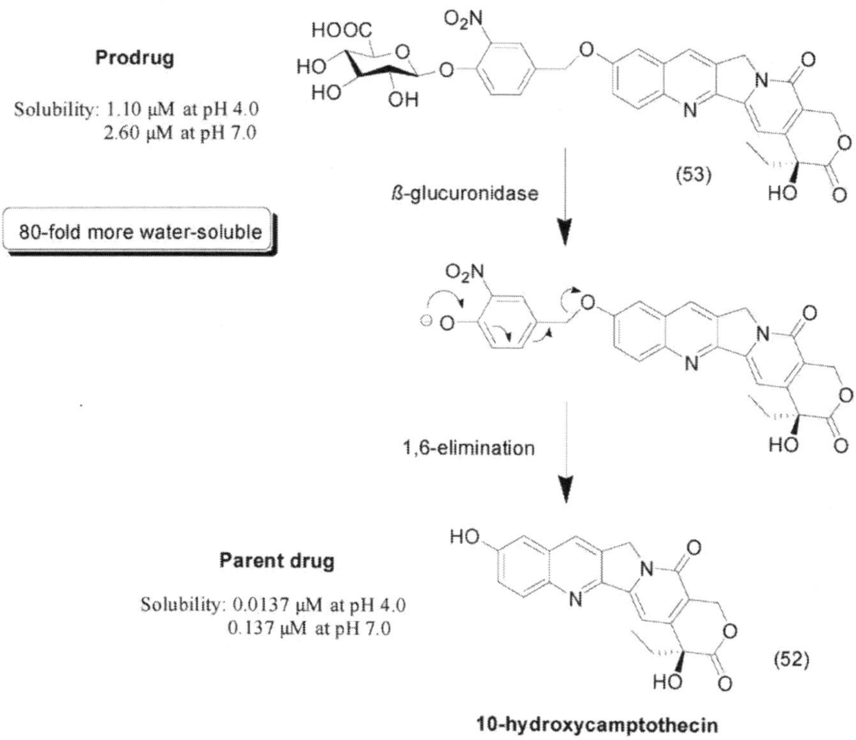

Figure 21. Prodrug release based on an enzymatic cleavage followed by a 1,6-elimination reaction [88].

Cadalene (**54**) is a flavonoid isolated from *Zelkova serrata Makino*. This molecule has a wide range of biological activities, including anticancer activities; however, its low solubility in water is a drawback [89]. Glycosylated cadalene derivatives linked by ether orethoxy spacers were synthesized as prodrugs. Prodrug **55** (Figure 22) exhibited the best solubility profile.*In vivo*, treatment with cadalene did not affect tumor volume, but treatment with the glycosylated prodrug reduced tumor volume by 50%

compared to the control group. The authors hypothesized that the better activity of the prodrug could be due to its better solubility compared to cadalene [90].

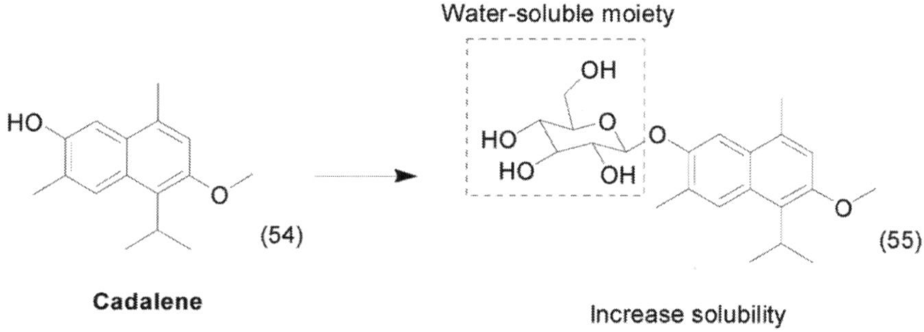

Figure 22. Cadalene prodrug with improved solubility in water [90].

7. IMINE PRODRUGS

Amphotericin B and nystatin prodrugs containing the pyridoxal phosphate as a water-soluble moiety were synthesized and evaluated. An imine group was used to allow the attachment of both subunits, and high water-soluble prodrugs **56** and**57** with solubilities up to 100 mg/mL were characterized (Figure 23) [91].

Another paper described the synthesis of a novel bisphosphonate doxorubicin prodrug to treat bone metastases. To improve the solubility of doxorubicin, the water-soluble group bisphosphonate was attached to the parental drug by a hydrazone acid-sensitive bond. Moreover, the bisphosphonate moiety acts as a bone-targeting ligand, improving the drug's selectivity. The prodrug **58** showed considerably higher stability at pH 7.4, with approximately 12% of the doxorubicin being released after 18 h of incubation (Figure 23) [92].

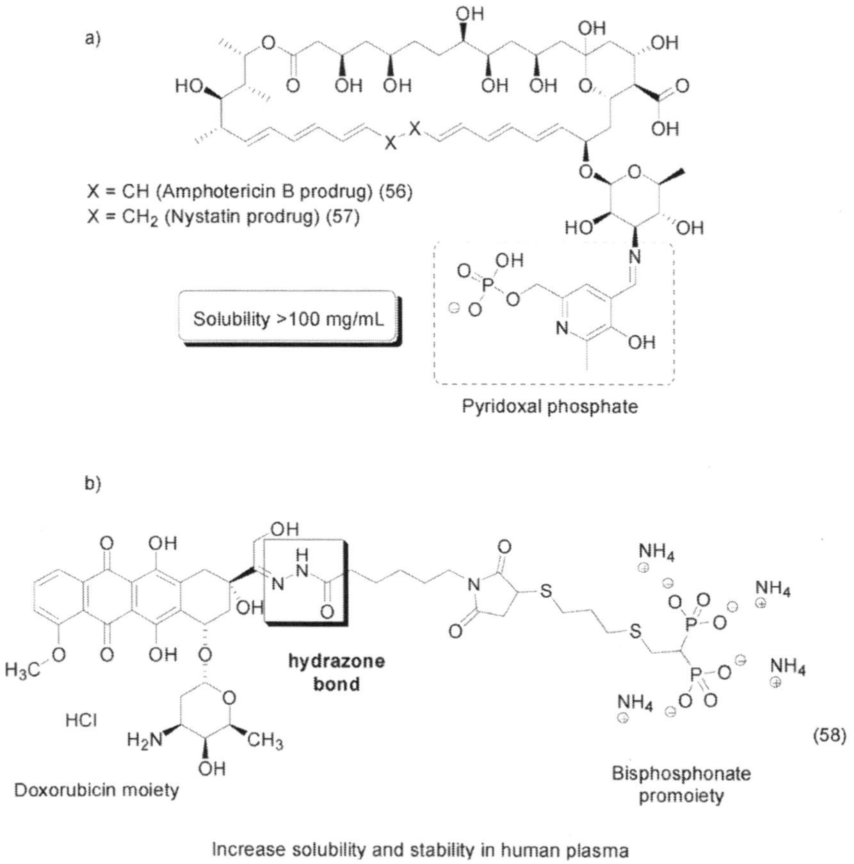

Figure 23. (**a**) Amphotericin B and nystatin prodrugs; (**b**) Bisphosphonate doxorubicin prodrug [91,92].

Synthesis of a soluble paclitaxel prodrug using an acid-sensitive hydrazone bond was reported by Moktan *et al.* The authors used a biopolymer elastin-like polypeptide (ELP) as a water-soluble group to improve the solubility of paclitaxel. ELP also enables thermal targeted delivery of paclitaxel, as hyperthermia above a specific transition temperature at the site of a tumor causes ELP to aggregate and accumulate, thus increasing the local concentration of the drug. The ELP paclitaxel prodrug **59** was able to increase solubility and cytotoxicity against a paclitaxel-resistant breast cancer cell line (Figure 24) [93].

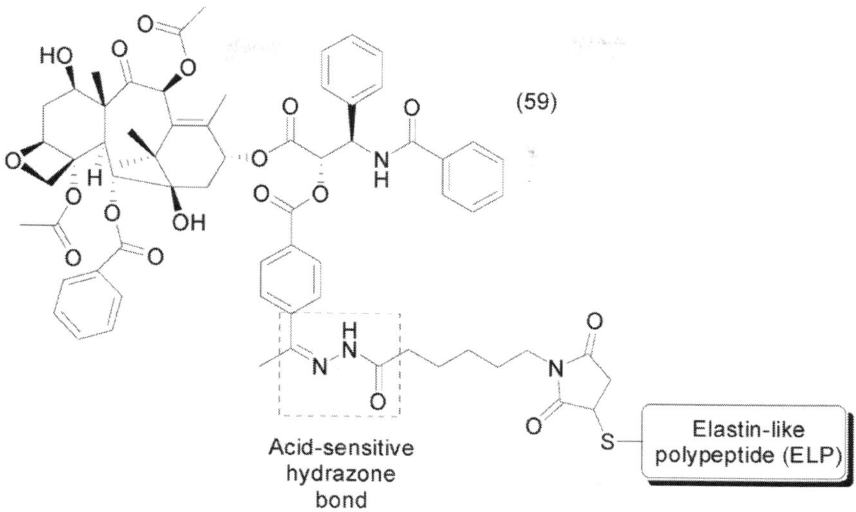

Figure 24. Paclitaxel-biopolymer-conjugated prodrug [93].

8. PHOSPHATE PRODRUGS

Phosphate ester compounds are relatively stable when free in metabolic surroundings because they have a low p*K*a, between 1 and 2. At a physiological pH of 7.0–7.4, the compounds are permanently deprotonated and therefore negatively charged but are readily activated upon complexation to various counterions in an enzyme's active site [94,95]. Intestinal alkaline phosphatases play an important role in drug metabolism by cleaving phosphate prodrugs and releasing the parental drugs [96]. Within the cell, the regeneration of the parental drug from the phosphate prodrug is difficult to determine. In most cases, this process is assumed to involve a specific enzyme or enzymes such as an esterase, an amidase, a phosphatase, or even a redox process [95,97,98]. Therefore, the phosphate-prodrug approach has been successfully used to enhance the solubility and bioavailability of the parental drug.

Degoey *et al.*, synthesized a series of prodrugs of lopinavir (LPV, **60**) and ritonavir (RTV, **61**), HIV protease inhibitors. The oxymethylphosphate and oxyethylphosphate prodrugs (**62/63** and **64/65**, respectively) had more than 700 and 1600-fold enhanced solubility in water compared to LPV and RTV, respectively (Figure 25). Pharmacokinetic studies in rats and dogs showed high plasma levels of the parent drugs after administration via the oral route [99].

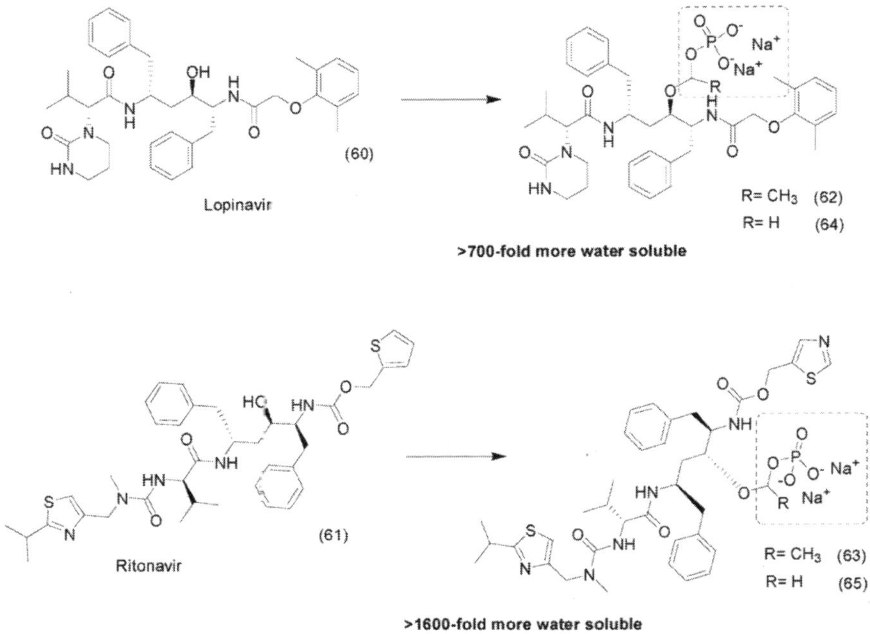

Figure 25. Structures of the prodrugs of lopinavir and ritonavir [99].

Another example of water soluble prodrug with phosphate group was synthesized by Flores-ramos *et al.*, using as prototype the compound α-6-chloro-2-(methylthio)-5-(naphthalen-1-yloxy)-1*H*-benzo[*d*]imidazole (**66**), a benzimidazole derivative. The solubility of the prodrug **67** was increased 50,000-fold when compared to its precursor by linking a disodium phosphate to the structure (Figure 26). *In vitro* and *in vivo* assays demonstrated that the prodrug have faciolicidal activity; however, administered dose *in vivo* was lower than the parent drug [100].

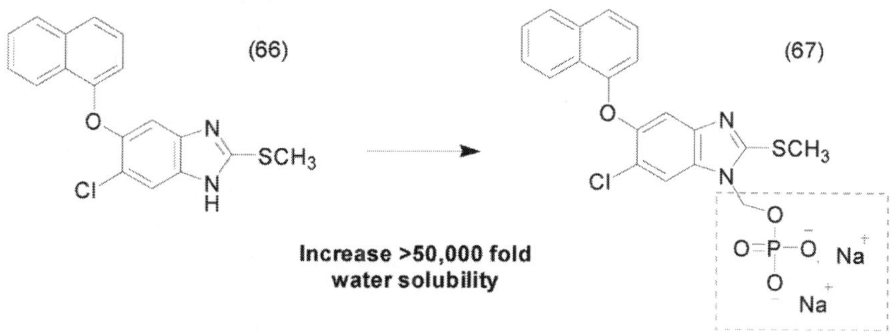

Figure 26. Benzimidazole phosphate prodrug [100].

The chalcone family is known to prevent CXCL12 chemokine from binding to its CXCR4 or CXCR7 receptors, preventing inflammatory reactions, but their poor solubility in water limits their use (9 μM/mL) [101]. To overcome this problem, chalcone prodrugs **68** were synthesized with different functional groups, such as phosphate, an L-seryl, and a sulfate. The phosphate prodrug **69** (Figure 27) showed a greater increase in solubility, to at least 3000 times greater solubility than the parent chalcone. These results allowed low-dose intranasal administration, with ≥50% inhibited eosinophil recruitment in the airways at a dose as low as 30 nmol/kg without any signs of toxicity [102].

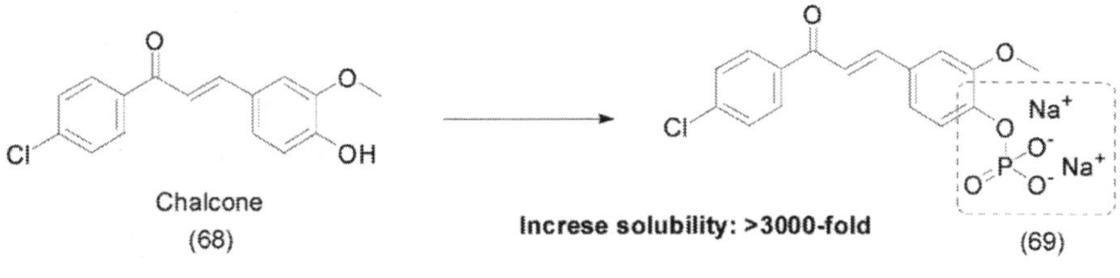

Figure 27. Chalcone-phosphate prodrugs [102].

Propofol (**70**) is another drug with high lipophilicity and is administered in an oil-in-water emulsion. To overcome this drawback, phosphate prodrugs of propofol have been generated using a phosphate group attached directly to the hydroxyl group of propofol with oxymethyl (compound **71**) and ethylenedioxy moieties (compound **72**) as spacers linked to a phosphate group (Figure 28). All these prodrugs demonstrated stability and enhanced solubility in water. Solubility of the ethyl dioxy phosphate derivative **72** was increased more than 70-fold compared to propofol. Furthermore, neither prodrug produced formaldehyde, a toxic subproduct, and thus both represent useful water-soluble prodrugs suitable for i.v. administration [103].

SB-3CT (**73**) is a selective inhibitor of matrix metalloproteinases 2 and 9 that is active in the treatment of traumatic brain injury (TBI) [63]. Its phosphate prodrug **74** (Figure 29) showed more than 2000-fold enhanced solubility in water. The prodrug was readily hydrolyzed to the active *p*-hydroxy SB-3CT (**75**), which crossed the blood-brain barrier and reached a therapeutic concentration in the brain. In an *in vivo* assay in TBI mice, the prodrug significantly decreased brain-lesion volume and improved neurological outcomes [104].

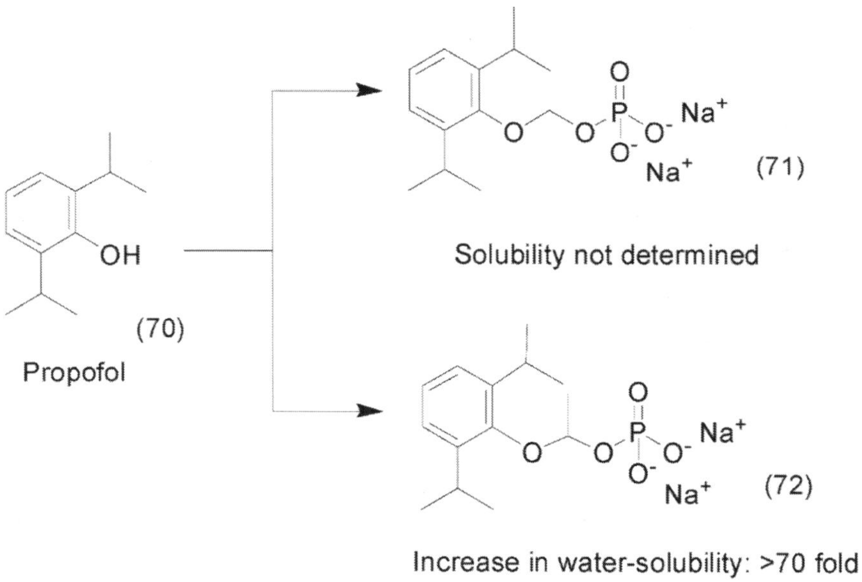

Figure 28. Propofol prodrug [103].

Compound **76** is a benzimidazole derivative with antibacterial activity that inhibits DNA gyrase and topoisomerase IV [105,106]. Due to its poor solubility in water, Dowd *et al.*, synthesized a benzimidazole phosphate-ester prodrug. The compound **77** was up to 30,000-fold more water soluble than the parent drug at physiological pH (Figure 30). Interestingly, the phosphate-ester prodrug was much less potent *in vitro* against all bacteria strains tested. Nevertheless, in the target-level assay, the prodrug exhibited similar activity to the parent drug against *S. aureus* gyrase and topoisomerase IV [107].

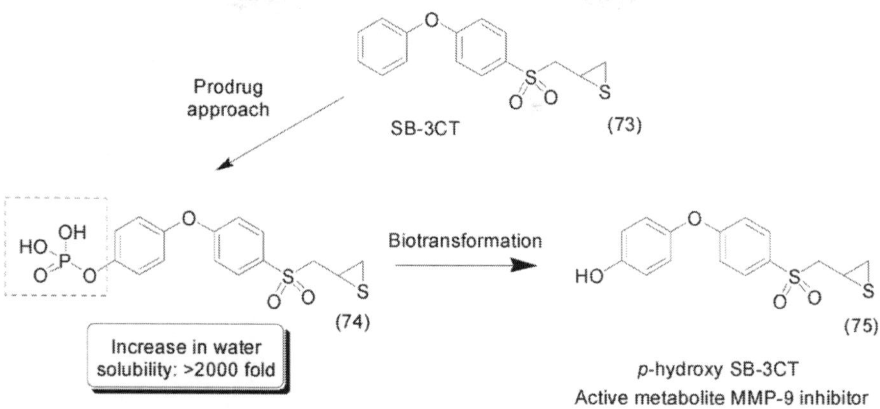

Figure 29. SB-3CT prodrug and its active metabolite [104].

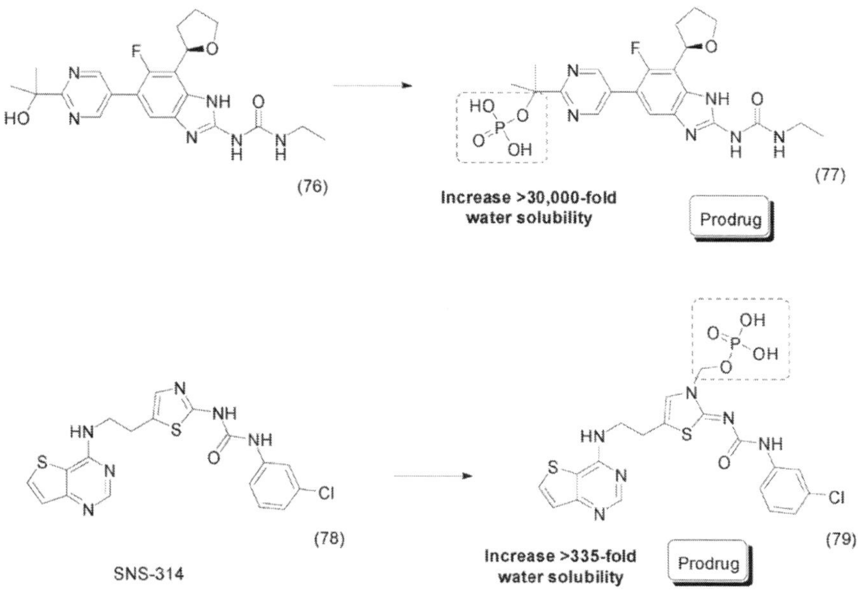

Figure 30. Benzimidazole and SNS-314 phosphate prodrugs [107,108].

Aurora serine/threonine kinases have been described as a promising anti-cancer target [109]. To overcome the poor solubility in water of SNS-314 (**78**), an aurora kinase inhibitor, a prodrug was designed with several moieties attached to the parent drug, such as acyl-oxymethylene, amine-containing acyl-oxymethylene and phosphate group. Prodrug **79**, a phosphate-ester derivative, was 335-fold more water soluble than the parent drug (Figure 30). Moreover, although the acyl-oxymethylene derivatives showed increased solubility (8–20-fold), they were much less effective than the phosphate-ester prodrug [108].

9. OTHER CHEMICAL PRODRUGS

Carbamazepine (CBZ) (**80**) is an effective anticonvulsant drug that causes less sedation and cognitive impairment than other anticonvulsants; however, its aqueous solubility is 120 µg/mL, hindering intravenous administration [110]. To overcome this problem, CBZ prodrugs **81** and **82** with improved solubility were synthesized (Figure 31). The aqueous chemical stability of **81** was superior to that of **82**. Compound **81** was dissolved in water, with apparent solubility in excess of 100 mg/mL (final pH value of 2.6). An *in vivo* pharmacokinetic assay with i.v. administration demonstrated that compound **81** was rapidly converted to CBZ [111].

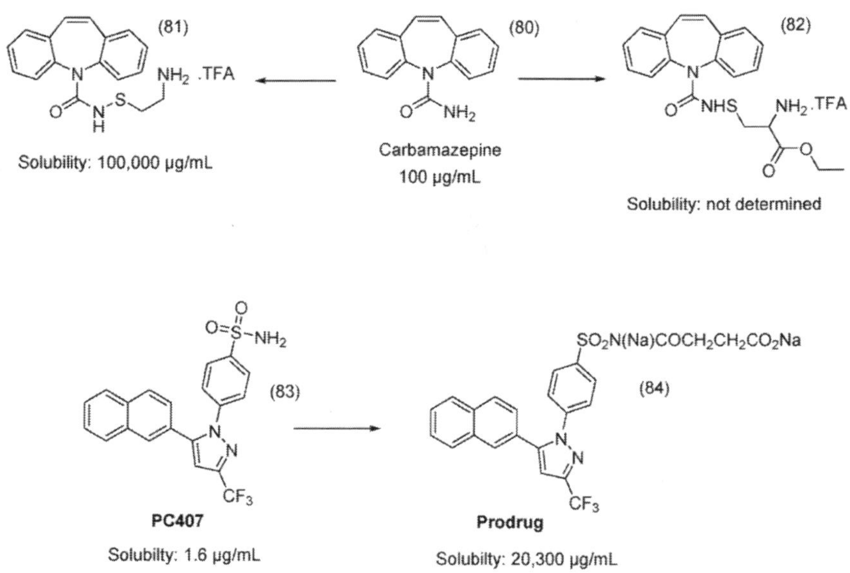

Figure 31. Carbamazepine and PC407 prodrugs [111,112].

The sulfonamide group was also utilized to develop a cyclooxygenase 2 (COX-2)-inhibitor prodrug of PC407 (**83**) for parenteral administration (Figure 31). The prodrug showed highly improved solubility, from 1.6 µg/mL to 20.3 mg/mL. Prodrug **84** showed analgesic activity *in vivo* and aqueous stability and is a promising candidate COX-2 inhibitor for injectable formulation [112].

Tetrahydrocurcumin (THCur) (**85**) is a metabolite of curcumin that shows antioxidant and anticancer activities [113,114,115]. A series of THCur prodrugs was synthesized using carboxymethylcellulose (CMC) as a water-soluble group linked by an azo bond. 4-Amino-THCur (**86**, Figure 32) has shown increased solubility in water up to 5 mg/mL and was released in the colon via azoreductase enzymes of colonic bacteria (up to 62% within 24 h). In human colon adenocarcinoma cell lines (HT-29), the prodrug was cytotoxic, with a IC_{50} of 28.67 ± 1.01 μg/mL, and selective to these cells, unlike curcumin, which was non-selective [116].

Famotidine (**87**) is a histamine H2-receptor blocker used to treat and prevent ulcers [117]. Vijayaraj *et al.*, synthesized a water-soluble prodrug of famotidine by introducing a sulphoxide group. This modification (compound **88**) increased solubility in water 6.7-fold compared to famotidine (Figure 32). The prodrug was stable at pH 1.2 and 7.4 and might be effective in ulcer therapy [118].

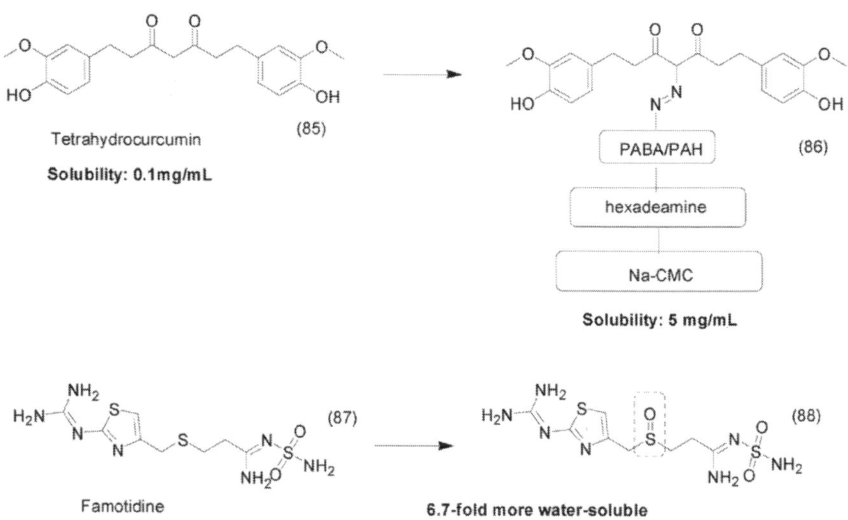

Figure 32. Tetrahydrocurcumin and famotidine prodrugs [116,118].

10. MARKETED PRODRUGS

From 2005 to 2015, several prodrugs were discovered and launched in the market (Table 1). Some of these drugs are summarized in Table 1 and will be discussed below.

Table 1. Prodrugs launched in the USA, 2005–2015 [a].

Year	Prodrug Name	Chemical Structure	Indication
2005	Nelarabine Arranon® GlaxoSmithKline plc	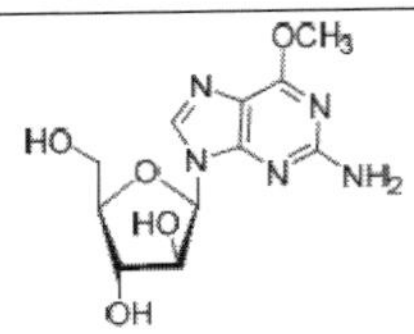	Lymphoblastic leukemia
2006	-	-	-
2007	Lisdexamfetamine dimesylate Vyvanse® New River, Inc.	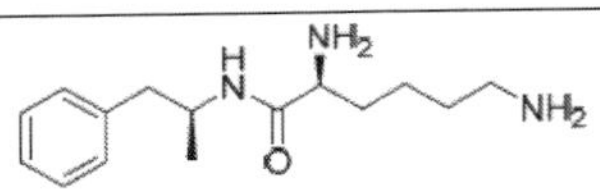	Attention–deficit/hyperactivity disorder
	Temsirolimus Torisel® Wyeth Pharm., Inc.	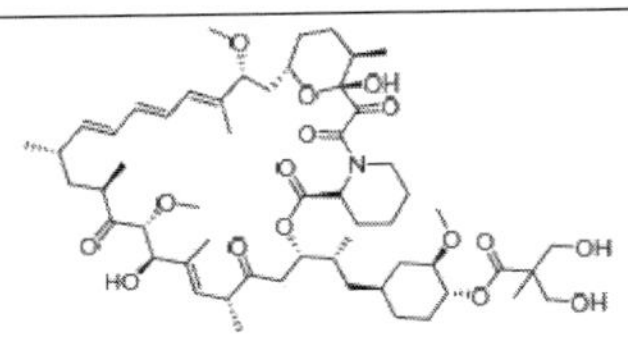	Advanced renal cell carcinoma
2008	Fesoterodinefumarate Toviaz® Pfizer, Inc.	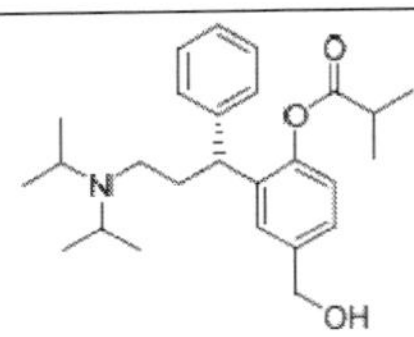	Overactive bladder disorder
	Fospropofoldisodium Lusedra® ElanPharm. plc	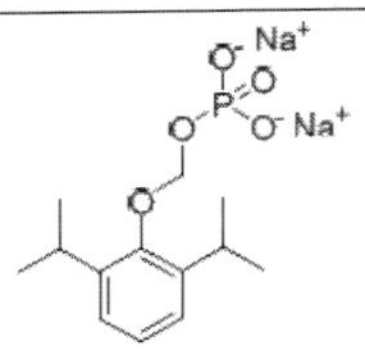	Monitored anesthesia care sedation. # discontinued
2009	Prasugrel Effient® Eli Lilly (developed with Daiichi Sankyo)	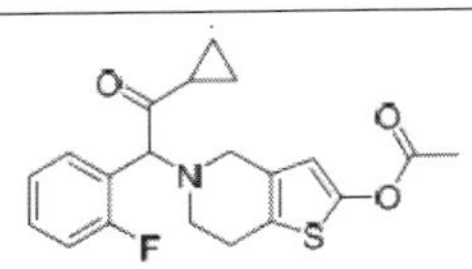	Prevention of thrombotic cardiovascular complications in acute coronary syndromes
	Romidepsin Istodax® Gloucester Pharm.	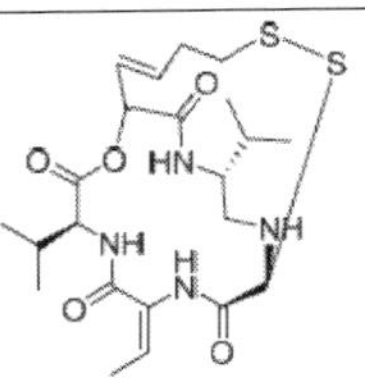	Cutaneous T cell lymphoma

Table 2. *Cont.*

Year	Prodrug Name	Chemical Structure	Indication
	Dabigatranetexilate mesylate Pradaxa® Boehringer Ingelheim GmbH	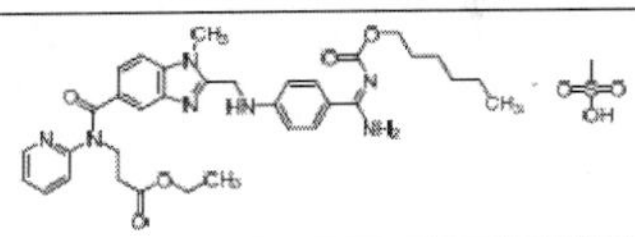	Thromboembolism acute coronary syndrome
2010	Fingolimod Gilenya® Novartis International AG	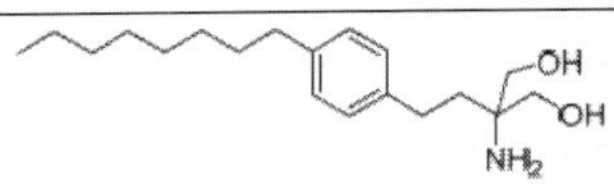	Multiple sclerosis [sphingosin-1-phosphate (S1P) agonist with cannabinoid antagonist]
	Ceftarolinefosamil Teflaro® Forest Laboratories, Inc.	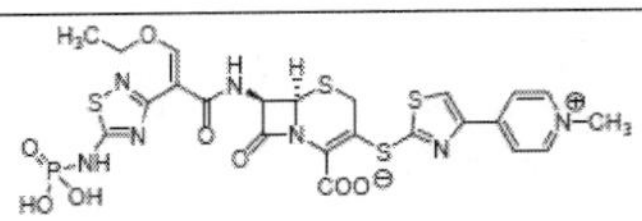	Bacterial skin infection
	Abiraterone acetate Zytiga® Janssen Biotech, Inc.	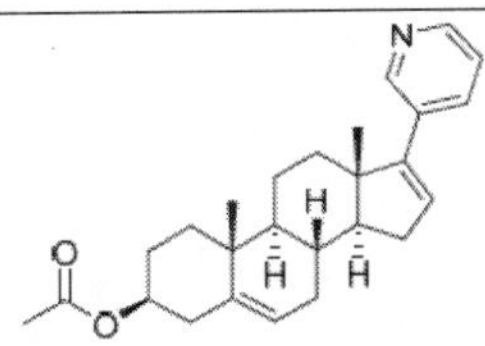	Hormone-refractory prostate cancer (17-α-hydrolase/C17,20lyase)
2011	Azilsartanmedoxomil Edarbi® Takeda Pharm.	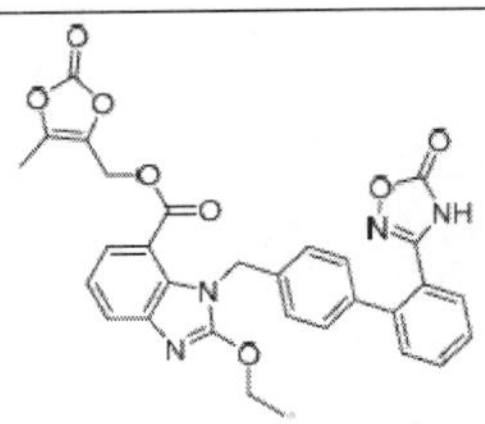	Hypertension (angiotensin II antagonist)
	Gabapentin encarbil Horizant® GlaxoSmithKline plc	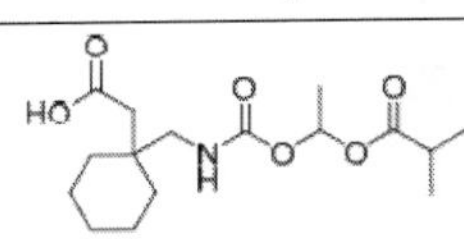	Restless leg syndrome (GABA and Ca channel modulator) new indication
2012	Tafluprost Zioptan® Merck & Co., Inc.	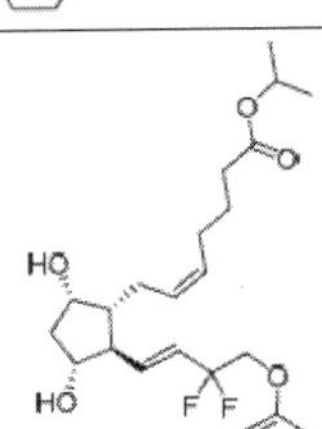	Elevated intraocular pressure (prostaglandin analog)

Table 3. *Cont.*

Year	Prodrug Name	Chemical Structure	Indication
	Dimethyl fumarate Tecfidera® Biogen Idec, Inc.	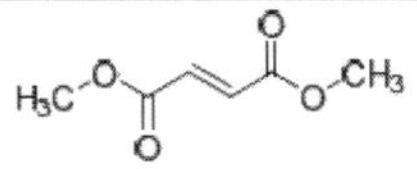	Relapsing multiple sclerosis
2013	Eslicarbazepineacetate Aptiom® Bial-Portela & Ca. S.A.	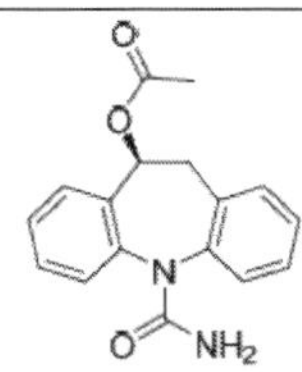	Anticonvulsant (partial-onset seizures)
	Sofosbuvir Sovaldi® Gilead Sciences, Inc.	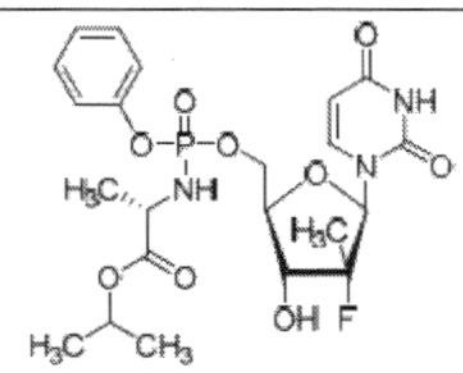	Treatment of hepatitis C virus (HCV) infection
2014	Droxidopa Northera® Lundbeck A/S	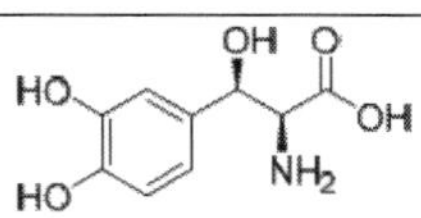	Neurogenic orthostatic hypotension
	Tedizolidphosphate Sivextro® Cubist Pharma., Inc.	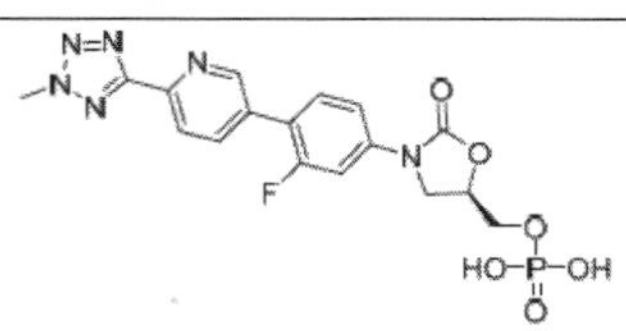	Acute bacterial skin and skin structure infections
2015	Isavuconazonium sulfate Cresemba® Astellas Pharma, Inc.	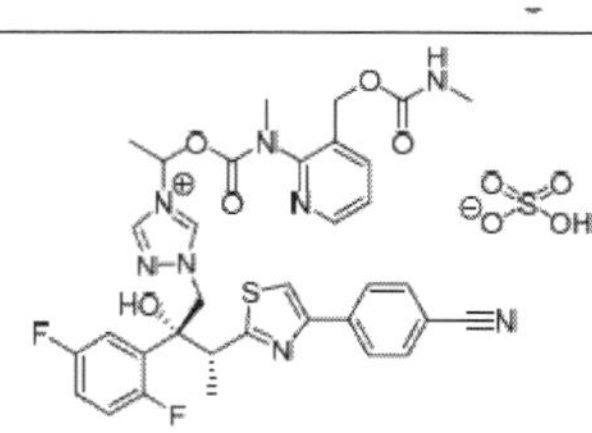	Invasive aspergillosis and invasive mucormycosis

[a] The table does not include the launch of reformulations or new indications of already-marketed prodrugs. Drugs approved till the first half 2015.

Eslicarbazepine (ESL) is an antiepileptic prodrug developed from oxcarbazepine and approved by the European Medicines Agency (EMA), the Food and Drug Administration and Health Canada as an adjunctive therapy in adults with partial-onset seizures. The development of the ESL acetate prodrug was based on the observation that the active *S*-enantiomer of ESL is produced after biotransformation at a rate 20-fold higher than the inactive *R*-enantiomer [119,120].

Fospropofol disodium, a phosphate ester prodrug of propofol, was approved by the FDA in 2008 as a sedative–hypnotic agent for monitored anesthesia-care sedation in patients who are undergoing diagnostic or therapeutic procedures. The aqueous solubility of fospropofol provides an advantage over propofol, which is available only as a lipid-containing oil-water emulsion. These advantages include lack of pain on injection, different pharmacokinetic profile due to *in vivo*conversion of fospropofol to propofol and rapid recovery from sedation [121,122].

Ceftaroline fosamil, a phosphoramidate prodrug launched in 2011 for the treatment of bacterial skin infections and bacterial pneumonia, showed 50-fold enhancement of solubility relative to the parental drug, thus enabling formulation of a marketable injectable solution [123].

Tedizolid phosphate, which has been marketed for the treatment of acute bacterial skin and skin-structure infections since 2014, is a phosphate-ester prodrug of the active compound tedizolid. The additional phosphate group provides improved aqueous solubility and thus bioavailability [124,125].

11. CONCLUSIONS

The prodrug approach has been a successful tool for improving solubility in water, as can be observed from several publications that showed up to 400,000-fold increased solubility compared to the parent drug, as described herein. This approach can make it possible to avoid discarding promising active prototypes or drugs with therapeutic uses limited by poor solubility. The rational selection of the adequate pro-moiety and the type of linkage, (e.g., ester, amide, carbamate and phosphate), may determine the prodrug selectivity, toxicity, and ideal bioconversion profile.

Furthermore, the prodrug approach could be viewed as an alternative in the early phases of drug discovery. This technique may be used to modulate pharmacokinetic properties (absorption, distribution, metabolism and excretion), as well as poor water solubility, a critical step in pre-clinical phases. The majority of prodrugs presented herein were esters and amides because esterases and amidases could activate them, releasing the parent drug. Amino acids were the most-used water-soluble pro-moieties and, as demonstrated by several reviewed authors, efficiently increase solubility in water. Nevertheless, several other chemical groups are represented herein to a lesser degree, such as glycol groups (e.g., polyethylene glycol and ethylene glycol) and glycosides.

Several prodrugs were approved in the last 10 years by the FDA, although not all prodrugs were designed to increase solubility. Nevertheless, many prodrugs were designed for this purpose, such as tedizolid phosphate, ceftaroline fosamil and fospropofol disodium. Therefore, the prodrug approach is an important tool in rational drug design to improve drug solubility in water.

ACKNOWLEDGMENTS

The authors thank the Programa de Apoio ao Desenvolvimento Científico da Faculdade de Ciências Farmacêuticas da UNESP (PADC-FCF UNESP) and Fundação de Amparo à Pesquisa do Estado de São Paulo (FAPESP 2014/02240-1; 2014/24811-0; 2014/06755-6 and 2014/14980-0) for financial support.

AUTHOR CONTRIBUTIONS

Authors D.H.J., G.F.S.F., D.E.C., T.R.F.M. and C.M.C. designed the study, analyzed and organized the literature papers. J.L.S. and C.M.C. helped rewrite the revised manuscript; J.L.S. and C.M.C. revised the manuscript and approved it in its final form. All authors edited and contributed to drafts of the manuscript. All authors approved the final form of the manuscript.

CONFLICTS OF INTEREST

The authors declare no conflict of interest.

REFERENCES

1. Rautio, J.; Kumpulainen, H.; Heimbach, T.; Oliyai, R.; Oh, D.; Järvinen, T.; Savolainen, J. Prodrugs: Design and clinical applications. Nat. Rev. Drug Discov. 2008, 7, 255–270.

2. Takagi, T.; Ramachandran, C.; Bermejo, M.; Yamashita, S.; Yu, L.X.; Amidon, G.L. A provisional biopharmaceutical classification of the top 200 oral drug products in the United States, Great Britain, Spain, and Japan. Mol. Pharm. 2006,3, 631–643.

3. Huttunen, K.; Raunio, H.; Rautio, J. Prodrugs: Design and clinical applications. Pharmacol. Rev. 2011, 63, 750–771.

4. Stella, V.J.; Nti-Addae, K.W. Prodrug strategies to overcome poor water solubility. Adv. Drug Deliv. Rev. 2007, 59, 677–694.

5. Abu-jaish, A.; Jumaa, S.; Karaman, R. Prodrug Overview. In Prodrugs Design: A New Era; Karaman, R., Ed.; Nova Publisher: Hauppauge, NY, USA, 2014; pp. 77–102.

6. Beaumont, K.; Webster, R.; Gardner, I.; Dack, K. Design of ester prodrugs to enhance oral absorption of poorly permeable compounds: Challenges to the discovery scientist. Curr. Drug Metab. 2003, 4, 461–485.

7. Testa, B. Prodrugs: Bridging pharmacodynamic/pharmacokinetic gaps. Curr. Opin. Chem. Biol. 2009, 13, 338–344.

8. Chung, M.-C.; Silva, A.T.D.A.; Castro, L.F.; Güido, R.V.C.; Nassute, J.C.; Ferreira, E.I. Latenciação e formas avançadas de transporte de fármacos. Rev. Bras. Ciênc. Farm. 2005, 41, 155–180.

9. Redasani, V.K.; Bari, S.B. Prodrug Design: Perspectives, Approaches and Applications in Medicinal Chemistry, 1st ed.; Academic Press: London, UK, 2015.

10. Zovko, M.; Zorc, B.; Novak, P.; Tepeš, P.; Cetina-Čižmek, B.; Horvat, M. Macromolecular prodrugs: XI. Synthesis and characterization of polymer–estradiol conjugate. Int. J. Pharm. 2004, 285, 35–41.

11. Ettmayer, P.; Amidon, G.L.; Clement, B.; Testa, B. Lessons Learned from Marketed and Investigational Prodrugs. J. Med. Chem. 2004, 47, 2393–2404.

12. Han, H.K.; Amidon, G.L. Targeted prodrug design to optimize drug delivery. AAPS PharmSci 2000, 2, E6.

13. Silva, A.T.D.A.; Chung, M.C.; Castro, L.F.; Güido, R.V.C.; Ferreira, E.I. Advances in prodrug design. Mini Rev. Med. Chem. 2005, 5, 893–914.

14. Williams, H.D.; Trevaskis, N.L.; Charman, S.A.; Shanker, R.M.; Charman, W.N.; Pouton, C.W.; Porter, C.J.H. Strategies to address low drug solubility in discovery and development. Pharmacol. Rev. 2013, 65, 315–499.

15. Müller, C.E. Prodrug approaches for enhancing the bioavailability of drugs with low solubility. Chem. Biodivers. 2009,6, 2071–2083.

16. Wenlock, M.C.; Austin, R.P.; Barton, P.; Davis, A.M.; Leeson, P.D. A comparison of physiochemical property profiles of development and marketed oral drugs. J. Med. Chem. 2003, 46, 1250–1256.

17. Chung, M.C.; Ferreira, E.I.; Leandro, J.; Giarolla, J.; Rando, D.G.; Almeida, A.E.; Bosquesi, P.L.; Menegon, R.F.; Blau, L. Prodrugs for the Treatment of Neglected Diseases. Molecules 2008, 616–677.

18. Lee, S.; Kim, S.M.; Lee, R.T. Thioredoxin and Thioredoxin Target Proteins: From Molecular Mechanisms to Functional Significance. Antioxid. Redox Signal. 2013, 18, 1165–1207.

19. Wipf, P.; Lynch, S.M.; Powis, G.; Birmingham, A.; Englund, E.E. Synthesis and biological activity of prodrug inhibitors of the thioredoxin-thioredoxin reductase system. Org. Biomol. Chem. 2005, 3, 3880–3882.

20. Vollmann, K.; Qurishi, R.; Hockemeyer, J.; Müller, C.E. Synthesis and properties of a new water-soluble prodrug of the adenosine A 2A receptor antagonist MSX-2. Molecules 2008, 13, 348–359.

21. De Lera Ruiz, M.; Lim, Y.; Zheng, J. Adenosine A2A receptor as a drug discovery target. J. Med. Chem. 2014, 57, 3623–3650.

22. Burbiel, J.C.; Mu, C.E. Multigram-Scale Syntheses, Stability, and Photoreactions of A 2A Adenosine Receptor Antagonists with 8-Styrylxanthine Structure: Potential Drugs for Parkinson's Disease. J. Org. Chem. 2004, 69, 3308–3318.

23. Sauer, R.; Maurinsh, J.; Reith, U.; Fülle, F.; Klotz, K.N.; Müller, C.E. Water-soluble phosphate prodrugs of 1-propargyl-8-styrylxanthine derivatives, A2A-selective adenosine receptor antagonists. J. Med. Chem. 2000, 43, 440–448.

24. McGuigan, C.; Balzarini, J. Aryl furano pyrimidines: The most potent and selective anti-VZV agents reported to date.Antiviral Res. 2006, 71, 149–153.

25. Diez-Torrubia, A.; Balzarini, J.; Andrei, G.; Snoeck, R.; de Meester, I.; Camarasa, M.J.; Velázquez, S. Dipeptidyl peptidase IV dependent water-soluble prodrugs of highly lipophilic bicyclic nucleoside analogues. J. Med. Chem.2011, 54, 1927–1942.

26. Han, H.; de Vrueh, R.L.; Rhie, J.K.; Covitz, K.M.; Smith, P.L.; Lee, C.P.; Oh, D.M.; Sadée, W.; Amidon, G. 5'-Amino acid esters of antiviral nucleosides, acyclovir, and AZT are absorbed by the intestinal PEPT1 peptide transporter.Pharm. Res. 1988, 15, 1154–1159.

27. Ohwada, J.; Ozawa, S.; Kohchi, M.; Fukuda, H.; Murasaki, C.; Suda, H.; Murata, T.; Niizuma, S.; Tsukazaki, M.; Ori, K.; et al. Synthesis and biological activities of a pH-dependently activated water-soluble prodrug of a novel hexacyclic camptothecin analog. Bioorg. Med. Chem. Lett. 2009, 19, 2772–2776.

28. Sugawara, M.; Huang, W.; Fei, Y.J.; Leibach, F.H.; Ganapathy, V.; Ganapathy, M.E. Transport of valganciclovir, a ganciclovir prodrug, via peptide transporters PEPT1 and PEPT2. J. Pharm. Sci. 2000, 89, 781–789.

29. Cao, F.; Jia, J.; Yin, Z.; Gao, Y.; Sha, L.; Lai, Y.; Ping, Q.; Zhang, Y. Ethylene Glycol-Linked Amino Acid Diester Prodrugs of Oleanolic Acid for PepT1-Mediated Transport: Synthesis, Intestinal Permeability and Pharmacokinetics.Mol. Pharm. 2012, 9, 2127–2135.

30. Cao, F.; Gao, Y.; Wang, M.; Fang, L.; Ping, Q. Propylene glycol-linked amino acid/dipeptide diester prodrugs of oleanolic acid for PepT1-mediated transport: Synthesis, intestinal permeability, and pharmacokinetics. Mol. Pharm.2013, 10, 1378–1387.

31. Zhou, G.B.; Kang, H.; Wang, L.; Gao, L.; Liu, P.; Xie, J.; Zhang, F.X.; Weng, X.Q.; Shen, Z.X.; Chen, J.; et al. Oridonin, a diterpenoid extracted from medicinal herbs, targets AML1-ETO fusion protein and shows potent antitumor activity with low adverse effects on t(8;21) leukemia in vitro and in vivo. Blood 2007, 109, 3441–3450.

32. Li, C.; Wang, E.; Cheng, Y.; Bao, J. Oridonin: An active diterpenoid targeting cell cycle arrest, apoptotic and autophagic pathways for cancer therapeutics. Int. J. Biochem. Cell Biol. 2011, 43, 701–704.

33. Shen, J.; Zhang, D.; Zhao, Z.; Jia, L.; Zheng, D.; Liu, G.; Hao, L.; Zhang, Q.; Tian, X.; Li, C.; et al. Synthesis, characterization, in vitro and in vivo evaluation of PEGylated oridonin conjugates. Int. J. Pharm. 2013, 456, 80–86.

34. Huang, G.M.; Sun, Y.; Ge, X.; Wan, X.; Li, C.B. Gambogic acid induces apoptosis and inhibits colorectal tumor growth via mitochondrial pathways. World J. Gastroenterol. 2015, 21, 6194–6205.

35. Tang, X.; Zhang, P.; Ye, H.; Zhang, C.; Shen, W.; Ping, Q. Water-soluble gambogic acid PEGylated prodrugs: Synthesis, characterization, physicochemical properties and in vitro hydrolysis. Pharmazie 2008, 63, 711–717. [PubMed]

36. Adams, J.D.; Flora, K.; Goldspiel, B.R.; Wilson, J.W.; Arbuck, S.G.; Finley, R. Taxol: A history of pharmaceutical development and current

pharmaceutical concerns. J. Natl. Cancer Inst. Monogr. 1993, 15, 141–147. [PubMed]

37. Skwarczynski, M.; Sohma, Y.; Noguchi, M.; Kimura, M.; Hayashi, Y.; Hamada, Y.; Kimura, T.; Kiso, Y. No auxiliary, no byproduct strategy for water-soluble prodrugs of taxoids: Scope and limitation of O–N intramolecular acyl and acyloxy migration reactions. J. Med. Chem. 2005, 48, 2655–2666.

38. Shimoda, K.; Kubota, N. Chemo-enzymatic synthesis of ester-linked docetaxel-monosaccharide conjugates as water-soluble prodrugs. Molecules 2011, 16, 6769–6777.

39. Estrelaa, J.M.; Ortegaa, A.; Obradora, E. Glutathione in Cancer Biology and Therapy. Crit. Rev. Clin. Lab. Sci. 2006, 43, 143–181.

40. Balendiran, G.K.; Dabur, R.; Fraser, D. The role of glutathione in cancer. Cell Biochem. Funct. 2004, 22, 343–352.

41. Gund, M.; Khanna, A.; Dubash, N.; Damre, A.; Singh, K.S.; Satyam, A. Water-soluble prodrugs of paclitaxel containing self-immolative disulfide linkers. Bioorg. Med. Chem. Lett. 2015, 25, 122–127.

42. Chen, J.; Du, W. Synthesis and Evaluation of Water-Soluble Etoposide Esters of Malic Acid as Prodrugs. Med. Chem. (Los. Angeles). 2013, 3, 740–747.

43. Ullah, N.; Huang, Z.; Sanaee, F.; Rodriguez-Dimitrescu, A.; Aldawsari, F.; Jamali, F.; Bhardwaj, A.; Islam, N.; Velázquez-Martínez, C. NSAIDs do not require the presence of a carboxylic acid to exert their anti-inflammatory effect—Why do we keep using it? J. Enzym. Inhib. Med. Chem. 2015, 24, 1–11.

44. Halen, P.; Murumkar, P.; Giridhar, R.; Yadav, M. Prodrug designing of NSAIDs. Mini Rev. Med. Chem. 2009, 9, 124–139.

45. Pawar, V.; Thosani, R.; Kanhed, A.; Giridhar, R.; Yadav, M.R. Potential of Piperazinylalkylester Prodrugs of 6-Methoxy-2-Naphthylacetic Acid (6-MNA) for Percutaneous Drug Delivery. AAPS PharmSciTech 2015, 16, 518–527.

46. Lobo, S.; Li, H.; Farhan, N.; Yan, G. Evaluation of diclofenac prodrugs for enhancing transdermal delivery. Drug Dev. Ind. Pharm. 2014, 40, 425–432.

47. Formica, J.V.; Regelson, W. Review of the biology of quercetin and related bioflavonoids. Food Chem. Toxicol. 1995, 33, 1061–1080.

48. Montenegro, L.; Carbone, C.; Maniscalco, C.; Lambusta, D.; Nicolosi, G.; Ventura, C.A.; Puglisi, G. In vitro evaluation of quercetin-3-O-acyl esters as topical prodrugs. Int. J. Pharm. 2007, 336, 257–262.

49. Kamat, J.P.; Devasagayam, T.P.A. Tocotrienols from palm oil as potent inhibitors of lipid peroxidation and protein oxidation in rat brain mitochondria. Neurosci. Lett. 1995, 195, 179–182.

50. Newaz, M.A.; Nawal, N.N.; Rohaizan, C.H.; Muslim, N.; Gapor, A. alpha-Tocopherol increased nitric oxide synthase activity in blood vessels of spontaneously hypertensive rats. Am. J. Hypertens. 1999, 12, 839–844.

51. Qureshi, A.A.; Dewitt, G.; Kramer, G.; Gapor, A. Lowering humans of serum cholesterol in hypercholesterolemic by tocotrienols. Am. J. Clin. Nutr. 1991, 53, 1021S–1026S. [PubMed]

52. Qureshi, A.A.; Bradlow, B.A.; Brace, L.; Manganello, J.; Peterson, D.M.; Pearce, B.C.; Wright, J.J.; Gapor, A.; Elson, C. Response of hypercholesterolemic subjects to administration of tocotrienols. Lipids 1995, 30, 1171–1177.

53. Qureshi, A.A.; Bradlow, B.A.; Salser, W.A.; Brace, L.D. Novel tocotrienols of rice bran modulate cardiovascular disease risk parameters of hypercholesterolomic humans. J. Nutr. Biochem. 1997, 8, 290–298.

54. Parker, B.A.; Pearce, B.C.; Clark, R.W.; Gordon, D.A.; Wright, J.J.K. Tocotrienols regulate cholesterol production in mammalian cells by post-transcriptional suppression of 3-hydroxy-3-methylglutaryl-coenzyme a reductase. J. Biol. Chem. 1993, 268, 11230–11238. [PubMed]

55. Raederstorff, D.; Elste, V.; Aebischer, C.; Weber, P. Effect of either gamma-tocotrienol or a tocotrienol mixture on the plasma lipid profile in hamsters. Ann. Nutr. Metab. 2002, 46, 17–23.

56. Guthrie, N.; Gapor, A.; Chambers, A.F.; Carroll, K.K. Inhibition of proliferation of estrogen receptor-negative MDA-MB-435 and -positive MCF-7 human breast cancer cells by palm oil tocotrienols and tamoxifen, alone and in combination. J. Nutr. 1997, 127, 544S–548S. [PubMed]

57. Nesaretnam, K.; Ambra, R.; Selvaduray, K.R.; Radhakrishnan, A.; Reimann, K.; Razak, G.; Virgili, F. Tocotrienol-rich fraction from palm oil affects gene expression in tumors resulting from MCF-7 cell inoculation in athymic mice.Lipids 2004, 39, 459–467.

58. Shah, S.J.; Sylvester, P. Gamma-tocotrienol inhibits neoplastic mammary epithelial cell proliferation by decreasing Akt and nuclear factor kappaB activity. Exp. Biol. Med. 2005, 230, 235–241.

59. Akaho, N.; Takata, J.; Fukushima, T.; Matsunaga, K.; Hattori, A.; Hidaka, R.; Fukui, K.; Yoshida, M.; Fujioka, T.; Karube, Y.; et al. Preparation and In Vivo Evaluation of a Water-Soluble Prodrug for 2R-γ-Tocotrienol and as a Two-Step Prodrug for 2,7,8-Trimethyl-2S-(β-carboxyethyl)-6-hydroxychroman (S-γ-CEHC) in Rat. Drug Metab. Dispos.2007, 35, 1502–1510.

60. Kim, S.H.; Markovitz, B.; Trovato, R.; Murphy, B.R.; Austin, H.; Willardsen, A.J.; Baichwal, V.; Morham, S.; Bajji, A. Discovery of a new HIV-1 inhibitor scaffold and synthesis of potential prodrugs of indazoles. Bioorg. Med. Chem. Lett.2013, 23, 2888–2892.

61. De Miranda, P.; Blum, M. Pharmacokinetics of acyclovir after intravenous and oral administration. J. Antimicrob. Chemother. 1983, 12, 29–37.

62. Diez-torrubia, A.; Cabrera, S.; Castro, S.D.; García-aparicio, C.; Mulder, G.; Meester, I.D.; Camarasa, M.; Balzarini, J.; Velázquez, S. Novel water-soluble prodrugs of acyclovir cleavable by the dipeptidyl-peptidase IV (DPP IV/CD26) enzyme. Eur. J. Med. Chem. 2013, 70, 456–468.

63. Zhou, J.; Tao, P.; Fisher, J.F.; Shi, Q.; Mobashery, S.; Schlegel, H.B. QM/MM Studies of the Matrix Metalloproteinase 2 (MMP2) Inhibition Mechanism of (S)-SB-3CT and its Oxirane Analogue. J. Chem. Theory Comput. 2010, 6, 3580–3587.

64. Gooyit, M.; Lee, M.; Schroeder, V.A.; Ikejiri, M.; Suckow, M.A.; Mobashery, S.; Chang, M. Selective water-soluble gelatinase inhibitor prodrugs. J. Med. Chem. 2011, 54, 6676–6690.

65. Lee, C.W.; Hong, D.H.; Han, S.B.; Jung, S.-H.; Kim, H.C.; Fine, R.L.; Lee, S.-H.; Kim, H.M. A novel stereo-selective sulfonylurea, 1-[1-(4-aminobenzoyl)-2,3-dihydro-1H-indol-6-sulfonyl]-4-phenyl-imidazolidin-2-one, has antitumor efficacy in in vitro and in vivo tumor models. Biochem. Pharmacol. 2002, 64, 473–480.

66. Lee, K.-C.; Venkateswararao, E.; Sharma, V.K.; Jung, S.-H. Investigation of amino acid conjugates of (S)-1-[1-(4-aminobenzoyl)-2,3-dihydro-1H-indol-6-sulfonyl]-4-phenyl-imidazolidin-2-one (DW2282) as water soluble anticancer prodrugs. Eur. J. Med. Chem. 2014, 80, 439–446.

67. Hirabayashi, H.; Takahashi, T.; Fujisaki, J.; Masunaga, T.; Sato, S.; Hiroi, J.; Tokunaga, Y.; Kimura, S.; Hata, T. Bone-specific delivery and sustained release of diclofenac, a non-steroidal anti-inflammatory drug, via bisphosphonic prodrug based on the Osteotropic Drug Delivery System (ODDS). J. Control. Release 2001, 70, 183–191.

68. Wang, D.; Miller, S.C.; Kopečková, P.; Kopeček, J. Bone-targeting macromolecular therapeutics. Adv. Drug Deliv. Rev.2005, 57, 1049–1076.

69. Murphy, M.B.; Hartgerink, J.D.; Goepferich, A.; Mikos, A.G. Synthesis and in vitro hydroxyapatite binding of peptides conjugated to calcium-binding moieties. Biomacromolecules 2007, 8, 2237–2243.

70. Ishizaki, J.; Waki, Y.; Takahashi-Nishioka, T.; Yokogawa, K.; Miyamoto, K.I. Selective drug delivery to bone using acidic oligopeptides. J. Bone Miner. Metab. 2009, 27, 1–8.

71. Sarig, S. Aspartic acid nucleates the apatite crystallites of bone: A hypothesis. Bone 2004, 35, 108–113.

72. Ouyang, L.; Zhang, J.; Pan, J.; Yan, L.; Guo, L. Synthesis and preliminary evaluation in vitro of novel naproxen-dendritic peptide conjugates. Drug Deliv. 2009, 16, 348–356.

73. Kaul, M.; Mark, L.; Zhang, Y.; Parhi, A.K.; LaVoie, E.J.; Pilch, D.S. An FtsZ-targeting prodrug with oral antistaphylococcal efficacy in vivo. Antimicrob. Agents Chemother. 2013, 57, 5860–5869.

74. Kaul, M.; Mark, L.; Zhang, Y.; Parhi, A.K.; Lavoie, E.J.; Pilch, D.S. Pharmacokinetics and in vivo antistaphylococcal efficacy of TXY541, a 1-methylpiperidine-4-carboxamide prodrug of PC190723. Biochem. Pharmacol. 2013, 86, 1699–1707.

75. Monneret, C. Histone deacetylase inhibitors. Eur. J. Med. Chem. 2005, 40, 1–13.

76. Dokmanovic, M.; Clarke, C.; Marks, P.A. Histone Deacetylase Inhibitors: Overview and Perspectives. Mol. Cancer Res.2007, 5, 981–989.

77. Thomas, M.; Clarhaut, J.; Tranoy-Opalinski, I.; Gesson, J.P.; Roche, J.; Papot, S. Synthesis and biological evaluation of glucuronide prodrugs of the histone deacetylase inhibitor CI-994 for application in selective cancer chemotherapy.Bioorg. Med. Chem. 2008, 16, 8109–8116.

78. Vignaroli, G.; Zamperini, C.; Dreassi, E.; Radi, M.; Angelucci, A.; Sanità, P.; Crespan, E.; Kissova, M.; Maga, G.; Schenone, S.; et al. Pyrazolo[3,4-d]pyrimidine prodrugs: Strategic optimization of the aqueous solubility of dual Src/Abl inhibitors. ACS Med. Chem. Lett. 2013, 4, 622–626.

79. Arora, A.; Scholar, E.M. Role of tyrosine kinase inhibitors in cancer therapy. J. Pharmacol. Exp. Ther. 2005, 315, 971–979.

80. Debele, T.; Peng, S.; Tsai, H.-C. Drug Carrier for Photodynamic Cancer Therapy. Int. J. Mol. Sci. 2015, 16, 22094–22136.

81. Macdonald, I.J.; Dougherty, T.J. Basic principles of photodynamic therapy. J. Porphyr. Phthalocyanines 2001, 5, 105–129.

82. Noguchi, M.; Skwarczynski, M.; Prakash, H.; Hirota, S.; Kimura, T.; Hayashi, Y.; Kiso, Y. Development of novel water-soluble photocleavable protective group and its application for design of photoresponsive paclitaxel prodrugs.Bioorg. Med. Chem. 2008, 16, 5389–5397.

83. Binderup, E.; Björkling, F.; Hjarnaa, P.V.; Latini, S.; Baltzer, B.; Carlsen, M.; Binderup, L. EB1627: A soluble prodrug of the potent anticancer cyanoguanidine CHS828. Bioorg. Med. Chem. Lett. 2005, 15, 2491–2494.

84. Erez, R.; Segal, E.; Miller, K.; Satchi-Fainaro, R.; Shabat, D. Enhanced cytotoxicity of a polymer-drug conjugate with triple payload of paclitaxel. Bioorg. Med. Chem. 2009, 17, 4327–4335.

85. Satchi-Fainaro, R.; Hailu, H.; Davies, J.W.; Summerford, C.; Duncan, R. PDEPT: Polymer-Directed Enzyme Prodrug Therapy. 2. HPMA Copolymer-β-lactamase and HPMA Copolymer-C-Dox as a Model Combination. Bioconjugate Chem. 2003, 14, 797–804.

86. Wani, M.C.; Ronman, P.E.; Lindley, J.T.; Wall, M.E. Plant antitumor agents. 18. Synthesis and biological activity of camptothecin analogs. J. Med. Chem. 1980, 23, 554–560.

87. Ling, Y.H.; Andersson, B.S.; Nelson, J.A. DNA topoisomerase I as a site of action for 10-hydroxycamptothecin in human promyelocytic leukemia cells. Cancer Biochem. Biophys. 1990, 11, 23–30. [PubMed]

88. Leu, Y.L.; Chen, C.S.; Wu, Y.J.; Chern, J.W. Benzyl ether-linked glucuronide derivative of 10-hydroxycamptothecin designed for selective camptothecin-based anticancer therapy. J. Med. Chem. 2008, 51, 1740–1746.

89. Kim, J.H.; Lee, H.J.; Yeon, S.C.; Choi, D.H.; Lee, S.S.; Kang, J.K.; Chae, C.H.; Paik, N.W.; Lee, K.H.; Cho, M.H. Antioxidative Effects of 7-Hydroxy-3-Methoxy-Cadalene Extracted from Zelkova serrata on Butanone-induced Oxidative Stress in A/J Mice. Phyther. Res. 2004, 18, 425–427.

90. Lee, H.Y.; Kwon, J.T.; Koh, M.; Cho, M.H.; Park, S.B. Enhanced efficacy of 7-hydroxy-3-methoxycadalene via glycosylation in in vivo xenograft study. Bioorg. Med. Chem. Lett. 2007, 17, 6335–6339.

91. Day, T.P.; Sil, D.; Shukla, N.M.; Anbanandam, A.; Day, V.W.; David, S.A. Imbuing aqueous solubility to amphotericin B and nystatin with a vitamin. Mol. Pharm. 2010, 8, 297–301.

92. Hochdörffer, K.; Abu Ajaj, K.; Schäfer-Obodozie, C.; Kratz, F. Development of novel bisphosphonate prodrugs of doxorubicin for targeting bone metastases that are cleaved pH dependently or by cathepsin B: Synthesis, cleavage properties, and binding properties to hydroxyapatite as well as bone matrix. J. Med. Chem. 2012, 55, 7502–7515.

93. Moktan, S.; Ryppa, C.; Kratz, F.; Raucher, D. A thermally responsive biopolymer conjugated to an acid-sensitive derivative of paclitaxel stabilizes microtubules, arrests cell cycle, and induces apoptosis. Investig. New Drugs 2012, 30, 236–248.

94. Schultz, C. Prodrugs of biologically active phosphate esters. Bioorg. Med. Chem. 2003, 11, 885–898.

95. Wiemer, A.; Wiemer, D. Prodrugs of phosphonates and phosphates: Crossing the membrane barrier. Top. Curr. Chem.2014, 360, 115–160.

96. Ghosh, K.; Mazumder Tagore, D.; Anumula, R.; Lakshmaiah, B.; Kumar, P.P.B.S.; Singaram, S.; Matan, T.; Kallipatti, S.; Selvam, S.; Krishnamurthy, P.; et al. Crystal structure of rat intestinal alkaline phosphatase—Role of crown domain in mammalian alkaline phosphatases. J. Struct. Biol. 2013, 184, 182–192.

97. Hecker, S.J.; Erion, M.D. Prodrugs of Phosphates and Phosphonates. J. Med. Chem. 2008, 51, 2328–2345.

98. Krise, J.P.; Stella, V.J. Prodrugs of Phosphates, Phosphonates, and Phosphinates. Adv. Drug Deliv. Rev. 1996, 19, 287–310.

99. Degoey, D.A.; Grampovnik, D.J.; Flosi, W.J.; Marsh, K.C.; Wang, X.C.; Klein, L.L.; Mcdaniel, K.F.; Liu, Y.; Long, M.A.; Kati, W.M.; Molla, A.; Kempf, D.J. Water-Soluble Prodrugs of the Human Immunodeficiency

Virus Protease Inhibitors Lopinavir and Ritonavir Water-Soluble Prodrugs of the Human Immunodeficiency Virus Protease Inhibitors Lopinavir and Ritonavir. J. Med. Chem. 2009, 52, 2964–2970.

100. Flores-ramos, M.; Ibarra-velarde, F.; Hernández-campos, A.; Vera-montenegro, Y.; Jung-cook, H.; Cantó-alarcón, G.J.; Misael, L.; Castillo, R. A highly water soluble benzimidazole derivative useful for the treatment of fasciolosis. Bioorg. Med. Chem. Lett. 2014, 24, 5814–5817.

101. Hachet-Haas, M.; Balabanian, K.; Rohmer, F.; Pons, F.; Franchet, C.; Lecat, S.; Chow, K.Y.C.; Dagher, R.; Gizzi, P.; Didier, B.; et al. Small Neutralizing Molecules to Inhibit Actions of the Chemokine CXCL12. J. Biol. Chem. 2008, 283, 23189–23199.

102. Gasparik, V.; Daubeuf, F.; Hachet-Haas, M.; Rohmer, F.; Gizzi, P.; Haiech, J.; Galzi, J.L.; Hibert, M.; Bonnet, D.; Frossard, N. Prodrugs of a CXC chemokine-12 (CXCL12) neutraligand prevent inflammatory reactions in an asthma model in vivo. ACS Med. Chem. Lett. 2012, 3, 10–14.

103. Kumpulainen, H.; Järvinen, T.; Mannila, A.; Leppänen, J.; Nevalainen, T.; Mäntylä, A.; Vepsäläinen, J.; Rautio, J. Synthesis, in vitro and in vivo characterization of novel ethyl dioxy phosphate prodrug of propofol. Eur. J. Pharm. Sci.2008, 34, 110–117.

104. Lee, M.; Chen, Z.; Tomlinson, B.N.; Gooyit, M.; Hesek, D.; Juárez, M.R.; Nizam, R.; Boggess, B.; Lastochkin, E.; Schroeder, V.A.; et al. Water-Soluble MMP-9 Inhibitor Reduces Lesion Volume after Severe Traumatic Brain Injury.ACS Chem. Neurosci. 2015, 10, 1658–1664.

105. Charifson, P.S.; Grillot, A.L.; Grossman, T.H.; Parsons, J.D.; Badia, M.; Bellon, S.; Deininger, D.D.; Drumm, J.E.; Gross, C.H.; LeTiran, A.; et al. Novel dual-targeting benzimidazole urea inhibitors of DNA gyrase and topoisomerase IV possessing potent antibacterial activity: Intelligent design and evolution through the judicious use of structure-guided design and structure-activity relationships. J. Med. Chem. 2008, 51, 5243–5263.

106. Grossman, T.H.; Bartels, D.J.; Mullin, S.; Gross, C.H.; Parsons, J.D.; Liao, Y.; Grillot, A.-L.; Stamos, D.; Olson, E.R.; Charifson, P.S.; Mani, N. Dual targeting of GyrB and ParE by a novel aminobenzimidazole class of antibacterial compounds. Antimicrob. Agents Chemother. 2007, 51, 657–666.

107. O'Dowd, H.; Shannon, D.E.; Chandupatla, K.R.; Dixit, V.; Engtrakul, J.J.; Ye, Z.; Jones, S.M.; O'Brien, C.F.; Nicolau, D.P.; Tessier, P.R.; et al. Discovery and Characterization of a Water-Soluble Prodrug of a Dual Inhibitor of Bacterial DNA Gyrase and Topoisomerase IV. ACS Med. Chem. Lett. 2015, 6, 822–826.

108. Oslob, J.D.; Heumann, S.A.; Yu, C.H.; Allen, D.A.; Baskaran, S.; Bui, M.; Delarosa, E.; Fung, A.D.; Hashash, A.; Hau, J.; et al. Water-soluble prodrugs of an Aurora kinase inhibitor. Bioorg. Med. Chem. Lett. 2009, 19, 1409–1412.

109. Mortlock, A.A.; Keen, N.J.; Jung, F.H.; Heron, N.M.; Foote, K.M.; Wilkinson, R.W.; Green, S. Progress in the development of selective inhibitors of aurora kinases. Curr. Top. Med. Chem. 2005, 5, 807–821.

110. Chen, X.Q.; Venkatesh, S. Miniature device for aqueous and non-aqueous solubility measurements during drug discovery. Pharm Res 2004, 21, 1758–1761.

111. Hemenway, J.N.; Nti-Addae, K.; Guarino, V.R.; Stella, V.J. Preparation, characterization and in vivo conversion of new water-soluble sulfenamide prodrugs of carbamazepine. Bioorg. Med. Chem. Lett. 2007, 17, 6629–6632.

112. Liu, H.-F.; Wan, N.; Huan, M.-L.; Jia, Y.-Y.; Yuan, X.-F.; Zhou, S.-Y.; Zhang, B.-L. Enhanced water-soluble derivative of PC407 as a novel potential COX-2 inhibitor injectable formulation. Bioorg. Med. Chem. Lett. 2014, 24, 4794–4797.

113. Lin, J.K.; Pan, M.H.; Lin-Shiau, S.Y. Recent studies on the biofunctions and biotransformations of curcumin. Biofactors2000, 13, 153–158.

114. Pan, M.H.; Huang, T.M.; Lin, J.K. Biotransformation of curcumin through reduction and glucuronidation in mice.Drug Metab. Dispos. 1999, 27, 486–494. [PubMed]

115. Kawakishi, S. Antioxidative of the P-Diketone Moiety in the Mechanism of Tetrahydrocurcumin. Science 1996, 52, 519–525.

116. Plyduang, T.; Lomlim, L.; Yuenyongsawad, S.; Wiwattanapatapee, R. Carboxymethylcellulose-tetrahydrocurcumin conjugates for colon-specific delivery of a novel anti-cancer agent, 4-amino tetrahydrocurcumin. Eur. J. Pharm. Biopharm. 2014, 88, 351–360.

117. Patel, A.R.; Vavia, P.R. Preparation and Evaluation of Taste Masked Famotidine Formulation Using Drug/β-cyclodextrin/Polymer Ternary Complexation Approach. AAPS PharmSciTech 2008, 9, 544–550.

118. Vijayaraj, S.; Omshanthi, B.; Anitha, S.; Sampath Kumar, K.P. Synthesis and Characterization of Novel Sulphoxide prodrug of Famotidine. Indian J. Pharm. Educ. Res. 2014, 48, 35–44.

119. Rocamora, R. A review of the efficacy and safety of eslicarbazepine acetate in the management of partial-onset seizures. Ther. Adv. Neurol. Disord. 2015, 8, 178–186.

120. Ben-Menachem, E. Eslicarbazepine Acetate: A Well-Kept Secret? Epilepsy Curr. 2010, 10, 7–8. [PubMed]

121. Mueller, S.W.; Moore, G.D.; Maclaren, R. Fospropofol Disodium for Procedural Sedation: Emerging Evidence of its Value? Clin. Med. Insights Ther. 2010, 2, 513–522.

122. Boules, R.; Szkiladz, A.; Nogid, A. Fospropofol disodium (lusedra) injection for anesthesia-care sedation: a clinical review. Pharm. Ther. 2012, 37, 395–422.

123. Ishikawa, T.; Matsunaga, N.; Tawada, H.; Kuroda, N.; Nakayama, Y.; Ishibashi, Y.; Tomimoto, M.; Ikeda, Y.; Tagawa, Y.; Iizawa, Y.; et al. TAK-599, a novel N-phosphono type prodrug of anti-MRSA cephalosporin T-91825: Synthesis, physicochemical and pharmacological properties. Bioorg. Med. Chem. 2003, 11, 2427–2437.

124. Flanagan, S.; Bartizal, K.; Minassian, S.L.; Fang, E.; Prokocimer, P. In vitro, in Vivo, and clinical studies of tedizolid to assess the potential for peripheral or central monoamine oxidase interactions. Antimicrob. Agents Chemother. 2013, 57, 3060–3066. [PubMed]

125. Rybak, J.M.; Roberts, K. Tedizolid Phosphate: A Next-Generation Oxazolidinone. Infect. Dis. Ther. 2015, 4, 1–14.

Chapter 3

Polymer-Based Prodrugs: Improving Tumor Targeting and the Solubility of Small Molecule Drugs in Cancer Therapy

Sonja Dragojevic, Jung Su Ryu and Drazen Raucher *

Department of Biochemistry, University of Mississippi Medical Center, 2500 North State Street Jackson, MS 39216, USA

ABSTRACT

The majority of anticancer drugs have poor aqueous solubility, produce adverse effects in healthy tissue, and thus impose major limitations on both clinical efficacy and therapeutic safety of cancer chemotherapy. To help circumvent problems associated with solubility, most cancer drugs are now formulated with co-solubilizers. However, these agents often also introduce severe side effects, thereby restricting effective treatment and patient quality of life. A promising approach to addressing problems in anticancer drug solubility and selectivity is their conjugation with polymeric carriers to form polymer-based prodrugs. These polymer-based prodrugs are macromolecular carriers, designed to increase the aqueous solubility of antitumor drugs, can enhance bioavailability. Additionally, polymer-based prodrugs approach exploits unique features of tumor physiology to passively facilitate intratumoral accumulation, and so improve chemodrug pharmacokinetics and pharmacological properties. This review introduces basic concepts of polymer-based prodrugs, provides an overview of currently emerging synthetic, natural, and genetically engineered polymers that now deliver anticancer drugs in preclinical or clinical trials, and highlights their major anticipated applications in anticancer therapies.

Keywords: polymer based prodrugs; anticancer drugs; macromolecules; drug delivery; cancer therapy; genetically engineered biopolymers; synthetic and natural polymers

1. INTRODUCTION

Cancer is one of the major fatal diseases, with current treatments of limited therapeutic efficacy, despite great progress in a range of approaches to tumor eradication, including chemotherapy. The mechanism of action of anti-cancer drugs relies on arresting the cell cycle and rapidly killing all proliferating cells. This killing includes non-cancerous cells, such as bone marrow, gut epithelia, lymphatic system, red blood cells and hair follicles. One of the main challenges in current chemotherapeutic treatments is drug toxicity to healthy organs due to lack of selectivity. To achieve therapeutic efficacy, it is often necessary to administer high doses of drugs because of the drugs' physicochemical properties, such as short circulation time in the blood plasma. The problems associated with the low specificity and poor pharmacokinetics and pharmacodynamics of chemotherapeutic drugs are well known; however, another frequently encountered difficulty is their poor aqueous solubility, which restricts bioavailability.

Intractable problems besetting drug solubility in aqueous solutions have imposed serious clinical constraints on ascertaining effective therapeutic plasma concentrations so as to achieve appropriately targeted pharmacological responses. At present, ensuring the presence of a therapeutic concentration of poorly soluble drugs in systemic circulation requires the administration of high concentrations of potent chemotherapeutics, with their solubility related to particle size. Smaller particles have surface area to volume ratio increases, which allows greater interaction with the solvent [1]. The most common methods for reducing particle size are micronization and nano suspension [2]. Micronization relies on mechanical stress, using a technique such as spray drying to break up drug aggregates. Nano suspension relies on forcing the drug suspension under pressure through a nano aperture valve, resulting in drug nanoparticles that can then be stabilized by surfactants. Other new techniques for reducing particle size are also forthcoming, such as sonocrystalization, supercritical fluid processes and wet milling [3,4].

Despite successes within each of these approaches, scientists continue to encounter major concerns related to particle size reduction. These concerns largely arise owing to the strong tendency for particles to agglomerate subsequent to size reduction. Such agglomerates are likely to reduce the drug's therapeutic activity.

A drug's chemical structure and property determine possible approaches for improving their solubility in aqueous solutions. As drugs may be weakly acidic, weakly basic or hydrophobic, appropriate approaches for increasing their rate of solubility may include pH adjustment, chemical modification, micellar solubilization, hydrotropy, solid dispersion, complexations, microemulsions and co-solvency [5]. Each method possesses advantages, but may also introduce challenges in formulation design, with consequent limitations in therapeutic efficacy. For example, drugs that can be dissolved in water by adjusting their pH may precipitate in blood upon intravenous administration, due to the strong buffering capacity of blood, with its pH between 7.2 and 7.4 [6], leading to emboli.

Similarly, another promising approach for improving drug solubility relies upon mixing hydrophobic drugs with co-solutes, which can open more hydrophilic moieties and so facilitate solubility. However, co-solutes often exert more toxic and immunogenic effects, limiting their effective use at doses required for eradicating cancer cells.

One typical example is Paclitaxel, a microtubule stabilizing agent that has been deemed one of the most significant advances in anticancer therapy [7,8]. This promising anti-cancer agent, however, has poor water solubility—but it does dissolve in organic solvents. Among the formulations used to increase Paclitaxel solubility for intravenous administration is a vehicle containing a blend of ethanol and Chremophor, or polyethoxyated castor oil. The side effects of Chremophor, however, include neurotoxicity, nephrotoxicity, vasodilatation lethargy, and hypersensitivity reactions mediated by histamine release, requiring premedication with corticosteroids and antihistamines [9]. To reduce these side effects' intensity and so increase Paclitaxel's safety, better strategies for improving its aqueous solubility are urgently needed. Moreover, even were micronization and/or co-solubilization to improve Paclitaxel's solubility, a method for selective delivery of this small molecule to tumor cells is equally important, to prevent its quick distribution to healthy cells in all organs immediately after administration.

Due to the side effects of both anticancer drugs and their co-solutes, a critical need exists for improvements in such drug pharmaceutical properties as solubility, stability, and distribution. Toward this end, many drug delivery systems have been developed, including: (1) lipid based drug delivery systems, such as liposomes, niosomes and solid lipid particles; (2) nanotechnology-based drug delivery systems, including micelles, fullerenes, dendrimers, carbon nanotubes, quantum dots, metal-based nanoparticles, and nanofibers; and (3) polymeric delivery systems, including polymeric micelles, polymeric vesicles, polymer protein or polymer small drug conjugates [10].

At administration, antitumor drugs are quickly distributed in the body, exhibiting fast renal clearance, short durations of action, and considerable dissipation, with only a fraction reaching cancer cells. By conjugating these small molecule agents with a high molecular weight polymer, an increase in their solubility and bioavailability can be attained [11] so as to slow renal elimination, lengthen construct duration in blood circulation and improve the construct's pharmacokinetic profile. A further contribution of such water-soluble macromolecular carriers lies in their ability to facilitate a more targeted drug delivery by exploiting the disorganized vasculature and poor lymphatic drainage that characterizes tumor physiology. This targeting occurs through a phenomenon collectively known as the enhanced permeation and retention effect (EPR) Figure 1. [12]. Exploiting the EPR effect for tumor targeting in this way permits polymer-drug conjugates to be applied at higher doses than possible for any free drug, as polymer-bound drugs will preferentially accumulate and release within tumors. With more drug delivered specifically to a tumor, drug toxicity and side effects will be reduced, and overall therapeutic outcomes improved.

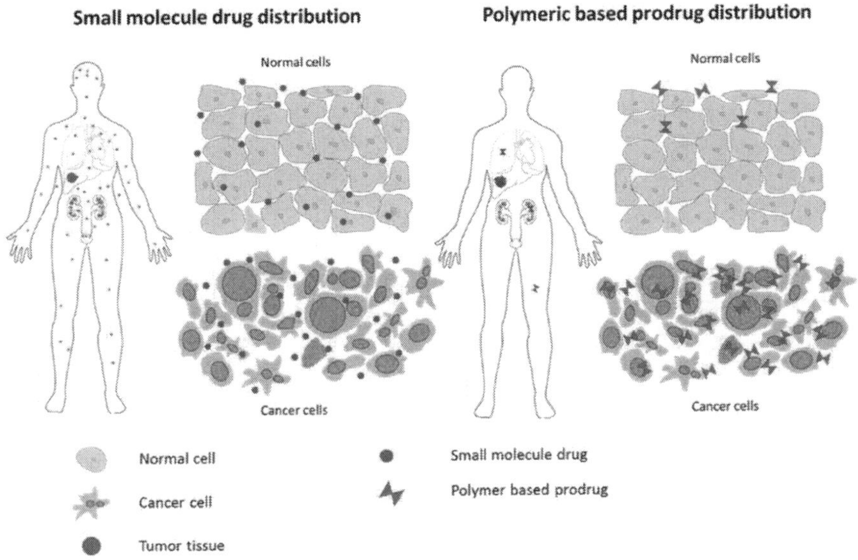

Figure 1. Small molecule drugs are distributed not only in tumor tissue, but are also widely distributed in all healthy tissues, resulting in adverse side effects (**left**); The polymeric macromolecular carrier is designed to spare healthy tissue from toxicity by preferentially accumulating within tumors, primarily by exploiting their disorganized tumor vasculature and poor lymphatic drainage through the enhanced permeation and retention (EPR) effect (**right**).

Advances in polymer science have resulted in the synthesis and design of polymers with unique properties. While once confined to roles as co-solutes or mechanical scaffolds, by which to stabilize and solubilize drugs or to control drug release, respectively, polymers now have sophisticated and advanced properties that can be engaged for the development of novel drug delivery vehicles. Prime examples of these advances can be seen in the recently developed synthetic and genetically engineered polymers that can respond to changes in such environmental conditions as pH or temperature. These new constructs currently underlie active research in the development of innovative approaches for active tumor cell targeting [13,14,15].

To optimize their potential as drug carriers, polymers must be well characterized, reproducibly synthesized in quantities, and of a purity acceptable for clinical application. They should also contain the appropriate functional groups needed for coupling or incorporation with specific drugs. Not least, they should be composed of biocompatible, non-immunogenic, and biodegradable materials, remain stable in circulation, but readily release their chemodrug cargo intratumorally or intracellularly. Polymer drug carriers should also maintain a low variability in particle size distribution to ensure the homogeneity of polymer-drug conjugates. By their potential to meet these requirements, synthetic, natural and genetically engineered protein polymers can each contribute urgently needed properties to contemporary drug delivery technology.

Synthetic polymers are human-made polymers, usually fabricated with organic solvents such as polyethylene glycol or*N*-(2-Hydroxypropyl) methacrylamide [16]. Although the structure of synthetic polymers can be easily modified to permit their wide use in drug delivery, problems associated with toxic degradation products or residual toxic components that arise during the process of chemical synthesis have been reported [17]. Natural polymers include polysaccharides, such as dextran and chitosan, polypeptides, and polynucleotides. The use of these polymers as drug-delivery agents is widely reported in cell culture and preclinical trials; however, they have not yet reached clinical trials despite offering greater biocompatibility and biodegradability than synthetic polymers. Genetically engineered polypeptides, developed for controlled and targeted drug delivery, offer polymers with uniform composition and precise molecular weight as compared with chemically synthesized products. These constructs are currently under further development within preclinical trials [15].

In summary, cancer therapies that rely on the systemic administration of anticancer agents face considerable obstacles, with poor solubility main among them. This insolubility, through the consequently lowered bioavailability of chemotherapeutic agents, has to date frustrated treatment efforts and led to the taxing outcome of moderate cancer treatment efficacy. The coupling of antitumor drugs with polymeric drug carriers, whether synthetic, natural or genetically engineered, can help overcome agent insolubility, rendering cancer treatment more effective. These polymers can potentially provide vehicles through which to improve application and target delivery of small, poorly water-soluble chemodrugs so as to improve therapeutic effectiveness and reduce the toll of treatment for cancer patients.

2. CLASSIFICATION OF POLYMERS

2.1. Synthetic Polymers

The advancement of modern polymer chemistry has allowed the modification of polymer structures, rendering "a tailored polymer for the need" possible. For this reason, a wide variety of synthetic polymers, characterized by variations in the main chain as well as in side chains, are currently available (Table 1).

2.1.1. Polyethylene Glycol (PEG)

One of the most well-known synthetic polymers, polyethylene glycol (PEG), is a linear polyether compound with ethylene glycol repeats. PEG terminates with hydroxyl group, and can readily be covalently conjugated to many peptides, proteins or drugs (Figure 2). For the coupling of PEG and drugs, agents such as dicyclohexyl carbodiimide (DCC), 1-ethyl-3-(3-dimethylaminopropyl) carbodiimide (EDC), or *N*-hydroxysuccinimide (NHS) esters are commonly used.

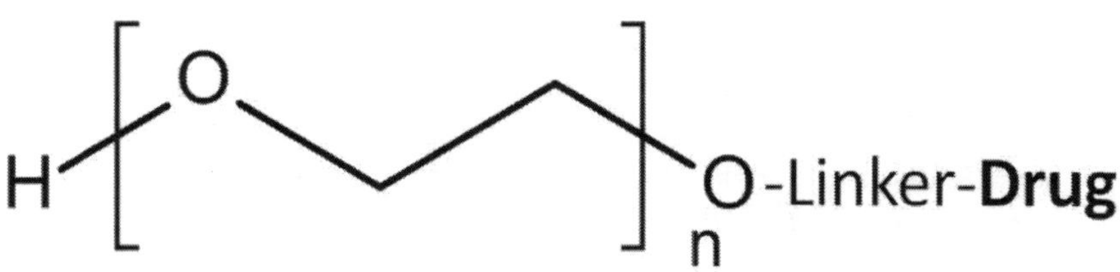

Figure 2. Structure of polyethylene glycol (PEG) and its drug conjugation.

The incorporation of PEG with a drug has the advantage of increasing solubility and therefore bioavailability of the resulting conjugates.

Additionally, immunogenic responses upon exposure to peptides/proteins may be diminished by the presence of PEG, which can prevent exposure of the peptide's epitopes and thereby mask peptide/protein recognition by the reticuloendothelial system (RES) [18,19]. The main problem with using PEG, however, lies in its poor drug-loading efficiency, with the limited number of reactive groups in PEG often restricting its broad applications as a drug carrier [20].

Table 1. Recent applications of synthetic polymers in prodrug delivery.

Polymer	Application	(co)Polymer	Prodrug	Linker(bond)	Ref.
PEG	Gene delivery	Polyethylenimine(PEI)-PEG	cDNA of herpes simplex virus thymidine kinase gene (HSVtk) and granulocyte-macrophage colony-stimulating factor (GM-CSF)	N.D.	[21]
	Tumor targeting	octreotide(Phe)-PEG	Paclitaxel		[22]
	Improved stability and intracellular drug release	methoxy PEG-b-(poly(2-(diisopropylamino) ethyl methacrylate-co-aminopropyl methacrylamide) (PEDPA)	Cis-aconityl-doxorubicin		[23]
	Tumor targeting	PEG	Fusion toxin consisting of the anti-EpCAM DARPin Ec1 and a domain I-deleted variant of ETA (ETA)	rhinovirus 3C model protease – cleavage linker	[24]
	Drug delivery	3,3'-dithiodipropionic acid functionalized PEG-b-poly(l-lysine) (mPEG-b-P(LL-DTPA))	Paclitaxel	Disulfide	[25]
	Multidrug resistance	D-α-tocopherol PEG succinate (TPGS)	Paclitaxel	Disulfide	[26]
	Improved the therapeutic efficacy	β-CD, PEG	Doxorubicin	hydrazone	[27]
	Tumor targeting	PEG	Paclitaxel		[28]
	Multidrug resistance	PEG-poly(D,L-lactide)	4-(N)-stearoyl Gemcitabine		[29]
	Theranosis	PEG-polylactic acid (PEG-PLA) l	Dicyanomethylene-4H-pyran-5-CPT	Disulfide	[30]
	Tumor targeting	PEG monomethyl ether (mPEG)	Artesunate	Ester	[31]
	Tumor targeting	PEG monomethyl ether	Camptothecin	Disulfide	[32]
	Tumor targeting	PEG	Camptothecin	Disulfide	[33]
	Tumor targeting	PEG2000	Paclitaxel	MMP2-cleavable linker	[34]
	Nanogel	PLGA-PEG-PLGA	PEGylated Taxol		[35]
	Drug delivery	mPEG-b-P(ATMC-co-DTC)	Doxorubicin	Hydrazone	[36]
HPMA	improving anticancer therapy	(mPEG₅₀₀₀-b-p(HPMAmLac₂-r-AzEMA)	Doxorubicin-glucuronide prodrug (DOX-propGA3)	Glucuronide (β-glucuronidase cleavable linker)	[37]
	Prevent metastasis	HPMA copolymer	E-selectin binding peptide (Esbp)-doxorubicinor (KLAKLAK)₂		[38]
	Drug delivery	HPMA copolymer	H1-S6A, F8A peptide	GFLG (Cathepsin cleavage linker)	[39]
	Tumor targeting	HPMA copolymer	Doxorubicin	GFLG and MMP cleavable linker	[40]
	Improved the therapeutic efficacy	HPMA copolymer	Doxorubicin, 5-FU	Hydrazone, GFLG	[41]
	Theranosis	Star polymer: poly(amido amine) (PAMAM) dendrimers and HPMA	Doxorubicin or TAMRA fluorescent dye	Hydrazone	[42]
	Improved Bioavailability	Star polymer: poly(amido amine) (PAMAM) dendrimers and HPMA	Pirarubicin	Hydrazone	[43,44]
	Drug delivery	HPMA copolymer	Iodine-125	Hydrazone	[45]
	Drug delivery	HPMA copolymer	Paclitaxel, Gemcitabine	GFLG	[46]
	Drug delivery	Starch + HPMA copolymer	Camptothecin		[47]
	Gene delivery	galactosylated 2-hydroxypropylmethacrylamide-s-3-guanidinopropyl methacrylamide (HPMAs–GPMA)	shRNA		[48]
	Drug delivery	HPMA	Indium-111, Yttrium-90		[49]
	Improved the therapeutic efficacy	multiblock poly HPMA	Gemcitabine, Paclitaxel and Doxorubicin	GFLG	[50]
	Theranosis	HPMA copolymer	Zinc protoporphyrin		[16,51]
SMA	Photodynamic therapy	SMA	Zinc protoporphyrin	Amide	[52]
PLGA	Improved Bioavailability	PLGA	Gemcitabine	Amide	[53]
PGG	Improved Bioavailability	PGG	Paclitaxel	Glutamic acid	[54,55]

In spite of these limitations, PEG has been used to deliver a variety of chemical, peptide and gene therapies. For example, Chen *et al.*, produced a Paclitaxel prodrug conjugated to poly (ethylene glycol)-b-poly (l-lysine) by connecting the drug to the polymer via a disulfide bond. This allowed the Paclitaxel to be released from the polymer conjugates when reduced by glutathione, widely utilized as the ideal trigger in redox-responsive delivery systems [25]. Other examples are found in the study of Cai *et al.* [56]. They modified PEG with functional dendrimers to fabricate linear-dendritic copolymers with multiple reactive groups, increasing the number of reactive groups in PEG-based polymers and improving the treatment efficacy of the conjugated drugs [56]. Xia *et al.* also developed a strategy conjugating PEG with multi-numbers of drugs, such as paclitaxel and cisplatin, to maximize the treatment efficiency of the drug-polymer conjugates [57].

2.1.2. The *N*-(2-hydroxypropyl) Methacrylamide (HPMA) Copolymer

After early uses of a HPMA monomer/homopolymer as a plasma expander, HPMA's conjugation with doxorubicin in 1994 clinical trials [58] to its extensive application and investigation for cancer treatment, especially as a copolymer. HPMA, like PEG, is a water soluble, biocompatible, non-immunogenic polymer. Unlike PEG, however, the large number of pendent functional groups in HPMA copolymers permits the conjugation of many drugs (Figure 3). Further, HPMA facilitates activating or releasing the compound from the polymer backbone within a targeted region by manipulation of their bond, a degradable linker. For example, Huang *et al.* used a GFLG cathepsin cleavable linker to deliver a c-Myc inhibitory peptide, the H1-S6A, F8A peptide. An increase in nuclear drug accumulation and an inhibition of tumor growth were achieved with a construct composed of a nuclear localization sequence (NLS) and the polymer complex (R8NLS-HPMAcoPolymer-GFLG-H1-S6A, F8A) [39]

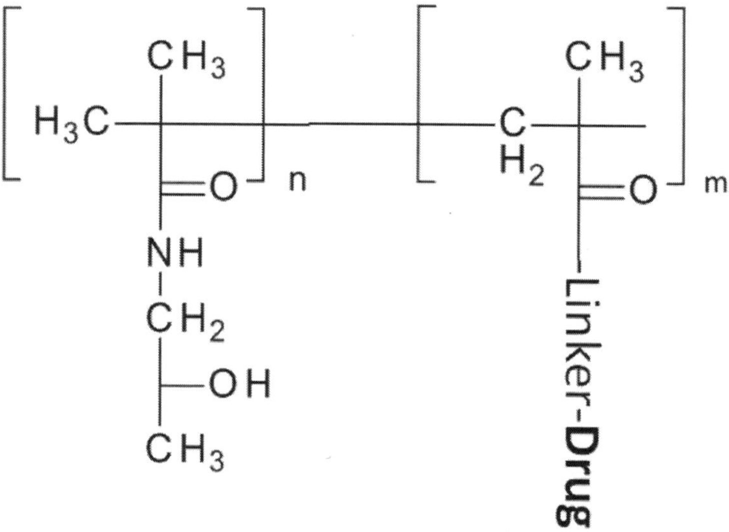

Figure 3. Structure of *N*-(2-hydroxypropyl) Methacrylamide (HPMA) and its drug conjugation.

In another study, Etrych *et al.* applied an acid-sensitive hydrazone linker to bridge the HPMA-poly (amide amine) (PAMAM) dendrimers copolymer [42]. In this study, a fluorescent dye, TAMRA, was also conjugated to this dendrimer to visualize the feasibility of the polymer carrier by *in vitro* optical imaging.

2.1.3. Poly (Styrene-Co-Maleic Acid/Anhydride) (SMA)

Poly (styrene-co-maleic acid/anhydride) (SMA) is an alternating copolymer composed of styrene and maleic anhydride (Figure 4). Since SMA forms micelles with a hydrophobic styrenic core and a hydrophilic maleic acid surface, a stable and rate-controllable release of micelles can allow surface modification for use in tissue targeting [59]. Although research on SMA is not as active as that for PEG or HPMA, studies have continued to aim at producing prodrugs using this polymer. In 1985, Maeda *et al.* developed a conjugate between SMA and neocarzinostatin (SMANCS) by using an amide bond between the peptide's terminal amino group and the SMA carboxylanhydride [60]. This work yielded a decreased clearance rate and an increase in the tumor concentration of neocarzinostatin. The enhanced bioavailability and tumor accumulation achieved by this conjugate led to the effective treatment of a variety of solid tumors [61,62]. Recently, Maeda *et al.* developed a styrene-maleic acid-copolymer conjugated with ZnPP (SMA–ZnPP) for photodynamic therapy [63]. In addition to increased tumor accumulation achieved through the EPR effect, light irradiation increased the complex's cytotoxicity up to six fold.

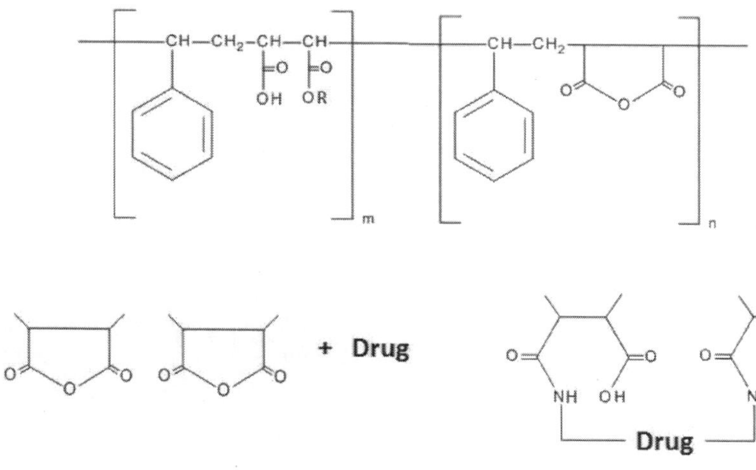

Figure 4. Structure of Poly (styrene-co-maleic acid/anhydride) (SMA) and its drug conjugation [61].

2.1.4. The Polyglutamic Acid Polymer

Bae *et al.* [64] developed a polymer nanogel based on poly (γ-glutamic acid) (γ-PGA) (Figure 5). Poly (γ-glutamic acid) (γ-PGA), a highly anionic polymer which exhibits excellent biocompatibility and non-cytotoxicity, is naturally synthesized in microbial species, especially *bacilli*. Bae and colleagues developed a thiolated γ-PGA nanogel for doxorubicin delivery that, when used to treat MCF 7 breast cancer cells, showed controlled drug release behavior and higher toxicity as compared to free doxorubicin. Their approach suggests that thiolated γ-PGA nanogel may be a promising drug delivery vehicle in anticancer therapy [64]. Furthermore, in the study of Yu *et al.* [54], the PGA was added with another glutamic acid to each glutamic acid in the polymer backbone producing a poly(L-γ-glutamyl-glutamine)(PGG). This modification shows additional hydrophilicity on a PGG-paclitaxel conjugate as well as improved paclitaxel-loading efficiency in comparison to PGA-paclitaxel conjugates.

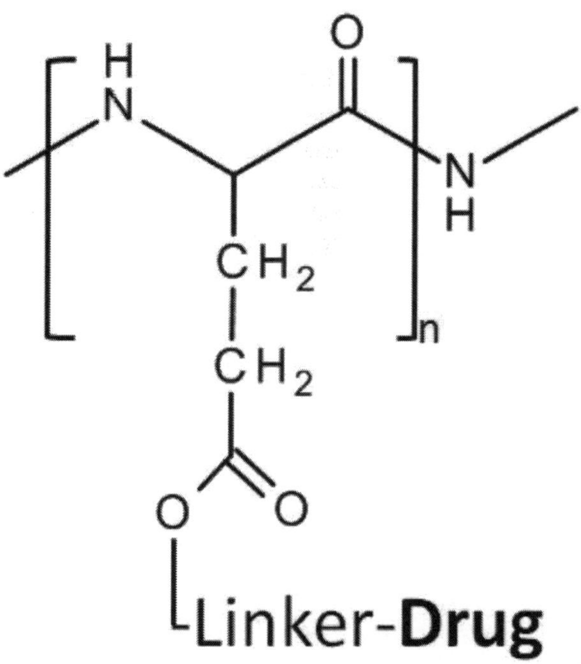

Figure 5. Structure of PGA and its drug conjugation.

Seth *et al.*, [65]used poly (γ-glutamic acid) (γ-PGA) in a mouse melanoma tumor model to deliver both the immuno-stimulating agent, toll-like receptor-7 (TLR-7) agonist-imiquimod and the chemotherapeutic agent paclitaxel. These water insoluble therapeutic agents formed crystalline microstructures that were dispersed in the IgPGA matrix. When the complex was administered by intra-tumoral injection, this combined treatment caused synergistic tumor regression as compared with effects observed when each of the compounds was used alone. This study demonstrates an excellent example and practical approach for the use of water soluble polymers, such as poly (γ-glutamic acid), in formulating micro-dispersions and delivering combinations of water insoluble drugs [65].

2.1.5. Poly (Lactic-Co-Glycolic Acid) (PLGA)

PLGA is a FDA approved biodegradable and biocompatible copolymer of poly lactic acid (PLA) and poly glycolic acid (Figure 6) [66]. This copolymer has been conjugated to gemcitabine via amide bonding [53]. Through the biodegradability of PLGA, the conjugated gemcitabine and released from the complex slowly and over time. Stability and *in vitro* efficacy testing have shown that PLGA-gemcitabine complex has advantages over free Gemcitabine.

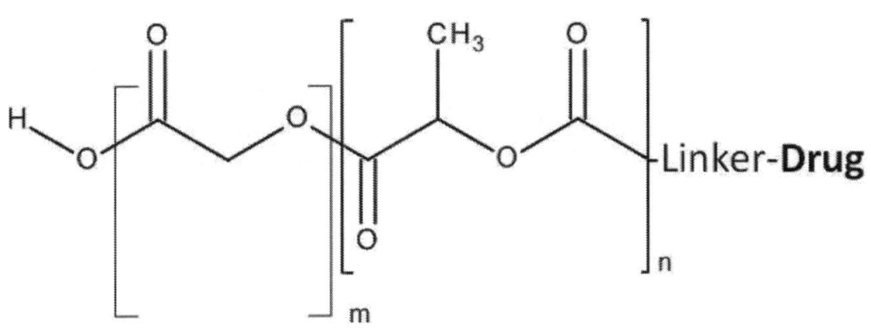

Figure 6. Structure of PLGA and its drug conjugation.

To further improve their drug delivery properties, synthetic polymers have often been combined with natural polymers. The synthetic polymer polyethylene glycol (PEG) and the natural polymer chitosan have been applied as conjugates in both*in vitro* and *in vivo* studies; however, the secondary stability and short half-life of chitosan-based micelles has hampered outcomes under *in vivo* conditions.

Emami *et al.* [67] used chitosan (see Natural Polymers, below) to develop a novel polymeric micelle with which to deliver paclitaxel. The new micelle, a tocopherol succinate–chitosan–polyethylene glycol–folic acid (TS–CS–PEG–FA), loaded with paclitaxel, yielded both an increased PTX accumulation in tumor tissue and increased toxicity to cancer cells as compared to free PTX. The increased tumor accumulation and favorable pharmacokinetics and tissue distribution observed in this study shows that chitosan-derived micelles have the potential to serve as an effective drug delivery vehicle for paclitaxel [67].

2.2. Natural Polymers

Natural polymers are commonly classified in three major classes: polysaccharides, polypeptides and polynucleotides. As polynucleotides are not used as drug carriers in anticancer therapy, only polysaccharides and polypeptides will be discussed below. For a discussion of polynucleotides in cancer therapy, however, interested readers are pointed to the work of Cheng *et al.* [68], Gorska *et al.* [69], and Xiang *et al.* [70].

Polysaccharides and polypeptides can be successfully conjugated with most anti-tumor drugs. Moreover, some of these polymers themselves possess antitumor characteristics. For example, chitosan, through such physical and chemical properties as small particle size and positive surface charge, exerts antitumor activity by disrupting the tumor cell membrane and thereby inducing apoptosis. The results of *in vitro* and *in vivo* studies that used chitosan as an anticancer agent against sarcomas and hepatocarinomas have furthered the view that chitosan can provide a promising antitumor agent, advancing its progress toward clinical trials [71,72].

Natural polymers can now be exploited as a foundation to which poorly soluble drugs can be attached and delivered to tumor sites. Through their considerable versatility the use of these polymers can occasion less drug loss and lower toxicity to healthy tissue.

2.2.1. Chitosan

Chitosan, a linear polymer composed of interchangeably organized, deacetylated β-(1-4)-linked D glucosamine units and of acetylated *N*-acetyl-D-glucosamine units (Figure 7), is widely used in drug delivery systems. The sole polysaccharide polycation used as a polymer, chitosan binds to negatively charged moieties[20]. Recently, Nogueira and colleagues produced chitosan nanoparticles loaded with the anticancer agent Methotrexate (MTX) [73]. This poorly soluble anticancer agent is extremely potent, but poses serious risks of adverse effects to normal cells, including kidney failure, neurotoxicity and mucositis [73,74,75].

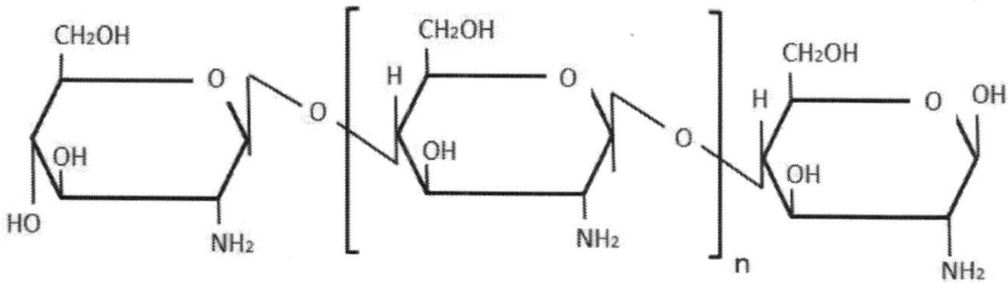

Figure 7. Chitosan structure, composed of deacetylated β-(1-4)-linked D-glucosamine units and of acetylated *N*-acetyl-D-glucosamine units.

To overcome chitosan's toxicity and poor solubility, Nogueira *et al.* [73] constructed chitosan-based nanoparticles encapsulating methotrexate (MTX–CS–NPs). Next, to make the chitosan-based nanoparticles pH responsive, these researchers added the amino acid based amphiphile surfactant, 77KS. The incorporation of this pH-sensitive moiety permitted a pH-dependent, controlled release of Methotrexate. *In vitro* experiments were conducted in MCF7 breast cancer cells, with promising results. Future *in vivo* studies will provide more information about the clinical potential of chitosan-based nanoparticles.

Another promising approach using grafted-chitosan polymeric micelles carrying two drugs was developed by Nam *et al.*[76]. The chitosan derivative O-carboxymethyl chitosan (OCMCh), with enhanced solubility, was conjugated first with α-tocopherol, forming α-tocopherol *O*-carboxymethyl (TOC). This chitosan polymer was then ligated with doxorubicin and an anti-human epidermal growth factor receptor 2 [HER2] target peptide, producing the final construct, HPTOC–DOX polymeric micelles. An *in vitro* study that employed this construct against the SK-BR-3 cell line showed a synergistic effect of tocopherol and dox. Furthermore, in an *in vivo* study with SK-BR-3 tumor bearing mice, the anti-HER targeting peptide enhanced not only cellular uptake, but also therapeutic efficacy.

The advantages of chitosan include that it is biodegradable and biocompatible, as well as able to efficiently transport polar drugs across an epithelial surface. In addition, chitosan oligomer derivatives of 3–6 kDa are considered relatively nontoxic. Chitosan's major disadvantage is its limited solubility at physiological pH, although this can be overcome by combining it with other polymers and/or by chemical modification.

2.2.2. Dextran

Dextran, structured of glucose molecules that form a complex branched glucan chain of variable length, is a water soluble polymeric saccharide (Figure 8). Dextran-based microspheres have been explored in *in vitro, in vivo* and most recently, in clinical trials of breast, colon, hepatic, and pancreatic tumors (please see Table 2. below). These microspheres can satisfy the vast majority of drug carrier requirements. Their primary use in cancer research is to improve the solubility of insoluble anti-tumor drugs.

The slow release systems associated with microspheres have made dextran microspheres (MS) the vehicle of choice for the delivery of mitomycin C (MMC), a promising, potent anticancer agent that functions through bioreductive activation [77]. Many acute as well as chronic toxicities are associated with MMC, however, severely limiting its clinical application. To overcome these impediments and achieve targeted drug delivery to the hypoxic regions of solid tumors, Cheung and colleagues developed a targeted approach composed of oxidized dextran microspheres loaded with MMC [77]. In an *in vitro* model of breast cancer cell-line, EMT6, these researchers applied a doxorubicin-MS therapy, followed by a MMC-MS therapy. This sequential application of two antitumor drugs resulted in a synergistically effective combination of cancer cell killing.

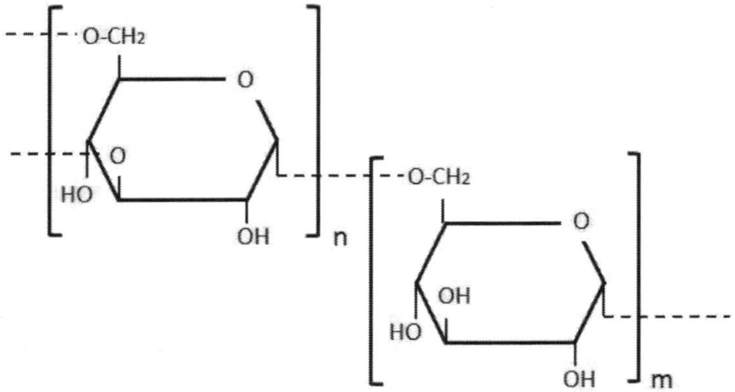

Figure 8. Dextran structure, composed of glucose molecules forming a complex branched glucan chain.

Table 2. Current clinical trials of investigational anticancer drugs based on polymer.

Polymer	Name	Drug	Status	Ref.
Dextran	OsteoDex	Alendronate	Phase I	[21,78,79]
	Somadex	Somatostatin	Phase I	[78]
PEG	NK105	Paclitaxel	Phase III	[21]
	NK102	SN-38	Phase II	[80,81]
	NC-6004	Cisplatin	Phase III	[21]
	NC-4016	Dachplatin	Phase I	[21]
	NC-6300	Doxorubicin	Phase I	[21]
poly-L-glutamate	paclitaxel poliglumex, CT-2103	Paclitaxel	Phase III	[82,83]
Cyclodextrin-PEG copolymer	CRLX101	Camptothecin	Phase II	[84]

Dextran microspheres exhibit biodegradability and biocompatibility, as well as being non-immunogenic and nontoxic, all important factors for future clinical applications. These microspheres, easily filtered and neutral, are above all water soluble. However, their clinical translation has been severely hampered by the vexing problem that, when conjugated to antitumor drugs, dextran is rendered immunogenic and non-biodegradable [20]. Other negative effects of dextran polymer use currently include platelet dysfunction, anaphylaxis, and pulmonary, as well as cerebral edema. Studies to determine whether or not these problems can be successfully resolved are now needed.

For nanoparticle assembly, it is important that an anticancer agent/drug and polymer have opposite ionic properties. Therefore, the incorporation of ionic macromolecules with oppositely charged ionic-molecules to form stable nanoparticles can be beneficial [85]. In order to combine advantages and circumvent disadvantages, Lee *et al.* [85] conjugated chitosan and dextran into self-organized nanoparticles for doxorubicin tumor targeted delivery.

2.2.3. Pullulan

Pullulan, a natural polysaccharide composed of maltotriose units structured as α-1-4; α-1-6-glucan (Figure 9.) and a liver specific biopolymer [86], is biodegradable, nontoxic, non-mutagenic, and noncarcinogenic. Its low immunogenicity and satisfactory solubility in aqueous and several organic solvents have together led to a range of applications in cancer drug delivery.

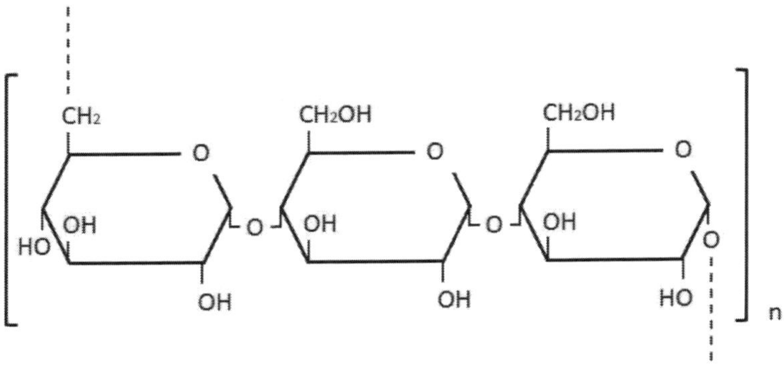

Figure 9. Pullulan structure, composed of maltotriose units.

Scomparin *et al.* [87] conducted an *in vitro* study with two pullulan bioconjugates. The aim of their study was to design an anticancer polymer therapeutic for targeted tumor cell therapy. The group synthesized the two pullulan derivatives by conjugating pullalan with (1) doxorubicin and (2) doxorubicin and folic acid, with the latter used as a moiety for targeting folic receptors overexpressed in some tumor cell lines. To achieve proper functionality, the pullulan was activated by such further chemical modifications as periodate oxidation and reductive conjugation with cysteamide. Both derivatives, the folic acid-free derivative [(NH_2 PEG)-Pull-(cyst-Dox)] and the folic acid-doxorubicin-coupled derivative [(FA-PEG)-Pull-(Cyst-Dox)], contained a similar doxorubicin concentration (*w/w*), *i.e.*, ~6%, with 4.3% (*w/w*) folic acid in (FA-PEG)-Pull-(Cyst-Dox). This group's *in vitro* study results showed reduced toxicity in the folic acid non-expressing receptor cell line, MCF7, but increased toxicity in the KB cell line, which has over-expressed folic acid receptors. This finding demonstrated the specificity of pullulan polymer bioconjugates and the improved pharmacokinetics of doxorubicin. Moreover, Scomparin and colleague's design for actively targeting tumors in folic acid over-expressing receptor tumor cell lines, offering a new approach for *in vivo* studies of this bioconjugate pullulan.

The synthesis of pullulan-stabilized gold nanoparticles (PAuNPs) was reported by Ganeshkumar *et al.* [86] for its use in delivering the anticancer drug 5-fluorouracil (5Fu) and folic acid. As described previously, folic acid can be used for active targeting of folate receptor over-expressing cell lines. 5-Fluorouracil was introduced in clinics in 1957 as a poorly-soluble, but potent drug for treating solid tumors. Despite its promise as a chemotherapeutic agent, it showed limited response rates of only 10%–30% [88]. Ganeshkumar and colleagues [86] designed their 2014 study in an effort to increase solubility, stability and specificity, as well as to minimize the side effects of this potent agent. *In vitro* cytotoxicity tests were conducted in free 5-Fu and 5-Fu@AuNPs, as well as in 5-Fu@AuNPs against HepG2 cells with over-expressed folate receptors. Biodistrubution studies in male Wistar rats showed no excessive toxicity in untargeted healthy cells. This group's encouraging results with 5-Fu@PAuNPs-Fa bioconjugates have set the groundwork for a novel approach in active liver cancer targeting.

Using a similar approach, Zhang *et al.* [89] constructed maleilated pullulan-doxorubicin conjugated with a folate polymeric prodrug for active tumor-targeted delivery, reported as FA-MP-DOX. The purpose of conjugating the construct with folates is to enhance cellular uptake by folate receptor mediated endocytosis. The folate-decorated maleilated pullulan-doxorubicin construct, used to treat ovarian carcinoma A2780 cells, was compared to free doxorubicin and to a maleilated pullulan doxorubicin construct. FA-MP-DOX showed higher toxicity in tumor cells, along with enhanced cellular uptake in ovarian carcinoma A278 cells as compared to free doxorubicin and the MP-DOX construct. These findings suggest that the FA-MP-DOX approach may have potential for clinical application in ovarian carcinoma treatment.

Genetically Engineered Polypeptides

Genetically engineered polypeptides have been gaining recognition in biomaterials and drug delivery fields for decades owing to custom design and simple, inexpensive manufacturing. These polypeptides are genetically encoded, providing control over their sequence and molecular weight (MW) to an extent impossible with synthetic polymer analogs. Moreover, these biopolymers are simple and inexpensive to manufacture as they are expressed in *E. coli* at high yields and can be easily purified; their custom design further provides opportunities to confer low immunotoxicity and immunogenicity. These polymers are thus now widely used in anticancer drug delivery to improve the solubility of anti-cancer therapeutics.

2.2.4. Elastin-Like Polypeptide (ELP)

An ELP is a thermally responsive pentapeptide $(VPGXaaG)_n$ amino acid repeat, where Xaa is any guest residue amino acid other than proline; the ELP polymer is easily tailored for a wide variety of applications. Depending on an ELP's primary structure (e.g., amino acid composition), it can form aggregates, hydrogels or micelles in response to stimuli (thermal or pH). As an ELP is a macromolecule, it can exploit the EPR effect to achieve passive targeting to solid tumors. Moreover, an ELP is designed to undergo an inverse transition in response to thermal stimuli. Below the transition temperature, the ELP is disordered and soluble in aqueous solution. At an increase in the solution's temperature above the transition temperature through application of a mild, external heat, the ELP forms aggregates and can be actively targeted to tumor tissue.

The ELP has now been extensively applied in targeted anticancer drug delivery [90], an area of research that has now expanded to include breast, prostate, ovarian, and brain tumor studies [91,92,93,94]. It has also been used in drug delivery for therapeutic angiogenesis in pre-eclampsia [95] and in biomaterials research [96].

For example, Bidwell *et al.* [97] conjugated the anticancer drug, doxorubicin, to an ELP composed of Tat, a cell penetrating peptide (CPP), ELP and GFLG, a tetrapeptide linker that is a substrate for the lysosomal cathepsin proteases and doxorubicin (Figure 10.). This therapeutic polymeric prodrug was used to treat sensitive MES/SA cells and resistant MES-SA/Dx5 cells. The study's findings showed that doxorubicin was equally cytotoxic in both cell lines and that ELP-bound doxorubicin accumulated in MES-SA/Dx5 cells, overcoming their drug resistance.

A similar approach was used by Moktan *et al.* [90] to deliver paclitaxel. This polymeric prodrug was composed of three domains: CPP, ELP and paclitaxel. The paclitaxel prodrug contained hydrazine, an acid sensitive linker that is cleaved at lysosomal pH. The study showed that ELP-delivered paclitaxel combined with mild hyperthermia effectively inhibits proliferation in MCF7 and in multidrug resistant MCF7 cells, providing *in vitro* proof of concept for the use of ELP-paclitaxel.

Furthermore, *in vivo* studies support that using ELP as a macromolecular carrier for the delivery of doxorubicin and paclitaxel is a promising approach for clinical application. Moktan *et al.* [92] conjugated an ELP with a CPP (SynB1) at its N-terminus and a 6-maleimidocaproyl hydrazine derivate of doxorubicin at its C-terminus to compare the efficacy of the construct with that of free doxorubicin. Results showed complete tumor inhibition with SynB1-ELP-Dox, as opposed to only moderate tumor inhibition with free dox. Furthermore, the delivery of dox using the ELP construct minimized the cardiac toxicity commonly present with the free form of doxorubicin.

MacKay developed a different approach to delivering Doxorubicin using ELP forming micelles. The ELP polypeptides were modified with short, cysteine-rich segments that were constructed so as to spontaneously self-assemble into 100 nm nanoparticles upon conjugation with doxorubicin. The conjugated doxorubicin nanoparticles exhibited a lower systemic toxicity, but induced almost complete tumor regression in a mouse cancer model [98].

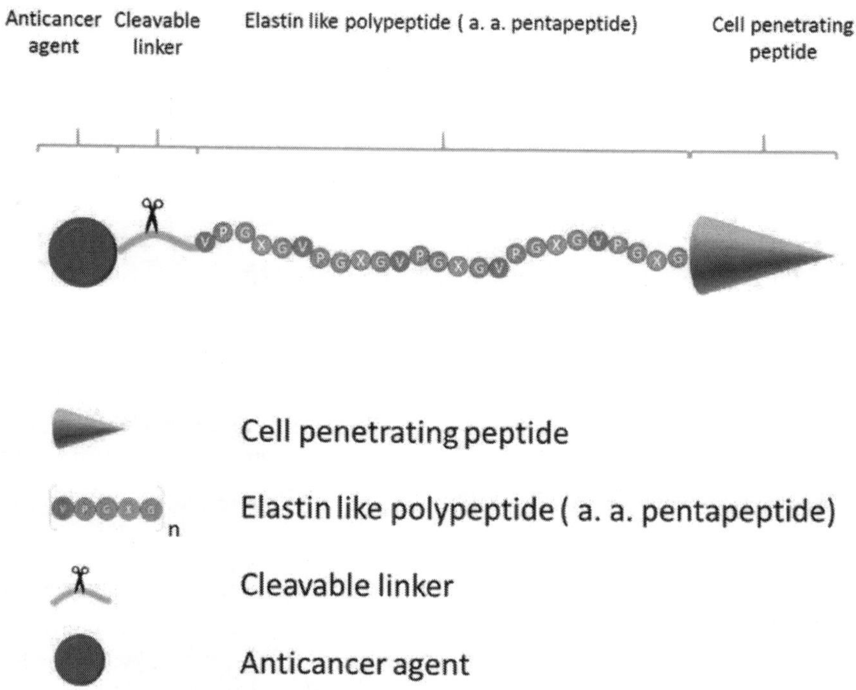

Figure 10. Schematic presentation of Elastin like polypeptide (ELP) polymeric prodrug. Construct consists of (1) Cell penetrating peptide, mediating uptake of ELP in tissues and cancer cells; (2) ELP, thermally responsive biopolymer which can be actively targeted to tumor tissue by external application of mild heat; (3) cleavable linker, releases anticancer agent in tumor tissues and cancer cells, and (4) anticancer agent.

2.2.5. Silk-Elastin Like Polypeptides

A silk-like polypeptide is a genetically engineered polypeptide that is composed of (Gly-Ala-Gly-Ala-Gly-Ser) tandem repeats. For years, silk-like polymers have posed a challenge in that a very low level of their expression is reported when they are derived from the commonly used *E. coli* host [99]. Utilizing substitute hosts that have higher expression levels such as yeast, insect cells, or transgenic plants may result in higher protein yield. However, these alternative routes may negatively affect protein purity. As a result, silk-like polypeptides are not as broadly used as elastin-like polypeptides, whose solubility facilitates numerous paths to more efficient drug delivery.

To make the silk-like polypeptide more suitable for drug delivery, it is often combined with an ELP. Xiao-Xia *et al.* [100] constructed a series of silk-elastin like polymers (SELP) which have the capacity to form micellar nanoparticles upon incubation with doxorubicin using heat as a stimuli. These nanoparticles effectively inhibited proliferation of HeLa cell lines, demonstrating the potential use of drug-loaded SELP for cancer treatment.

3. POLYMER DRUG CONJUGATES IN CLINICAL TRIALS

Despite the variety of novel drug targets and sophisticated chemistries available, to the best of our knowledge there have been no FDA-approved small molecule prodrugs based on polymer for cancer treatment yet, while only a limited number of polymer-drug conjugates has been placed in the stage of clinical trials. Currently less than 10 polymer–drug conjugates are on-going in clinical trials (Table 2).

3.1. Dextran Conjugates

Through GuaDex technology, a DexTech patented technology platform, dextran is modified and used as the backbone for construction of new candidate medications. OsteoDex and Somadex, which are made from alendronate and somatostatin by DexTech, are being tested in clinical trials to treat castration resistant prostate cancer (CRPC) [78].

3.2. HPMA Conjugates

FCE28068, a conjugate of doxorubicin with HPMA copolymer via GFLG linker, showed 5-fold decreased toxicity compared to doxorubicin in a phase I study [101]. In a phase II trial with breast, non-small cell lung, and colorectal cancer patients, FCE28068 had a total of 6 breast and NSCL cancer patients showing partial responses out of 62 patients without severe toxicities [101]. No subsequent reports have been released.

3.3. PEG Conjugates

3.3.1. EZN-2208/BEL-0222

SN38 (10-hydroxy-7-ethyl-camptothecin) is a potent topoisomerase I inhibitor and the active moiety of irinotecan (CPT-11). PEG-SN38 is thus a PEGylated conjugate of SN38 for better solubility and an improved pharmacokinetic profile [102]. In addition, the ester bond between the polymer and the drug can be cleaved in PBS, releasing approximately 57% of the conjugated SN-38 from the micelles [80]. PEG-SN38 has been shown to down-modulate HIF-1α involved in tumor invasion, migration, angiogenesis and production of vascular endothelial growth factor (VEGF). PEG-SN38 has produced positive responses in clinical trials for metastatic breast cancer and pediatric neuroblastoma and currently is in a phase III trial [103]

3.3.2. NC-6300

NC-6300 consists of a PEG polyaspartate block copolymer, an acid-labile hydrazone linker and epirubicin. Preclinical studies using Hep3B/Luc orthotopic hepatic tumor models showed a significant improvement in the survival rate in the group administered NC-6300 at 10 mg/kg compared with that in the group with EPI at 10 mg/kg [81]. Currently, Nano Carrier is conducting a phase I clinical trial in Japan [21].

4. CONCLUSIONS AND FUTURE DIRECTION

Poor solubility and selectivity of anti-cancer drugs limits efficacy of the current cancer treatments. Conjugating drugs to polymers improves drug solubility, pharmacokinetic and pharmacodynamics, as well as increasing their tumor specificity. Thus, polymers provide a remarkable tool through which to increase the therapeutic efficacy of drugs. Despite decades of significant progress in polymer-based drug delivery, the number of FDA approved polymer-based drugs remains very small.

Major objectives in designing new polymer based prodrugs include: (1) improving solubility of the drugs in aqueous solution or lipids; (2) conjugating the drug to the polymer without decreasing its potency; (3) binding the drug tightly to a polymer in circulation, but releasing it inside the tumor cells; and (4) ensuring that the polymer is non-immunogenic, biodegradable and can be cost-effectively produced in sufficient quantities. Successfully addressing these challenges will yield novel and efficient polymer-based prodrugs and establish their significant role in current cancer therapy.

ACKNOWLEDGEMENTS

We would like to express our great appreciation to Jo Anne Fordham of the UMMC Center for Bioethics and Medical Humanities for editorial support. We would further like to acknowledge Rebecca Mackey for helpful discussion.

AUTHOR CONTRIBUTIONS

Sonja Dragojevic wrote sections Natural polymers, Genetically engineered polymers, and created associated figures, as well as figure in the Introduction. Jung Su Ryu wrote sections Synthetic polymers, Polymer Drug Conjugates in Clinical Trials, and created associated figures and tables. Drazen Raucher coordinated writing of all sections and he wrote Introduction and Conclusions and Future Direction sections. All authors were involved in the editing process.

CONFLICTS OF INTEREST

The authors declare no conflict of interest.

REFERENCES

1. Williams, H.D.; Trevaskis, N.L.; Charman, S.A.; Shanker, R.M.; Charman, W.N.; Pouton, C.W.; Porter, C.J. Strategies to address low drug solubility in discovery and development. Pharmacol. Rev. 2013, 65, 315–499.

2. Bikiaris, D.N. Solid dispersions, part II: New strategies in manufacturing methods for dissolution rate enhancement of poorly water-soluble drugs. Expert Opin. Drug Deliv. 2011, 8, 1663–1680.

3. Sheth, P.; Sandhu, H.; Singhal, D.; Malick, W.; Shah, N.; Serpil Kislalioglu, M. Nanoparticles in the pharmaceutical industry and the use of supercritical fluid technologies for nanoparticle production. Curr. Drug Deliv. 2012, 9, 269–284.

4. Cooper, E.R. Nanoparticles: A personal experience for formulating poorly water soluble drugs. J. Control. Release2010, 141, 300–302.

5. Jouyban, A. Review of the cosolvency models for predicting solubility of drugs in water-cosolvent mixtures. J. Pharm. Pharm. Sci. 2008, 11, 32–58.

6. Venkatesh, S.; Li, J.; Xu, Y.; Vishnuvajjala, R.; Anderson, B.D. Intrinsic solubility estimation and pH-solubility behavior of cosalane (NSC 658586), an extremely hydrophobic diprotic acid. Pharm. Res. 1996, 13, 1453–1459.

7. Adams, J.D.; Flora, K.; Goldspiel, B.; Wilson, J.; Arbuck, S.; Finley, R. Taxol: A history of pharmaceutical development and current pharmaceutical concerns. J. Natl. Cancer Inst. Monogr. 1992, 141–147.

8. Arbuck, S.G.; Christian, M.; Fisherman, J.; Cazenave, L.; Sarosy, G.; Suffness, M.; Adams, J.; Canetta, R.; Cole, K.; Friedman, M. Clinical development of Taxol. J. Natl. Cancer Inst. Monogr. 1992, 11–24.

9. Dorr, R.T. Pharmacology and toxicology of Cremophor EL diluent. Ann. Pharmacother. 1994, 28, S11–S14.

10. Duncan, R.; Vicent, M.J. Polymer therapeutics-prospects for 21st century: The end of the beginning. Adv. Drug Deliv. Rev. 2013, 65, 60–70.

11. Giang, I.; Boland, E.L.; Poon, G.M. Prodrug applications for targeted cancer therapy. AAPS J. 2014, 16, 899–913.

12. Maeda, H. Macromolecular therapeutics in cancer treatment: The EPR effect and beyond. J. Control. Release 2012, 164, 138–144.

13. John, J.V.; Johnson, R.P.; Heo, M.S.; Moon, B.K.; Byeon, S.J.; Kim, I. Polymer-Block-Polypeptides and Polymer-Conjugated Hybrid Materials as Stimuli-Responsive Nanocarriers for Biomedical Applications. J. Biomed. Nanotechnol. 2015, 11, 1–39.

14. Taghizadeh, B.; Taranejoo, S.; Monemian, S.A.; Salehi Moghaddam, Z.; Daliri, K.; Derakhshankhah, H.; Derakhshani, Z. Classification of stimuli-responsive polymers as anticancer drug delivery systems. Drug Deliv. 2015, 22, 145–155.

15. Ryu, J.S.; Raucher, D. Elastin-like polypeptide for improved drug delivery for anticancer therapy: Preclinical studies and future applications. Expert Opin. Drug Deliv. 2015, 12, 653–667.

16. Pan, H.; Sima, M.; Yang, J.; Kopeček, J. Synthesis of Long-Circulating, Backbone Degradable HPMA Copolymer–Doxorubicin Conjugates and Evaluation of Molecular-Weight-Dependent Antitumor Efficacy. Macromol. Biosci. 2013,13, 155–160.

17. Duncan, R.; Kopeček, J. Soluble synthetic polymers as potential drug carriers. In Polymers in Medicine; Springer: Berlin, Heidelberg, Germany, 1984; Volume 57, pp. 51–101.

18. Jang, H.-J.; Shin, C.Y.; Kim, K.-B. Safety Evaluation of Polyethylene Glycol (PEG) Compounds for Cosmetic Use.Toxicol. Res. 2015, 31, 105–136.

19. Pang, X.; Du, H.L.; Zhang, H.Q.; Zhai, Y.J.; Zhai, G.X. Polymer-drug conjugates: Present state of play and future perspectives. Drug Discov. Today 2013, 18, 1316–1322.

20. Rohini, N.A.; Joseph, A.; Mukerji, A. Polymeric prodrugs: Recent achievements and general strategies. J. Antivir. Antiretrovir. 2013.

21. Alekseenko, I.V.; Snezhkov, E.V.; Chernov, I.P.; Pleshkan, V.V.; Potapov, V.K.; Sass, A.V.; Monastyrskaya, G.S.; Kopantzev, E.P.; Vinogradova, T.V.; Khramtsov, Y.V.; et al. Therapeutic properties of a vector carrying the HSV thymidine kinase and GM-CSF genes and delivered as a complex with a cationic copolymer. J. Transl. Med. 2015, 13, 78.

22. Huo, M.; Zhu, Q.; Wu, Q.; Yin, T.; Wang, L.; Yin, L.; Zhou, J. Somatostatin receptor-mediated specific delivery of paclitaxel prodrugs for efficient cancer therapy. J. Pharm. Sci. 2015, 104, 2018–2028.

23. Huang, P.; Wang, W.; Zhou, J.; Zhao, F.; Zhang, Y.; Liu, J.; Liu, J.; Dong, A.; Kong, D.; Zhang, J. Amphiphilic polyelectrolyte/prodrug nanoparticles constructed by synergetic electrostatic and hydrophobic interactions with cooperative pH-sensitivity for controlled doxorubicin delivery. ACS Appl. Mater. Interfaces 2015, 7, 6340–6350.

24. Stefan, N.; Zimmermann, M.; Simon, M.; Zangemeister-Wittke, U.; Pluckthun, A. Novel prodrug-like fusion toxin with protease-sensitive bioorthogonal PEGylation for tumor targeting. Bioconjug. Chem. 2014, 25, 2144–2156.

25. Lv, S.; Tang, Z.; Zhang, D.; Song, W.; Li, M.; Lin, J.; Liu, H.; Chen, X. Well-defined polymer-drug conjugate engineered with redox and pH-sensitive release mechanism for efficient delivery of paclitaxel. J. Control. Release 2014, 194, 220–227.

26. Bao, Y.; Guo, Y.; Zhuang, X.; Li, D.; Cheng, B.; Tan, S.; Zhang, Z. D-alpha-tocopherol polyethylene glycol succinate-based redox-sensitive paclitaxel prodrug for overcoming multidrug resistance in cancer cells. Mol. Pharm. 2014, 11, 3196–3209.

27. Wang, Y.; Wang, H.; Chen, Y.; Liu, X.; Jin, Q.; Ji, J. pH and hydrogen peroxide dual responsive supramolecular prodrug system for controlled release of bioactive molecules. Colloids Surf. B Biointerfaces 2014, 121, 189–195.

28. Nam, J.P.; Park, J.K.; Son, D.H.; Kim, T.H.; Park, S.J.; Park, S.C.; Choi, C.; Jang, M.K.; Nah, J.W. Evaluation of polyethylene glycol-conjugated novel polymeric anti-tumor drug for cancer therapy. Colloids Surf. B Biointerfaces2014, 120, 168–175.

29. Daman, Z.; Ostad, S.; Amini, M.; Gilani, K. Preparation, optimization and in vitro characterization of stearoyl-gemcitabine polymeric micelles: A comparison with its self-assembled nanoparticles. Int. J. Pharm. 2014, 468, 142–151.

30. Wu, X.; Sun, X.; Guo, Z.; Tang, J.; Shen, Y.; James, T.D.; Tian, H.; Zhu, W. In vivo and in situ tracking cancer chemotherapy by highly photostable NIR fluorescent theranostic prodrug. J. Am. Chem. Soc. 2014, 136, 3579–3588.

31. Meng, H.; Xu, K.; Xu, Y.; Luo, P.; Du, F.; Huang, J.; Lu, W.; Yu, J.; Liu, S.; Muir, B. Nanocapsules based on mPEGylated artesunate prodrug and its cytotoxicity. Colloids Surf. B 2014, 115, 164–169.

32. Xu, Z.; Wang, D.; Xu, S.; Liu, X.; Zhang, X.; Zhang, H. Preparation of a camptothecin prodrug with glutathione-responsive disulfide linker for anticancer drug delivery. Chem. Asian J. 2014, 9, 199–205.

33. Henne, W.A.; Kularatne, S.A.; Hakenjos, J.; Carron, J.D.; Henne, K.L. Synthesis and activity of a folate targeted monodisperse PEG camptothecin conjugate. Bioorg. Med. Chem. Lett. 2013, 23, 5810–5813.

34. Zhu, L.; Wang, T.; Perche, F.; Taigind, A.; Torchilin, V.P. Enhanced anticancer activity of nanopreparation containing an MMP2-sensitive PEG-drug conjugate and cell-penetrating moiety. Proc. Natl. Acad. Sci. USA 2013, 110, 17047–17052.

35. Wei, J.; Wang, H.; Zhu, M.; Ding, D.; Li, D.; Yin, Z.; Wang, L.; Yang, Z. Janus nanogels of PEGylated Taxol and PLGA-PEG-PLGA copolymer for cancer therapy. Nanoscale 2013, 5, 9902–9907.

36. Jiang, T.; Li, Y.M.; Lv, Y.; Cheng, Y.J.; He, F.; Zhuo, R.X. Amphiphilic polycarbonate conjugates of doxorubicin with pH-sensitive hydrazone linker for controlled release. Colloids Surf. B 2013, 111, 542–548.

37. Talelli, M.; Morita, K.; Rijcken, C.J.; Aben, R.W.; Lammers, T.; Scheeren, H.W.; van Nostrum, C.F.; Storm, G.; Hennink, W.E. Synthesis and characterization of biodegradable and thermosensitive polymeric micelles with covalently bound doxorubicin-glucuronide prodrug via click chemistry. Bioconjug. Chem. 2011, 22, 2519–2530.

38. Shamay, Y.; Raviv, L.; Golan, M.; Voronov, E.; Apte, R.N.; David, A. Inhibition of primary and metastatic tumors in mice by E-selectin-targeted polymer-drug conjugates. J. Control. Release 2015, 217, 102–112.

39. Zhong, J.; Li, L.; Zhu, X.; Guan, S.; Yang, Q.; Zhou, Z.; Zhang, Z.; Huang, Y. A smart polymeric platform for multistage nucleus-targeted anticancer drug delivery. Biomaterials 2015, 65, 43–55.

40. Peng, Z.H.; Kopecek, J. Enhancing Accumulation and Penetration of HPMA Copolymer-Doxorubicin Conjugates in 2D and 3D Prostate Cancer Cells via iRGD Conjugation with an MMP-2 Cleavable Spacer. J. Am. Chem. Soc. 2015, 137, 6726–6729.

41. Yang, Q.; Yang, Y.; Li, L.; Sun, W.; Zhu, X.; Huang, Y. Polymeric nanomedicine for tumor-targeted combination therapy to elicit synergistic genotoxicity against prostate cancer. ACS Appl. Mater. Interfaces 2015, 7, 6661–6673.

42. Chytil, P.; Koziolova, E.; Janouskova, O.; Kostka, L.; Ulbrich, K.; Etrych, T. Synthesis and Properties of Star HPMA Copolymer Nanocarriers Synthesised by RAFT Polymerisation Designed for Selective Anticancer Drug Delivery and Imaging. Macromol. Biosci. 2015, 15, 839–850.

43. Nakamura, H.; Koziolova, E.; Etrych, T.; Chytil, P.; Fang, J.; Ulbrich, K.; Maeda, H. Comparison between linear and star-like HPMA conjugated pirarubicin (THP) in pharmacokinetics and antitumor activity in tumor bearing mice.Eur. J. Pharm. Biopharm. 2015, 90, 90–96.

44. Nakamura, H.; Etrych, T.; Chytil, P.; Ohkubo, M.; Fang, J.; Ulbrich, K.; Maeda, H. Two step mechanisms of tumor selective delivery of N-(2-hydroxypropyl)methacrylamide copolymer conjugated with pirarubicin via an acid-cleavable linkage. J. Control. Release 2014, 174, 81–87.

45. Sedlacek, O.; Kucka, J.; Mattova, J.; Parizek, M.; Studenovsky, M.; Zadinova, M.; Pouckova, P.; Hruby, M. Multistage-targeted pH-

responsive polymer conjugate of Auger electron emitter: optimized design and in vivo activity. Eur. J. Pharm. Sci. 2014, 63, 216–225.

46. Zhang, R.; Yang, J.; Sima, M.; Zhou, Y.; Kopecek, J. Sequential combination therapy of ovarian cancer with degradableN-(2-hydroxypropyl)methacrylamide copolymer paclitaxel and gemcitabine conjugates. Proc. Natl. Acad. Sci. USA2014, 111, 12181–12186.

47. Mohanraj, S.; Murugan, D.; Rengarajan, A.; Rajiv, S. Anticancer activity of starch/poly[N-(2-hydroxypropyl)methacrylamide]: Biomaterial film to treat skin cancer. Int. J. Biol. Macromol. 2014, 70, 116–123.

48. Wu, Y.; Ji, J.; Yang, R.; Zhang, X.; Li, Y.; Pu, Y.; Li, X. Galactosylated 2-hydroxypropyl methacrylamide-s-3-guanidinopropyl methacrylamide copolymer as a small hairpin RNA carrier for inhibiting human telomerase reverse transcriptase expression. J. Gene Med. 2014, 16, 109–121.

49. Buckway, B.; Frazier, N.; Gormley, A.J.; Ray, A.; Ghandehari, H. Gold nanorod-mediated hyperthermia enhances the efficacy of HPMA copolymer-90Y conjugates in treatment of prostate tumors. Nucl. Med. Biol. 2014, 41, 282–289.

50. Larson, N.; Yang, J.; Ray, A.; Cheney, D.L.; Ghandehari, H.; Kopecek, J. Biodegradable multiblock poly(N-2-hydroxypropyl)methacrylamide gemcitabine and paclitaxel conjugates for ovarian cancer cell combination treatment. Int. J. Pharm. 2013, 454, 435–443.

51. Nakamura, H.; Liao, L.; Hitaka, Y.; Tsukigawa, K.; Subr, V.; Fang, J.; Ulbrich, K.; Maeda, H. Micelles of zinc protoporphyrin conjugated to N-(2-hydroxypropyl)methacrylamide (HPMA) copolymer for imaging and light-induced antitumor effects in vivo. J. Control. Release 2013, 165, 191–198.

52. Kayat, J.; Mehra, N.K.; Gajbhiye, V.; Jain, N.K. Drug targeting to arthritic region via folic acid appended surface-engineered multi-walled carbon nanotubes. J. Drug Target. 2015, 1–10.

53. Khare, V.; Kour, S.; Alam, N.; Dubey, R.D.; Saneja, A.; Koul, M.; Gupta, A.P.; Singh, D.; Singh, S.K.; Saxena, A.K.; et al. Synthesis, characterization and mechanistic-insight into the anti-proliferative potential of PLGA-gemcitabine conjugate. Int. J. Pharm. 2014, 470, 51–62.

54. Van, S.; Das, S.K.; Wang, X.; Feng, Z.; Jin, Y.; Hou, Z.; Chen, F.; Pham, A.; Jiang, N.; Howell, S.B. Synthesis, characterization, and biological

evaluation of poly (L-γ-glutamyl-glutamine)-paclitaxel nanoconjugate. Int.J. Nanomed. 2010, 5, 825–837.

55. Yang, D.; Liu, X.; Jiang, X.; Liu, Y.; Ying, W.; Wang, H.; Bai, H.; Taylor, W.D.; Wang, Y.; Clamme, J.-P. Effect of molecular weight of PGG–paclitaxel conjugates on in vitro and in vivo efficacy. J. Control. Release 2012, 161, 124–131.

56. Cai, L.; Xu, G.; Shi, C.; Guo, D.; Wang, X.; Luo, J. Telodendrimer nanocarrier for co-delivery of paclitaxel and cisplatin: A synergistic combination nanotherapy for ovarian cancer treatment. Biomaterials 2015, 37, 456–468.

57. Xiao, H.; Song, H.; Yang, Q.; Cai, H.; Qi, R.; Yan, L.; Liu, S.; Zheng, Y.; Huang, Y.; Liu, T. A prodrug strategy to deliver cisplatin (IV) and paclitaxel in nanomicelles to improve efficacy and tolerance. Biomaterials 2012, 33, 6507–6519.

58. Duncan, R.; Vicent, M.J. Do HPMA copolymer conjugates have a future as clinically useful nanomedicines? A critical overview of current status and future opportunities. Adv. Drug Deliv. Rev. 2010, 62, 272–282.

59. Larson, N.; Greish, K.; Bauer, H.; Maeda, H.; Ghandehari, H. Synthesis and evaluation of poly(styrene-co-maleic acid) micellar nanocarriers for the delivery of tanespimycin. Int. J. Pharm. 2011, 420, 111–117.

60. Mu, Y.; Kamada, H.; Kaneda, Y.; Yamamoto, Y.; Kodaira, H.; Tsunoda, S.; Tsutsumi, Y.; Maeda, M.; Kawasaki, K.; Nomizu, M.; et al. YBioconjugation of laminin peptide YIGSR with poly(styrene co-maleic acid) increases its antimetastatic effect on lung metastasis of B16-BL6 melanoma cells. Biochem. Biophys. Res. Commun. 1999, 255, 75–79.

61. Maeda, H. SMANCS and polymer-conjugated macromolecular drugs: Advantages in cancer chemotherapy. Adv. Drug Deliv. Rev. 2001, 46, 169–185.

62. Maeda, H.; Ueda, M.; Morinaga, T.; Matsumoto, T. Conjugation of poly(styrene-co-maleic acid) derivatives to the antitumor protein neocarzinostatin: Pronounced improvements in pharmacological properties. J. Med. Chem. 1985,28, 455–461.

63. Fang, J.; Tsukigawa, K.; Liao, L.; Yin, H.; Eguchi, K.; Maeda, H. Styrene-maleic acid-copolymer conjugated zinc protoporphyrin as a candidate drug for tumor-targeted therapy and imaging. J. Drug Target. 2015, 1–9.

64. Bae, H.H.; Cho, M.Y.; Hong, J.H.; Poo, H.; Sung, M.H.; Lim, Y.T. Bio-derived poly(gamma-glutamic acid) nanogels as controlled anticancer drug delivery carriers. J. Microbiol. Biotechnol. 2012, 22, 1782–1789.

65. Seth, A.; Heo, M.B.; Lim, Y.T. Poly (gamma-glutamic acid) based combination of water-insoluble paclitaxel and TLR7 agonist for chemo-immunotherapy. Biomaterials 2014, 35, 7992–8001.

66. Makadia, H.K.; Siegel, S.J. Poly Lactic-co-Glycolic Acid (PLGA) as Biodegradable Controlled Drug Delivery Carrier.Polymers 2011, 3, 1377–1397.

67. Emami, J.; Rezazadeh, M.; Hasanzadeh, F.; Sadeghi, H.; Mostafavi, A.; Minaiyan, M.; Rostami, M.; Davies, N. Development and in vitro/in vivo evaluation of a novel targeted polymeric micelle for delivery of paclitaxel. Int. J. Biol. Macromol. 2015, 80, 29–40.

68. Cheng, C.J.; Bahal, R.; Babar, I.A.; Pincus, Z.; Barrera, F.; Liu, C.; Svoronos, A.; Braddock, D.T.; Glazer, P.M.; Engelman, D.M.; et al. MicroRNA silencing for cancer therapy targeted to the tumour microenvironment. Nature2015, 518, 107–110.

69. Gorska, A.; Swiatkowska, A.; Dutkiewicz, M.; Ciesiolka, J. Modulation of p53 expression using antisense oligonucleotides complementary to the 5′-terminal region of p53 mRNA in vitro and in the living cells. PLoS ONE2013, 8, e78863.

70. Xiang, D.; Shigdar, S.; Qiao, G.; Wang, T.; Kouzani, A.Z.; Zhou, S.F.; Kong, L.; Li, Y.; Pu, C.; Duan, W. Nucleic acid aptamer-guided cancer therapeutics and diagnostics: The next generation of cancer medicine. Theranostics 2015, 5, 23–42.

71. Qi, L.; Xu, Z. In vivo antitumor activity of chitosan nanoparticles. Bioorg. Med. Chem. Lett. 2006, 16, 4243–4245.

72. Qi, L.; Xu, Z.; Chen, M. In vitro and in vivo suppression of hepatocellular carcinoma growth by chitosan nanoparticles. Eur. J. Cancer 2007, 43, 184–193.

73. Nogueira, D.R.; Scheeren, L.E.; Macedo, L.B.; Marcolino, A.I.; Pilar Vinardell, M.; Mitjans, M.; Rosa Infante, M.; Farooqi, A.A.; Rolim, C.M. Inclusion of a pH-responsive amino acid-based amphiphile in methotrexate-loaded chitosan nanoparticles as a delivery strategy in cancer therapy. Amino Acids 2015. in press.

74. Banerjee, T.; Mitra, S.; Kumar Singh, A.; Kumar Sharma, R.; Maitra, A. Preparation, characterization and biodistribution of ultrafine chitosan nanoparticles. Int. J. Pharm. 2002, 243, 93–105.

75. Rubino, F.M. Separation methods for methotrexate, its structural analogues and metabolites. J. Chromatogr. B 2001,764, 217–254.

76. Nam, J.P.; Lee, K.J.; Choi, J.W.; Yun, C.O.; Nah, J.W. Targeting delivery of tocopherol and doxorubicin grafted-chitosan polymeric micelles for cancer therapy: In vitro and in vivo evaluation. Colloids Surf. B 2015, 133, 254–262.

77. Cheung, R.Y.; Rauth, A.M.; Ronaldson, P.T.; Bendayan, R.; Wu, X.Y. In vitro toxicity to breast cancer cells of microsphere-delivered mitomycin C and its combination with doxorubicin. Eur. J. Pharm. Biopharm. 2006, 62, 321–331.

78. DexTch Medical. Available online: http://dextechmedical.com (accessed on 30 November 2015).

79. ClinicalTrials.gov. Available online: http://clinicaltrials.gov (NCT01595087) (accessed on 30 November 2015).

80. Cabral, H.; Kataoka, K. Progress of drug-loaded polymeric micelles into clinical studies. J. Control. Release 2014, 190, 465–476.

81. Matsumura, Y. The drug discovery by nanomedicine and its clinical experience. Jpn. J. Clin. Oncol. 2014, 44, 515–525.

82. CTI BioPharma. Available online: http://www.ctibiopharma.com (accessed on 30 November 2015).

83. Langer, C.J.; O'Byrne, K.J.; Socinski, M.A.; Mikhailov, S.M.; Lesniewski-Kmak, K.; Smakal, M.; Ciuleanu, T.E.; Orlov, S.V.; Dediu, M.; Heigener, D.; et al. Phase III trial comparing paclitaxel poliglumex (CT-2103, PPX) in combination with carboplatin versus standard paclitaxel and carboplatin in the treatment of PS 2 patients with chemotherapy-naive advanced non-small cell lung cancer. J. Thorac. Oncol. 2008, 3, 623–630.

84. Cerlulean. Available online: http://www,ceruleanrx.com (accessed on 30 November 2015).

85. Lee, K.D.; Choi, S.H.; Kim da, H.; Lee, H.Y.; Choi, K.C. Self-organized nanoparticles based on chitosan-folic acid and dextran succinate-doxorubicin conjugates for drug targeting. Arch. Pharm. Res. 2014, 37, 1546–1553.

86. Ganeshkumar, M.; Ponrasu, T.; Raja, M.D.; Subamekala, M.K.; Suguna, L. Green synthesis of pullulan stabilized gold nanoparticles for cancer targeted drug delivery. Spectrochim. Acta A Mol. Biomol. Spectrosc. 2014, 130, 64–71.

87. Scomparin, A.; Salmaso, S.; Bersani, S.; Satchi-Fainaro, R.; Caliceti, P. Novel folated and non-folated pullulan bioconjugates for anticancer drug delivery. Eur. J. Pharm. Sci. 2011, 42, 547–558.

88. Malet-Martino, M.; Martino, R. Clinical studies of three oral prodrugs of 5-fluorouracil (capecitabine, UFT, S-1): A review. Oncologist 2002, 7, 288–323.

89. Zhang, H.; Li, F.; Yi, J.; Gu, C.; Fan, L.; Qiao, Y.; Tao, Y.; Cheng, C.; Wu, H. Folate-decorated maleilated pullulan-doxorubicin conjugate for active tumor-targeted drug delivery. Eur. J. Pharm. Sci. 2011, 42, 517–526.

90. Moktan, S.; Ryppa, C.; Kratz, F.; Raucher, D. A thermally responsive biopolymer conjugated to an acid-sensitive derivative of paclitaxel stabilizes microtubules, arrests cell cycle, and induces apoptosis. Investig. New Drugs 2012, 30, 236–248.

91. Mikecin, A.M.; Walker, L.R.; Kuna, M.; Raucher, D. Thermally targeted p21 peptide enhances bortezomib cytotoxicity in androgen-independent prostate cancer cell lines. Anticancer Drugs 2014, 25, 189–199.

92. Moktan, S.; Perkins, E.; Kratz, F.; Raucher, D. Thermal targeting of an acid-sensitive doxorubicin conjugate of elastin-like polypeptide enhances the therapeutic efficacy compared with the parent compound in vivo. Mol. Cancer Ther.2012, 11, 1547–1556.

93. Massodi, I.; Bidwell, G.L., 3rd; Davis, A.; Tausend, A.; Credit, K.; Flessner, M.; Raucher, D. Inhibition of ovarian cancer cell metastasis by a fusion polypeptide Tat-ELP. Clin. Exp. Metastasis 2009, 26, 251–260.

94. Bidwell, G.L., 3rd; Perkins, E.; Hughes, J.; Khan, M.; James, J.R.; Raucher, D. Thermally targeted delivery of a c-Myc inhibitory polypeptide inhibits tumor progression and extends survival in a rat glioma model. PLoS ONE 2013, 8, e55104.

95. Bidwell, G.L., 3rd; George, E.M. Maternally sequestered therapeutic polypeptides—A new approach for the management of preeclampsia. Front. Pharmacol. 2014, 5, 201.

96. Amruthwar, S.S.; Janorkar, A.V. In vitro evaluation of elastin-like polypeptide-collagen composite scaffold for bone tissue engineering. Dent. Mater. 2013, 29, 211–220.

97. Bidwell, G.L., 3rd; Davis, A.N.; Fokt, I.; Priebe, W.; Raucher, D. A thermally targeted elastin-like polypeptide-doxorubicin conjugate overcomes drug resistance. Investig. New Drugs 2007, 25, 313–326.

98. Andrew MacKay, J.; Chen, M.; McDaniel, J.R.; Liu, W.; Simnick, A.J.; Chilkoti, A. Self-assembling chimeric polypeptide-doxorubicin conjugate nanoparticles that abolish tumours after a single injection. Nat. Mater. 2009, 8, 993–999.

99. Chow, D.; Nunalee, M.L.; Lim, D.W.; Simnick, A.J.; Chilkoti, A. Peptide-based Biopolymers in Biomedicine and Biotechnology. Mater. Sci. Eng. R Rep. 2008, 62, 125–155.

100. Xia, X.X.; Wang, M.; Lin, Y.; Xu, Q.; Kaplan, D.L. Hydrophobic drug-triggered self-assembly of nanoparticles from silk-elastin-like protein polymers for drug delivery. Biomacromolecules 2014, 15, 908–914.

101. Seymour, L.W.; Ferry, D.R.; Kerr, D.J.; Rea, D.; Whitlock, M.; Poyner, R.; Boivin, C.; Hesslewood, S.; Twelves, C.; Blackie, R.; et al. Phase II studies of polymer-doxorubicin (PK1, FCE28068) in the treatment of breast, lung and colorectal cancer. Int. J. Oncol. 2009, 34, 1629–1636.

102. Patnaik, A.; Papadopoulos, K.P.; Tolcher, A.W.; Beeram, M.; Urien, S.; Schaaf, L.J.; Tahiri, S.; Bekaii-Saab, T.; Lokiec, F.M.; Rezai, K.; et al. Phase I dose-escalation study of EZN-2208 (PEG-SN38), a novel conjugate of poly(ethylene) glycol and SN38, administered weekly in patients with advanced cancer. Cancer Chemother. Pharmacol. 2013, 71, 1499–1506.

Chapter 4

Positively Charged Nanostructured Lipid Carriers and Their Effect on the Dissolution of Poorly Soluble Drugs

Kyeong-Ok Choi, Jaehyeog Choe, Seokjin Suh and Sanghoon Ko *

Department of Food Science and Technology, Sejong University, 261 Neungdong-ro, Gwangjin-gu, Seoul 143-747, Korea

ABSTRACT

The objective of this study is to develop suitable formulations to improve the dissolution rate of poorly water soluble drugs. We selected lipid-based formulation as a drug carrier and modified the surface using positively charged chitosan derivative (HTCC) to increase its water solubility and bioavailability. Chitosan and HTCC-coated lipid particles had higher zeta-potential values than uncoated one over the whole pH ranges and improved encapsulation efficiency. *In vitro* drug release showed that all NLC formulations showed higher *in vitro* release efficiency than drug particle at pH 7.4. Furthermore, NLC formulation prepared with chitosan or HTCC represented good sustained release property. The results indicate that chitosan and HTCC can be excellent formulating excipients of lipid-based delivery carrier for improving poorly water soluble drug delivery.

Keywords: poorly water soluble drug; indomethacin; nanostructured lipid nanocarrier; *N*-(2-hydroxy)propyl-3-trimethyl ammonium chitosan chloride; *in vitro* release

1. INTRODUCTION

Recently, poorly water soluble drugs have received increasing interests in the pharmaceutical industry. According to the Biopharmaceutics Classification System (BCS), drugs are categorized into four classes (class I, II, III, and IV) based on their solubility and membrane permeability [1]. For BCS class II drugs, dissolution is a rate-limiting step to their gastrointestinal absorption and bioavailability. The bioavailability of a class II pharmaceutical compound with adequate cell membrane permeability depends on the dissolution of the compound. This has encouraged the pharmaceutical industry to look for new strategies to enhance the dissolution rate of such drugs in order to improve their bioavailability. When a pharmaceutical compound has high permeability but a water solubility of less than 1 μg/mL, conventional drug formulations may be unacceptable. Thus, the development and selection of suitable formulations are crucial to producing successful products for oral administration of BCS class II drugs.

Orally administered drugs that are rapidly absorbed from the gastrointestinal tract and have a short half-life are quickly eliminated from systemic circulation. This is one of the factors that affect the oral bioavailability of some drugs. Various types of sustained release drug formulations have been suggested to allow the slow release of drugs, resulting in increased residence time in the gastrointestinal tract. However, clinical trials often fail to show evidence of acceptable efficacy.

Currently, several strategies have been suggested for improving the solubility of BCS class II drugs, including crystal modification, particle size reduction, pH modification, amorphization, and lipid-based formulations [2]. Lipid-based formulations have been extensively studied, and have shown various advantages such as the ability to release the drug in a controlled and targeted manner; high and enhanced drug content and stability; favorable biodegradability and biocompatibility; versatility of excipients and formulations; and low risk profile [3]. In particular, nanostructured lipid nanocarriers (NLC) are considered an alternative lipid-based carrier system to liposomal vesicles, emulsions, and solid lipid nanoparticles (SLN). However, they also have some drawbacks that need to be overcome for successful delivery to the target site with controlled release of incorporated drugs. Short gastric retention time (GRT), unpredictable short gastric emptying time (GET), and lipolysis in the upper intestine are responsible for reduced bioavailability of drugs upon oral administration [4]. Surface modification of a drug delivery carrier via introduction of a positive charge is hypothesized to overcome the drawbacks of conventional formulations, improving system stability as well as allowing effective, site-specific drug delivery. In general, positively charged delivery carriers regulate the release rate of drugs depending on biological conditions and enhance drug uptake via electrostatic interactions with sialic acid of mucin in the GI tract [5,6].

In pharmaceutical industry, chitosan is often used as a surface modification agent for various drug delivery systems to improve their performance [7,8]. Chitosan and its derivatives have been extensively studied with respect to their potential to improve the performance of drug carriers by enhancing drug loading, absorption, via specific targeting, controlled and sustained release, among others. In particular, the coating of a drug delivery carrier with cationically modified chitosan derivative was demonstrated to have improved the delivery performance of the cationic primary amino group of chitosan [9]. Based on previous reports, we hypothesized that the impregnation of cationically modified chitosan derivative could improve the dissolution rate of poorly water soluble drugs by coating the surface of conventional NLC formulations. The aim of this study was to improve the solubility of poorly water soluble drugs and investigate the possibility of the application of cationically modified chitosan derivative as a surface modification agent to improve the drug delivery performance of lipid nanocarrier. A model drug, indomethacin was incorporated in NLC formulations to improve its water solubility and bioavailability.

2. RESULTS AND DISCUSSION

2.1. Synthesis and Characterization of HTCC

HTCC synthesis was confirmed by ^{1}H-NMR and FT-IR spectra. The ^{1}H-NMR spectra of unmodified chitosan and HTCC are shown in Figure 1. A strong resonance at 3.2 ppm was observed in the HTCC spectrum, indicative of a trimethyl group of the quaternary ammonium side chain; however, this was not observed in the chitosan spectrum [10]. The resonances at around 4.8 ppm and 3.1 ppm were attributable to the H-1 and H-2 of glucosamine unit (GlcN), respectively, in both chitosan and HTCC spectra [11]. Furthermore, the resonance at around 4.6 ppm corresponded to the H-1 of the *N*-acetylglucosamine unit (GlcNAc). The resonance at around 2.0 ppm represents three protons in the acetyl group [12]. The resonances of H-3, 4, 5, and 6 appeared in the range of 3.5–4.0 ppm [12]. The FT-IR spectra shown in Figure 2 also confirmed the successful introduction of GTMAC into the amino group of chitosan. Bands at around 3268, 2882, 1730, 1646, 1580, 1413, 1320, and 1413 cm^{-1} represent N-H, C-H, carbonyl group, primary amide, secondary amide, C-H of secondary amide, and C-N stretching, respectively, which are general characteristic vibrations of GlcN and GlcNAc. A new band at 1475 cm^{-1} in the HTCC spectrum indicates the introduction of trimethylammonium group on the chitosan backbone [13].

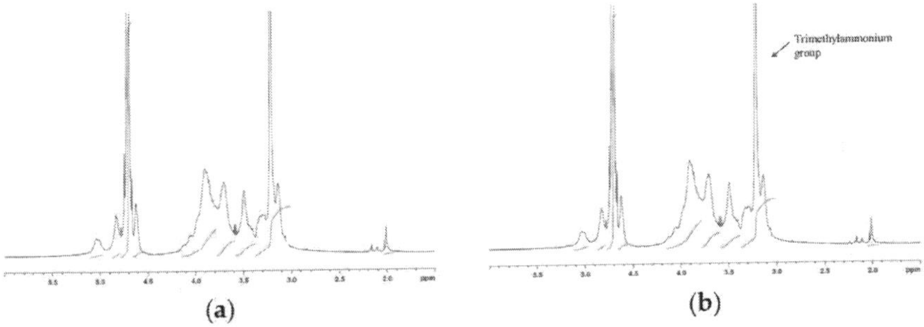

Figure 1. ^{1}H-NMR spectra of (**a**) chitosan and (**b**) HTCC at 25 °C.

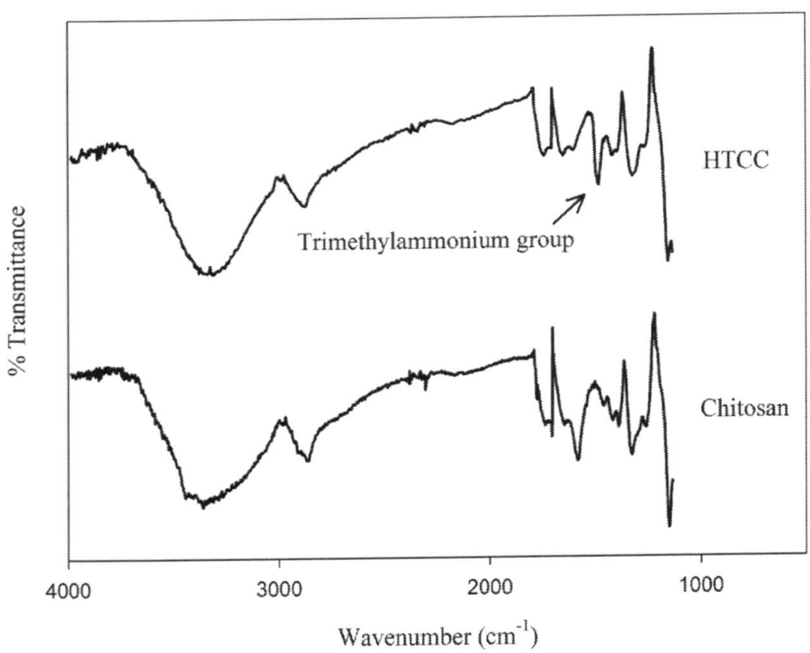

Figure 2. FT-IR spectra of chitosan and HTCC.

Zeta-potential profiles of chitosan and HTCC as a function of pH are shown in Figure 3. HTCC showed a highly positive zeta-potential value of +50.98 mV at pH 2. Chitosan also revealed a similar value to HTCC because the amino groups in the chitosan molecules are fully protonated at low pH. However, the zeta-potential value of chitosan decreased below zero at increasing pH. The zeta-potential value of HTCC also decreased with increasing pH; however, this change was more gradual and retained a positive value over the whole pH range. From these results, we concluded that chitosan could be successfully modified with GTMAC to provide permanent positive value, regardless of changes in pH.

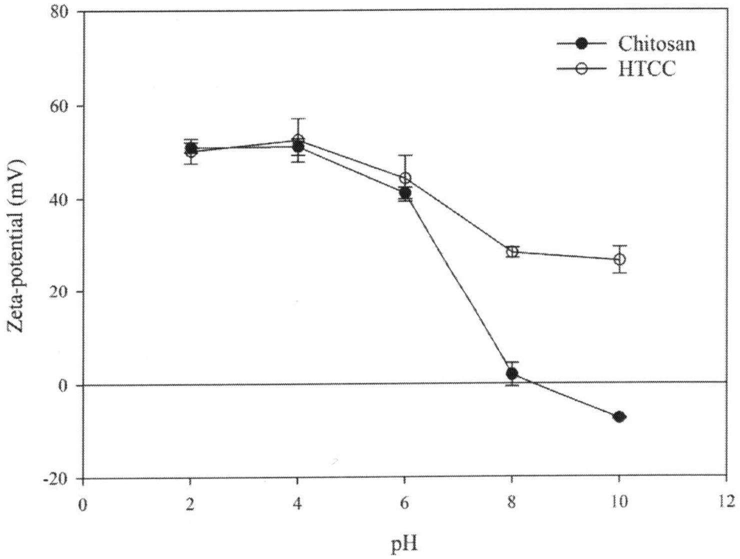

Figure 3. Zeta-potential profiles of chitosan and HTCC as a function of pH.

2.2. Appearance and Encapsulation Efficiency of Nanostructured Lipid Carriers

Different formulations of NLC dispersion were prepared using a hot emulsification method based on phase inversion technique, with the aid of ultrasonic treatment. Compositions of lipid and aqueous phases are represented in Table 1. Blank LNC, indomethacin-loaded NLC, chitosan-coated NLC, and HTCC-coated NLC formulations were abbreviated to NLC_B, NLC_I, NLC/CS_I, and NLC/HTCC_I, respectively. NLC_B formulation was translucent and whitish in appearance. NLC_I formulation was translucent and yellowish, while the NLC/CS_I and NLC/HTCC_I formulations were yellowish, turbid, and milky after preparation. All indomethacin-loaded formulations had a yellowish tinge because indomethacin particle (white) turns bright yellow upon dissolution in lipid phase. After storage for a month, visual observation indicated particle aggregation and precipitation with growing turbidity in NLC_B and NLC_I formulations. In agreement with previous reports that suggested that macromolecular biopolymers stabilize lipid colloids upon storage, chitosan and HTCC seemed to stabilize the surface of the lipid particle to become resistant to inter-particle aggregation [14].

Table 1. Compositions to prepare indomethacin loaded NLC formulations.

Formulation	Solid Lipid (%)	Liquid Lipid (%)	Tween 80 (%)	Coating Material (%)	Viscosity of Aqueous Phase (mPa·s)	Indomethacin (mg)
NLC_B	1	1	1	-	0.36	-
NLC_I	1	1	1	-	0.36	74
NLC/CS_I	1	1	1	0.1	4.8	74
NLC/HTCC_I	1	1	1	0.1	4.2	74

NLC_B, blank NLC formulation; NLC_I, indomethacin loaded NLC formulation; NLC/CS_I, indomethacin loaded NLC formulation coated with chitosan; NLC/HTCC_I, indomethacin loaded NLC formulation coated with HTCC. Viscosity values were measured at 80 °C.

The encapsulation efficiency of three different formulations of indomethacin-loaded NLC was evaluated and the results are represented in Table 1. Encapsulation efficiency of NLC_I formulation was low, despite the high solubility of indomethacin in lipid phase. On the other hand, more drugs were incorporated in NLC/CS_I and NLC/HTCC_I formulations than in the NLC_I formulation. It is assumed that some of the drug molecules would exist at the lipid surface or surfactant layer, so that, the surface-bound drug fraction could easily diffuse out upon dilution and storage. However, for the NLC/CS_I and NLC/HTCC_I formulations, surface coverage with the coating material could prevent the diffusion of surface-bound drug molecules.

2.3. Particle Size Distribution and Zeta-Potential Profiles

Mean particle sizes and polydispersity indexes (PDI) of NLC/CS and NLC/HTCC formulations are listed in Table 2. The mean particle size increased significantly with the addition of chitosan and HTCC, while the PDI value decreased. An increase in mean particle size of NLC/CS and NLC/HTCC formulations seems to relate to their viscosity of the continuous phase. Prior to the preparation of NLC formulation, viscosity values of chitosan and HTCC solutions were 4.8 and 4.2 mPa·s at 80 °C, respectively, which were approximate 4 times higher than that of deionized water. The viscosity of NLC/CS and NLC/HTCC formulations increased after phase inversion. Subsequent ultrasonic treatment increased the viscosities of NLC/CS_I and NLC/HTCC_I formulations up to 87 and 82 mPa·s, respectively. In cases of NLC_B and NLC_I formulations, viscosity increased up to 3.9 and 4.8 mPa·s, respectively, after ultrasonic treatment.

Table 2. Particle size distribution and encapsulation efficiency of indomethacin loaded NLC formulations.

Formulation	Particle Size Distribution		Encapsulation Efficiency (%)	Drug Loading (%)
	Mean Particle Size (nm)	Polydispersity Index		
NLC_B	255.6 ± 2.4 [b]	0.268 ± 0.002 [a]	-	-
NLC_I	277.0 ± 22.9 [b]	0.270 ± 0.015 [a]	56.8 ± 0.5 [c]	4.2 ± 0.04 [c]
NLC/CS_I	360.2 ± 34.1 [a]	0.247 ± 0.026 [b]	72.2 ± 1.9 [b]	5.3 ± 0.13 [b]
NLC/HTCC_I	351.9 ± 22.4 [a]	0.238 ± 0.02 [b]	74.4 ± 0.7 [a]	5.5 ± 0.05 [a]

NLC_B, blank NLC formulation; NLC_I, indomethacin loaded NLC formulation; NLC/CS_I, indomethacin loaded NLC formulation coated with chitosan; NLC/HTCC_I, indomethacin loaded NLC formulation coated with HTCC. Letters in the same column indicates significant difference at $p < 0.05$.

An increase in viscosity of the continuous phase inhibits the mobility and flexibility of surfactant molecules, which retards the adsorption of the surfactant on the surface of newly formed lipid particles. An increase in viscosity of aqueous medium also makes it harder to achieve sufficient acoustic cavitation, which is a powerful driving force for particle reduction during ultrasonic treatment.

The zeta-potential profiles of the NLC formulations as a function of pH are shown in Figure 4. The formulations of NLC_B and NLC_I showed almost zero mV at pH 2 and 4, which gradually decreased to negative value with increasing pH. The carboxyl groups of oleic acid chains are responsible for the zeta-potentials of NLC_B and NLC_I formulations because carboxyl groups are protonated at low pH and deprotonated at high pH to give a negative zeta-potential value. On the other hand, the formulations of NLC/CS_I and NLC/HTCC_I showed a highly positive zeta-potential value at low pH owing to protonation of the amino group of chitosan and the quaternized ammonium group of HTCC. The zeta-potential values of NLC/CS_I and NLC/HTCC_I also decreased with increasing pH owing to deprotonation of the carboxyl group of oleic acid chain, but more gradually than those of NLC_B and NLC_I formulations.

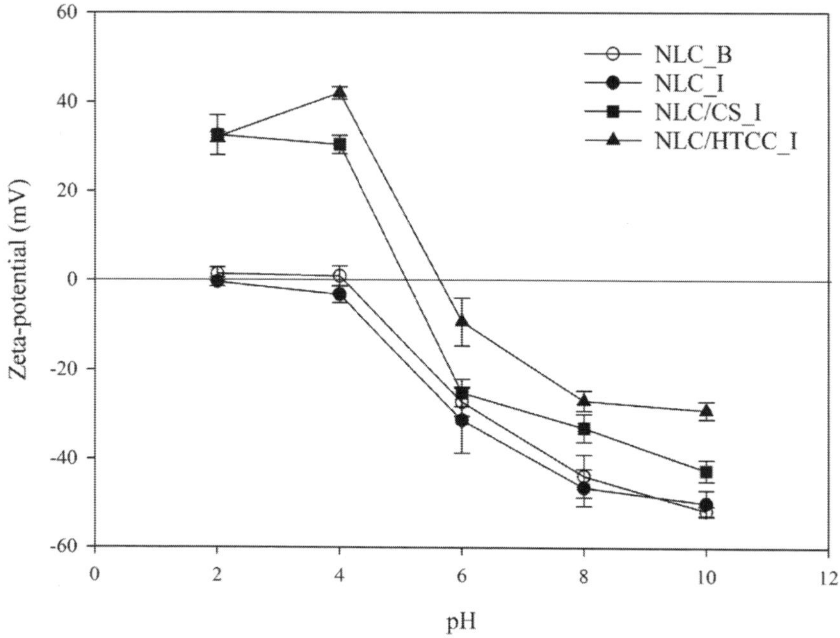

Figure 4. Zeta-potential profiles of different NLC formulations as a function of pH. NLC_B, blank NLC formulation; NLC_I, indomethacin loaded NLC formulation; NLC/CS_I: indomethacin loaded NLC formulation coated with chitosan; NLC/HTCC_I, indomethacin loaded NLC formulation coated with HTCC.

2.4. Differential Scanning Calorimetry

Melting behavior and relative crystallinity of NLC formulations were compared with the bulk lipid material as shown inTable 3. Melting peaks of glyceryl monostearate component for bulk lipid, NLC/HTCC_I, and the other formulations were observed at 58.4, 52.5, and around 47 °C, respectively. As crystallinity of NLC formulations increased, melting point increased as well. It is known that higher crystallinity (highly ordered lipid structure) affects negatively the performance of lipid based carriers such as less encapsulation efficiency and initial burst release *in vitro* [15]. However, in this study, interestingly NLC/HTCC_I formulation showed relatively high crystallinity compared to the other formulations, which seems not to affect the encapsulation efficiency and *in vitro* drug release. Although NLC/HTCC_I formulation was higher in crystallinity, it still provides enough space for entrapment of guest molecules. In addition, higher encapsulation efficiency and less initial drug burst were observed in NLC/HTCC_I formulation compared to the other formulations during *in vitro*test.

Table 3. Melting parameters and relative crystallinity of indomethacin loaded NLC formulations.

Formulation	Melting Temp. (°C)	Enthalpy (J/mg)	Relative Crystallinity (%)
NLC_B	47.4 ± 0.9 [b]	31.9 ± 1.6 [b]	30.0 ± 069 [b]
NLC_I	46.8 ± 0.5 [b]	31.8 ± 4.3 [b]	30.0 ± 0.0 [b]
NLC/CS_I	47.8 ± 2.1 [b]	31.0 ± 2.3 [b]	29.2 ± 2.1 [b]
NLC/HTCC_I	52.4 ± 0.5 [a]	39.0 ± 0.5 [a]	36.7 ± 0.4 [a]

NLC_B, blank NLC formulation; NLC_I, indomethacin loaded NLC formulation; NLC/CS_I, indomethacin loaded NLC formulation coated with chitosan; NLC/HTCC_I, indomethacin loaded NLC formulation coated with HTCC. Different letters in the same column indicate a significant difference ($p < 0.05$).

Endothermic curves of bulk lipid, indomethacin, and their mixture are shown in Figure 5a. Melting peaks of bulk lipid and indomethacin appeared at 58.4 and 162.7 °C, respectively. The mixture of lipid phase and indomethacin showed relatively low melting point and low enthalpy compared to individual lipid or indomethacin since melting and subsequent cooling induced the recrystallization between lipid and indomethacin resulting in the change in the crystalline state [16]. Melting peaks of NLC formulations are shown in Figure 5b. The melting points and enthalpy values were relatively low compared to bulk lipid owing to the same reason in the mixture. Melting peak did not appear in NLC/HTCC_I formulation, but tiny peaks were observed in NLC_I (116.9 °C) and NLC/CS_I (114.4 °C) formulations.

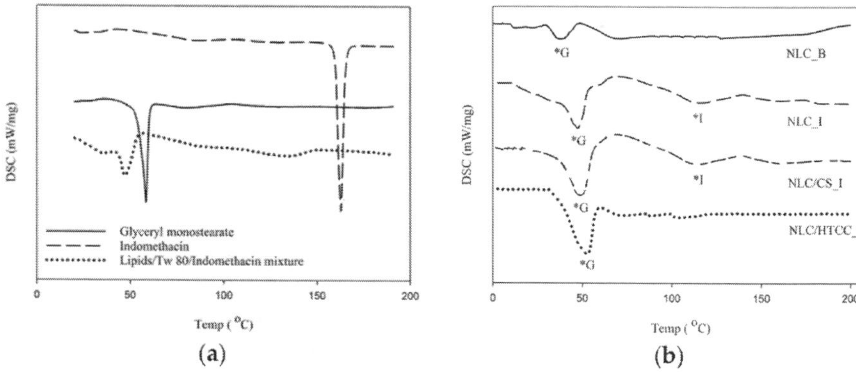

Figure 5. Endothermic peaks of (**a**) bulk glyceryl monostearate, indomethacin, and their mixture and (**b**) NLC formulations; *G, endothermic peak of glyceryl monostearate; *I, endothermic peak of indomethacin.

2.5. In vitro Drug Release

Figure 6 shows *in vitro* release profiles of three different indomethacin loaded NLC formulations. The release rate of indomethacin from three different NLC formulations varied depending on the formulations for the carriers. For the all formulations tested in the PBS buffer, indomethacin released fast at the initial period of the *in vitro* release test but subsequently released slowly over time. The order of drug release rate in the formulations was found to decrease in the following order NLC_I > NLC/CS_I > NLC/HTCC_I > drug for 32 h. NLC formulations coated with chitosan and HTCC showed intermediate drug release rate; the cationic surface modification of lipid carrier using chitosan and HTCC provided sustained release characteristics for the indomethacin loaded lipid carriers. It is expected that chitosan and HTCC coated lipid carrier could enhance dissolution rate of drug thanks to the sustained release behavior. In addition, the mucoadhesive property of chitosan and HTCC contributes to the increase in bioavailability of drug since residence time was prolonged in the gastrointestinal track.

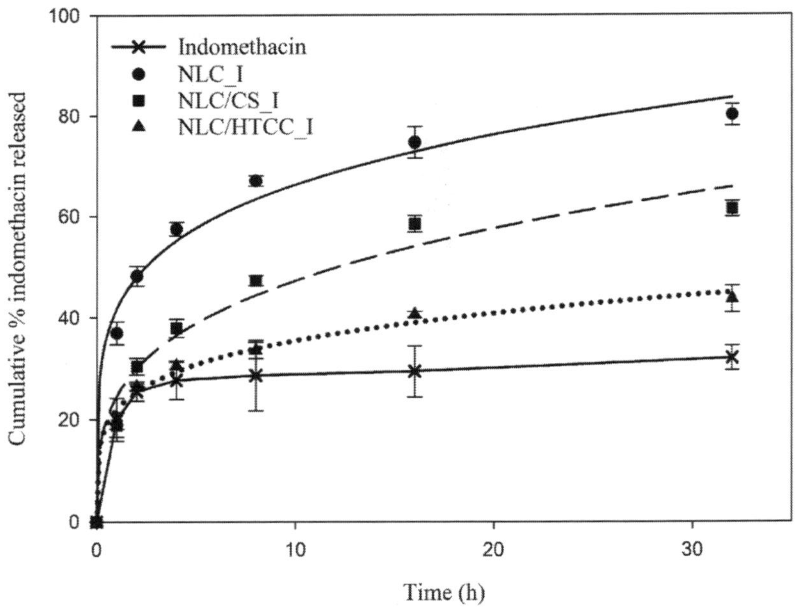

Figure 6. *In vitro* release profiles of three different indomethacin loaded NLC formulations. NLC_I, indomethacin loaded NLC formulation; NLC/CS_I: indomethacin loaded NLC formulation coated with chitosan; NLC/HTCC_I, indomethacin loaded NLC formulation coated with HTCC. Each formulation was fitted according to Korsmeyer-Peppas regression model except pure indomethacin.

In vitro release profiles of indomethacin loaded NLC formulations were fitted using various non-linear regression models such as zero order, first order, Higuchi, and Korsmeyer-Peppas equations as shown in Figure 6. In addition, the regression coefficient (R^2) and release exponent (n) values calculated using the non-linear regression models for the formulations are listed in Table 4. All NLC formulations were fitted well with Korsmeyer-Peppas model since R^2 values of NLC_I, NLC/CS_I and NLC/HTCC_I were 0.955, 0.9372, and 0.9645, respectively. The n value (<0.43) of Korsmeyer-Peppas equation indicated that the release behavior of all formulations followed Fickian diffusion mechanism.

Table 4. Regression coefficient (R^2) and release exponent (n) values calculated using the non-linear regression models for *in vitro* release profiles of indomethacin loaded NLC formulations.

Formulation	Zero Order R^2	First Order R^2	Higuchi R^2	Korsmeyer-Peppas R^2 (n)
NLC_I	0.7319	0.8545	0.8822	0.9555 (0.2202)
NLC/CS_I	0.7405	0.8126	0.8894	0.9372 (0.3312)
NLC/HTCC_I	0.7894	0.8263	0.9172	0.9645 (0.2200)

NLC_B, blank NLC formulation; NLC_I, indomethacin loaded NLC formulation; NLC/CS_I, indomethacin loaded NLC formulation coated with chitosan; NLC/HTCC_I, indomethacin loaded NLC formulation coated with HTCC. n value indicates release exponent.

2.6. Static In vitro Digestion

In vitro drug release characteristics of indomethacin with the presence of digestive enzymes were measured in static, simulated gastric and intestinal fluids for 4 h (2 h each), and the results are shown in Figure 7. For NLC_I and NLC/CS_I formulations, release rates of indomethacin from lipid particles were relatively fast in the simulated gastric fluid and slow in the intestinal fluid. We also observed that more than 60% of entrapped indomethacin was released from lipid nanoparticles in the gastric fluid. NLC/HTCC_I formulation showed a relatively slow release rate of indomethacin (approximately 30% within 2 h), and similar release characteristics to the other formulations in the simulated intestinal fluid. From the theoretical point of view, rapid release of drug entrapped in lipid nanoparticle should take place in the intestinal fluid in which lipolysis occurs. In this study, however, significant amounts of indomethacin were released in the gastric fluid. Thus, we assumed that a significant amount of indomethacin was distributed on the surface of the lipid particle or the interface between hydrophobic region of lipid and the surfactant layer during crystallization of the lipid particle. This surface-bound indomethacin could simply diffuse out into aqueous medium during digestion. For

NLC/HTCC_I formulation, HTCC with positively charged alkyl chains on the surface of the lipid particle seemed to prevent the diffusion of indomethacin molecules, thus retarding the release of the drug in the gastric fluid.

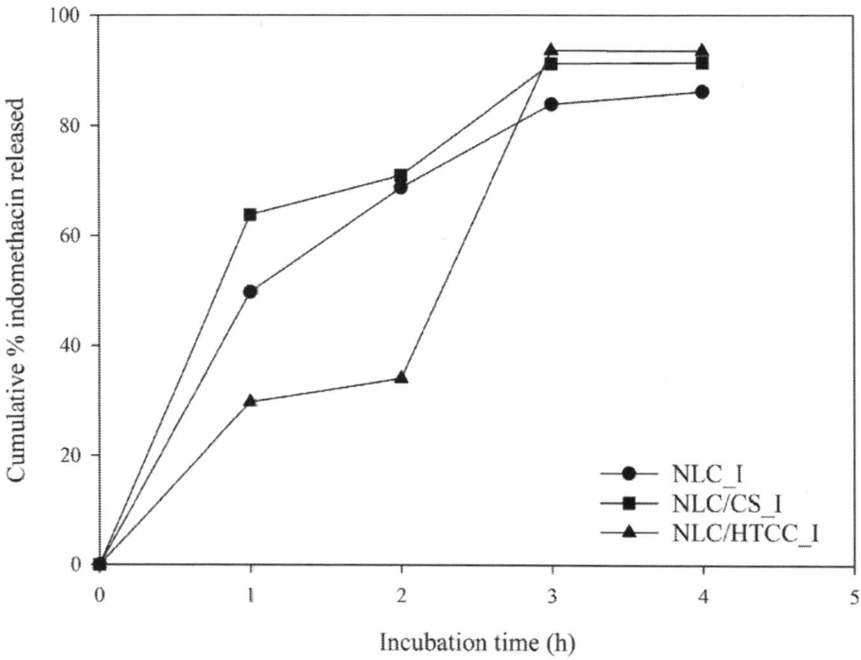

Figure 7. *In vitro* digestion characteristics of three different indomethacin loaded NLC formulations. NLC_I, indomethacin loaded NLC formulation; NLC/CS_I: indomethacin loaded NLC formulation coated with chitosan; NLC/HTCC_I, indomethacin loaded NLC formulation coated with HTCC.

3. EXPERIMENTAL SECTION

3.1. Synthesis and Characterization of HTCC

Chitosan (Mw 600,000–800,000, DD 90%) and glycidyl trimethylammonium chloride (GTMAC) were purchased from Sigma-Aldrich (St. Louis, MO, USA) and used to synthesize HTCC. *N*-(2-Hydroxy)propyl-3-trimethyl ammonium chitosan chloride (HTCC) was prepared by reacting chitosan with GTMAC in acetic aqueous solution (2% *v*/*v*) [17]. Chitosan powder (6 g) was dissolved in 500 mL of acetic aqueous solution under magnetic stirring at 60 °C. Following complete dissolution of chitosan particles, the temperature was raised to 85 °C with constant stirring. GTMAC was added to the chitosan solution dropwise in five portions at 3-h intervals (7.3 mL each). The total reaction time was 18 h from the first addition of GTMAC. The reaction mixture was then cooled to room temperature and diluted with ethanol. HTCC was precipitated by pouring the reaction mixture in acetone. The precipitate was isolated by centrifugation and washed by ethanol, acetone and finally, diethyl ether. The final product was dried and kept in a desiccator until use.

FT-IR spectra were recorded on a Nicolet 380 FT-IR Spectrometer (Thermo Fisher Scientific Inc., Waltham, MA, USA). ^{1}H-NMR spectra were obtained using a Bruker Avance II 500 spectrophotometer (Karlsruhe, Germany) at room temperature. The data was collected with 128 scans. NMR samples were prepared by dissolving chitosan and HTCC powder in D_2O containing a drop of DCl. Zeta potentials were measured using a zeta-potential analyzer (Delsa nano C, Beckman Coulter, Inc., Fullerton, CA, USA) at a pH range of 2–10.

3.2. Solubility Test of Indomethacin in Lipid Phase

Liquid oleic acid and solid glyceryl monostearate were used as components of lipid matrix and Tween 80 was used as a surfactant. They were purchased from Sigma-Aldrich. Indomethacin was used as a model drug and purchased from Sigma-Aldrich. Chitosan and HTCC prepared as described above, were used for surface modification of lipid nanoparticles.

The solubility of indomethacin in lipid phase consisting of oleic acid, glyceryl monostearate, and Tween 80 (1:1:1, *w/w/w*) was determined prior to encapsulation. Indomethacin (100 mg) was dissolved in 1 g of lipid phase at 80 °C for 20 min and cooled to room temperature. The lipid phase was centrifuged at 8000 rpm for 2 min at 40 °C to remove undissolved drug particles. The supernatant was separated carefully from the sediment and dissolved in DMSO at 60 °C. After cooling, the solution was centrifuged at 8000 rpm for 30 min at 4 °C to separate the oil phase. The DMSO layer was carefully transferred to a cuvette and the absorbance was measured at a wavelength of 320 nm on a spectrophotometer (DU 730, Beckman Coulter, Inc., Fullerton, CA, USA). A calibration curve was prepared from the indomethacin standard solutions (3, 5, 10, 20, 30, 40, 50 μg/mL in DMSO). The concentration of indomethacin in the lipid phase was calculated from calibration curve (R^2= 0.9993).

3.3. Preparation of Nanostructured Lipid Carriers

Four different formulations of nanostructured lipid carrier were prepared by a two-step emulsification method. In the first step, coarse emulsion was prepared by phase inversion emulsification. Lipid phase (3%) and 48.5 mL of aqueous phase were heated to 80 °C separately, and the aqueous phase was added dropwise to the lipid phase at a constant rate (5 mL/min) using a peristaltic pump (Longer Precision Pump Co. Ltd., Hebei, China) under magnetic stirring. For NLC/CS_I and NLC/HTCC_I formulations, 0.1% chitosan and HTCC were dissolved directly in the aqueous phase. The coarse emulsion was further stirred for 10 min. In the second step, the coarse emulsion was treated with a probe ultrasonic processor (VCX 750, Sonics & Materials, Inc., Newtown, CT, USA) at 35% amplitude for 20 min. The hot nanoemulsion was cooled to room temperature in order to induce crystallization.

3.4. Particle Size Measurement

Hydrodynamic particle diameters were measured using a particle size analyzer (Delsa nano C, Beckman Coulter, Inc.). Mean particle size and polydispersity index were recorded based on intensity-based distribution. Each sample was measured in triplicate with three different sets at 25 °C and a scattering angle of 165°.

3.5. Zeta-Potential Measurement

Zeta-potential was measured to determine surface charge distributions as a function of pH. Prior to measurement, 10 mL of LNC dispersions was washed with ethanol and water three times by repeated centrifugation and resuspension. The final precipitates were redispersed in 10 mL of deionized water by ultrasonic treatment at 30% amplitude for 10 min. Each sample was diluted in deionized water and subjected to a pH gradient from 2 to 10. Then, the diluted sample was injected into a flow cell and the zeta-potential was measured at different cell positions, as follows: 0.7, 0.35, 0, −0.35 and −0.7. Each sample was measured in triplicate with three different sets at 25 °C and a scattering angle of 35° using a zeta-potential analyzer.

3.6. Encapsulation Efficiency

Encapsulation efficiency and drug loading were determined by measuring the concentration of indomethacin loaded in the lipid carrier. A 1 mL aliquot of NLC dispersion was added to 4 mL of DMSO and stirred at 60 °C in order to completely dissolve the lipid particle and collect the released indomethacin. The mixture was then cooled and filtered through a 0.2 µm syringe filter. An HPLC system (Agilent Technologies, Santa Clara, CA, USA) equipped with an auto sampler and U*V*/*V*is detector was used to determine the concentration of indomethacin. The chromatograms of indomethacin were recorded using a Unisol reversed-phase C18 column (4.6 mm × 250 mm, 5 µm, 100 Å) (Agela Technologies, Wilmington, DE, USA) with a U*V*/*V*is detector at 320 nm. The flow rate was 1 mL/min and the mobile phase consisted of acetonitrile and 0.02 M acetate buffer (60:40, *v*/*v*). A calibration curve (R^2 = 0.9998) was prepared from indomethacin standard solutions (10, 50, 100, 300, 500, 700 and 1000 µg/mL). Encapsulation efficiency was calculated using the following equation [18].

$$\text{Encapsulation efficiency (\%)} = \frac{\text{Total drug content enprapped}}{\text{Initial drug content}} \times 100 \qquad (1)$$

3.7. Differential Scanning Calorimetry

Melting behaviors and relative crystallinity of the NLC formulations were determined by differential scanning calorimetry (DSC 200 F3, Netzsch-Gerätebau GmbH, Selb, Germany). For the DSC measurement, exactly 3 mg of the lyophilized NLC formulations were weighed in a standard aluminum pan, and hermetically sealed with a standard aluminum lid. An empty pan was used as a reference. All samples were heated and cooled from 0 to 200 °C at a scan rate of 5 °C/min. Melting temperature and relative crystallinity of the formulations were determined from the melting curve based on the temperature at which endothermic peak appeared and the integration of the peak, respectively. Crystallinity was calculated by the manufacturer's instructions (Netzsch-Gerätebau GmbH). In this study, crystallization enthalpy of all NLC formulations was negligible.

$$\text{Relative crystallinity (\%)} = \frac{\text{Integration of melting enthalpy of NLC}}{\text{Integration of melting enthalpy of bulk lipid}} \times 100 \qquad (2)$$

3.8. In vitro Drug Release Test

In vitro release test of indomethacin loaded NLC carriers was carried out for NLC_I, NLC/CS_I and NLC/HTCC_I formulations. Briefly, 5 mL of aqueous NLC formulations equivalent to 4.2, 5.3 and 5.6 mg indomethacin for NLC_I, NLC/CS_I and NLC/HTCC_I, respectively, were introduced in 10 mM PBS (pH 7.4, Sigma-Aldrich) solution in 250 mL flask. Additionally, drug release test on non-encapsulated indomethacin (7.4 mg) was performed as a control group. The flasks with 100 mL of *in vitro* release test solutions were incubated at 37 °C on a temperature controlled hot plate with constant stirring at 300 rpm for 36 h. Five mL aliquot was collected from each flask at predetermined time intervals using a syringe with a 0.45 µm pre-filter. After collecting each aliquot, 5 mL of fresh PBS buffer solution was added to the flask. The collected sample solution was filtered using 0.22 µm syringe filter again and transferred to a cuvette. Absorbance was measured at a wavelength of 298 nm using a spectrophotometer. A calibration curve (R^2 = 0.9966) was plotted using the absorbance data measured at indomethacin concentrations (1, 3, 5, 10, 30, 50, and 100 µg/mL in 10mM NaOH). *In vitro* drug release tests were performed in triplicate for each sample. *In vitro* release profiles were fitted using several non-linear models such as zero-order (cumulative percentage of drug released *vs.* time), first-order (log cumulative percentage of drug remaining *vs.* time), Higuchi (cumulative percentage of released *vs.* square root of time), and Korsmeyer-Peppas (log cumulative percentage of drug released *vs.* log time) equations.

3.9. Static in Vitro Digestion Test

The indomethacin release properties of lipid nanoparticles were determined by a static *in vitro* digestion model described by Minekus *et al.* [17]. Digestion was initiated by mixing 10 mL of NLC dispersion with 9 mL of simulated gastric fluid (SGF) containing electrolytes and pepsin (EC 3.4.23.1) (6.9 mM KCl, 9.0 mM KH_2PO_4, 12.5 mM $NaHCO_3$, 11.8 mM NaCl, 0.4 mM $MgCl_2(H_2O)_6$, 0.5 mM $(NH_4)_2CO_3$, 0.15 mM $CaCl_2(H_2O)_2$, 2000 U/mL pepsin). The pH of the mixture was adjusted to 3 by 1N HCl solution, and the resulting mixture was made up to 20 mL using deionized water. The mixture was incubated at 37 °C under constant magnetic stirring at 300 rpm for 2 h. After digestion in SGF, 10 mL of gastric chyme was mixed with 9 mL of simulated intestinal fluid (SIF) containing appropriate electrolyte concentrations, pancreatin (EC 232-468-9,4 USP specifications), and bile extract (bile extract porcine, Sigma-Aldrich) (6.8 mM KCl, 0.8 mM KH_2PO_4, 85 mM $NaHCO_3$, 38.4 mM NaCl, 0.33 mM $MgCl_2(H_2O)_6$, 0.6 mM $CaCl_2(H_2O)_2$, 800 U/mL pancreatin, 300 mg bile extract porcine), and incubated for a further 2 h. During the digestion, 1 mL of reaction mixture was withdrawn and the same volume of fresh SGF or SIF solution was added every 1 h. Free indomethacin molecules were collected by ultrafiltration (Vivaspin20 1000K device, MWCO 1,000 K, Sartorius Stedim Lab Ltd., Gloucestershire, UK). The amount of free indomethacin concentration was measured by HPLC described in the previous section.

4. CONCLUSIONS

The present research paper proposed cationically modified lipid-based drug carriers (NLCs) for improving water solubility and bioavailability of poorly soluble drug. Indomethacin was selected as a model drug and incorporated in NLCs. Positively charged chitosan derivative, HTCC was synthesized to modify the surface of NLCs cationically. NLC/CS_I and NLC/HTTC_I formulations showed higher encapsulation efficiency for indomethacin than NLC_I formulation. NLC/CS_I and NLC/HTCC_I formulations are considered as effective carriers for sustained drug release. Thus, it is concluded that positively charged NLCs such as NLC/CS_I and NLC/HTCC_I formulations could improve bioavailability of poorly soluble drugs.

ACKNOWLEDGMENTS

This research was supported by Basic Science Research Program through the National Research Foundation of Korea (NRF) funded by the Ministry of Science, ICT & Future Planning (No. 2014M3A7B4051125).

AUTHOR CONTRIBUTIONS

K.-O. Choi prepared nanostructured lipid carriers and drafted the manuscript. J. Choe prepared and characterized HTCC. S. Suh performed static *in vitro* release test. S. Ko designed the study, interpreted the results and drafted the manuscript.

CONFLICTS OF INTEREST

The authors declare no conflict of interest.

REFERENCES

1. Arrunategui, L.B.; Silva-Barcellos, N.M.; Bellavinha, K.R.; Ev, L.S.; de Souza, J. Biopharmaceutics classification system: importance and inclusion in biowaiver guidance. Braz. J. Pharm. Sci. 2015, 51, 143–154.

2. Kawabata, Y.; Wada, K.; Nakatani, M.; Yamada, S.; Onoue, S. Formulation design for poorly water-soluble drugs based on biopharmaceutics

classification system: Basic approaches and practical applications. Int. J. Pharm. 2011, 420, 1–10.

3. Shrestha, H.; Bala, R.; Arora, S. Lipid-based drug delivery systems. Int. J. Pharm. 2014, 2014.

4. Nayak, A.K.; Maji, R.; Das, B. Gastroretentive drug delivery systems: A review. Asian J. Pharm. Clin. Res. 2010, 3, 2–10. [Google Scholar]

5. Kobayashi, K.; Wei, J.; Iida, R.; Ijiro, K.; Niikura, K. Surface engineering of nanoparticles for therapeutic applications.Polym. J. 2014, 46, 460–468.

6. Mishra, G.P.; Bagui, M.; Tamboli, V.; Mitra, A.K. Recent applications of liposomes in ophthalmic drug delivery. J. Drug Deliv. 2011, 2011.

7. Wang, Y.; Li, P.; Kong, L. Chitosan-modified PLGA nanoparticles with versatile surface for improved drug delivery.AAPS PharmSciTech 2013, 14, 585–592.

8. Ramalingam, P.; Ko, Y. Enhanced oral delivery of curcumin from N-trimethyl chitosan surface-modified solid lipid nanoparticles: Pharmacokinetic and brain distribution evaluations. Pharm. Res. 2015, 32, 389–402.

9. Sashiwa, H.; Aiba, S. Chemically modified chitin and chitosan as biomaterials. Prog. Polym. Sci. 2004, 29, 887–908.

10. Wu, J.; Su, Z.G.; Ma, G.H. A thermo- and pH-sensitive hydrogel composed of quaternized chitosan/glycerophosphate. Int. J. Pharm. 2006, 315, 1–11.

11. Hirai, A.; Odani, H.; Nakajima, A. Determination of degree of deacetylation of chitosan by 1H NMR spectroscopy.Polym. Bull. 1991, 26, 87–94.

12. Varum, K.M.; Antohonsen, M.W.; Grasdalen, H.; Smidsrød, O. Determination of the degree of N-acetylation and the distribution of N-acetyl groups in partially N-deacetylated chitins (chitosans) by high-field n.m.r. spectroscopy.Carbohydr. Res. 1991, 211, 17–23.

13. Yang, X.; Zhang, C.; Qiao, C.; Mu, X.; Li, T.; Xu, J.; Shi, L.; Zhang, D. A simple and convenient method to synthesizeN-[(2-hydroxyl)-propyl-3-trimethylammonium] chitosan chloride in an ionic liquid. Carbohydr. Polym. 2015, 130, 325–332.

14. Desbrieres, J.; Bousquet, C.; Babak, V. Surfactant-chitosan interactions and application to emulsion stabilization. Cell. Chem. Technol. 2010, 44, 395–406. [Google Scholar]

15. Montasser, I.; Shahgaldian, P.; Perret, F.; Coleman, A. Solid lipid nanoparticle-based calix[n]arenes and calix-resorcinarenes as building blocks: Synthesis, formulation and characterization. Int. J. Mol. Sci. 2013, 14, 21899–21942.

16. Choi, K.O.; Aditya, N.P.; Ko, S. Effect of aqueous pH and electrolyte concentration on structure, stability and flow behavior of non-ionic surfactant based solid lipid nanoparticles. Food Chem. 2014, 147, 239–244.

17. Lim, S.H.; Hudson, S.M. Synthesis and antimicrobial activity of a water-soluble chitosan derivative with a fiber-reactive group. Carbohydr. Res. 2004, 339, 313–319.

18. Papadimitriou, S.; Bikiaris, D. Novel self-assembled core–shell nanoparticles based on crystalline amorphous moieties of aliphatic copolyesters for efficient controlled drug release. J. Control. Release 2009, 138, 177–184.

Chapter 5

Determination of Solubility Parameters of Ibuprofen and Ibuprofen Lysinate

Teja Kitak [1], Aleksandra Dumičić [2], Odon Planinšek [1], Rok Šibanc [1] and Stanko Srčič [1,*]

[1]*Faculty of Pharmacy, University of Ljubljana, Aškerčeva cesta 7, 1000 Ljubljana, Slovenia*

[2]*Zenitva k.s., U Kabelovny 130, 102 37 Prague 10, Czech Republic*

ABSTRACT

In recent years there has been a growing interest in formulating solid dispersions, which purposes mainly include solubility enhancement, sustained drug release and taste masking. The most notable problem by these dispersions is drug-carrier (in)solubility. Here we focus on solubility parameters as a tool for predicting the solubility of a drug in certain carriers. Solubility parameters were determined in two different ways: solely by using calculation methods, and by experimental approaches. Six different calculation methods were applied in order to calculate the solubility parameters of the drug ibuprofen and several excipients. However, we were not able to do so in the case of ibuprofen lysinate, as calculation models for salts are still not defined. Therefore, the extended Hansen's approach and inverse gas chromatography (IGC) were used for evaluating of solubility parameters for ibuprofen lysinate. The obtained values of the total solubility parameter did not differ much between the two methods: by the extended Hansen's approach it was $\delta t = 31.15$ MPa$^{0.5}$ and with IGC it was $\delta_t = 35.17$ MPa$^{0.5}$. However, the values of partial solubility parameters, *i.e.*, δ_d, δ_p and δ_h, did differ from each other, what might be due to the complex behaviour of a salt in the presence of various solvents.

Keywords: Hansen's solubility parameters; extended Hansen's approach; inverse gas chromatography; ibuprofen; ibuprofen lysinate

1. INTRODUCTION

One of the main challenges in the fields of formulation development is the formulation of new and outstanding products that will meet the required regulatory and market criteria. The limits that often prevent the success of such goals may occur at various steps of product development. In this work we focused on how to overcome the limits which often occur during pre-formulation studies and are closely related to the solubility of a drug with excipients. One way of determining whether the drug is miscible with a selected excipient is the use of a parameter, called solubility parameter. Initially, the idea of solubility parameter was introduced by Hildebrand and Scott and was based on an estimation of the solubility behavior of substances [1,2]. Hildebrand defined the solubility parameter, δ_t, as square root of the cohesive energy density (CED) of a substance. CED is an energy required to separate the atoms or molecules from each other and it's a direct criterion of attractiveness [3]:

$$\delta_t = (\mathrm{CED})^{0.5} = \left(\frac{\Delta H_v - RT}{V_m}\right)^{0.5} \quad (1)$$

where ΔH_v is the heat of vaporization, R is the gas constant, T is the temperature and V_m is the molar volume.

This concept was originally developed for simple liquid mixtures and can be used in various practical applications for this reason. However, it can't meet the needs of more complex systems, as are typical pharmaceutical substances [4]. A number of approximations and extensions of Hildebrand's concept have been made due to better understanding of the complex interactions between substances. The most established is the so-called Hansen approach [5,6,7]. The basis of his theory lies within dividing total cohesive energy, E_{coh}, on components arising from dispersion forces, E_d, permanent dipole-dipole interactions, E_p, and hydrogen bonds, E_h [8]:

$$E_{coh} = E_d + E_p + E_h \quad (2)$$

If the equation above is divided by molar volume, V_m, and squared, it gives us a new three dimensional approach in which Hildebrand's total solubility parameter squared is equal to the sum of the squares of Hansen's components [8]:

$$\delta_t^2 = \delta_d^2 + \delta_p^2 + \delta_h^2 \quad (3)$$

where δ_d, δ_p and δ_h are Hansen's solubility parameters which represent the three main types of interactions in organic matter.

The most commonly studied interactions in the pharmaceutical industry are interactions related to solubility, where the greatest interest lies in solubility interactions between solvent and polymer, solvent and drug and also polymer and drug. For this purpose, a number of approaches based on solubility parameter theory or "like dissolves like", have been developed.

Van Krevelen and Hoftyzer used the parameter $\Delta\bar{\delta}$ to describe mixing between two substances [3]:

$$\Delta\bar{\delta} = \left[(\delta_{d2} - \delta_{d1})^2 + (\delta_{p2} - \delta_{p1})^2 + (\delta_{h2} - \delta_{h1})^2\right]^{0.5} \quad (4)$$

The threshold value at which there is a high probability of miscibility between two substances, was determined at $\Delta\bar{\delta} \leq 5$ MPa0.5 [3,9]. Primarily, miscibility was only considered for a solution with a constant concentration of polymer as a solute [3,9,10]. However, recent studies have brought the usefulness of the parameter $\Delta\bar{\delta}$ to completely different level, by determining the degree of mixing between active substance and various excipients [11].

On the other hand, Bagley [12] presented the theory of thermodynamic similarity between δ_d and δ_p, and introduced new, combined solubility parameter, δ_v:

$$\delta_v = \left(\delta_d^2 + \delta_p^2\right)^{0.5} \quad (5)$$

By using this approach one can project the three-dimensional solubility parameter space into a two-dimensional plot. Plot δ_v as a function of δ_h is called Bagley's plot. The distance, $R_{a(v)}$, in the Bagley's plot, can be expressed by the following equation:

$$R_{a(v)} = \left[4\left(\delta_{v2} - \delta_{v1}\right)^2 + \left(\delta_{h2} - \delta_{h1}\right)^2\right]^{0.5} \tag{6}$$

and represents the level of solubility/miscibility between substances. Bagley plots have proved especially useful in the case of determining solubility between polymers and solvents [3], but other studies [11,13] also prove its usefulness in determining miscibility between an active substance and an excipient or polymer. In the case of miscibility, the value of the parameter $R_{a(v)}$ should be in the range of $R_{a(v)} \leq 5.6$ MPa$^{0.5}$.

It seems like the most often used approach in determining miscibility between two components, is Greenhalgh's approach. In order to define miscibility, Greenhalgh *et al.* [14] used the difference in values of total solubility parameters, δ_t:

$$\Delta\delta_t = \left|\delta_{t2} - \delta_{t1}\right| \tag{7}$$

The results of many combinations of active substances and excipients gave a clear trend of solubility at $\Delta\delta_t < 7$ MPa$^{0.5}$. They were also able to estimate the difference of total solubility parameters at which the mixing is not likely to occur, $\Delta\delta_t > 10$ MPa$^{0.5}$ [14]. Numerous studies have confirmed the validity of Greenhalgh's approach [11,13,15,16,17], however, on the other hand some studies [18,19] have pointed out the deficiency of comparisons between total solubility parameters.

The total solubility parameter, δ_t, for solvents, polymers and active substances, can be determined by many different ways, using different methods, all being applied in the existing literature. In order to obtain greater insight on interactions between substances, it is crucial to divide the total solubility parameter into the individual contributions (δ_d, δ_p and δ_h), but there are only few methods for determination of all three partial solubility parameters. These correlations are still under research. In particular, the biggest limitation is the determination of partial solubility parameters of a drug, since most methods, used in the case of solvents and polymers, cannot be applied to drug molecules [20]. In this work we focused on determination of solubility parameters, using two different approaches: solely by calculation and an experimental approach.

1.1. Calculation Methods

There are number of approaches for the calculation of solubility parameters of a substance by using group contribution methods. All of them share the same theory that the total cohesive energy density of a molecule is an additive property and it is the sum of contributions from individual functional groups within the molecule [21]. The main advantage is the fast and easy way of determination of dispersive, polar and hydrogen bonding solubility parameters by having an adequate knowledge of molecular structure. Furthermore, numerous references demonstrate the credibility of solubility parameter calculations for various excipients, with a predominance of polymers, and several drugs [3,8,11,13,14,17,21,22]. Group contribution methods may be accurate and time-saving methods, however, their usage is limited. One main disadvantage is the absence of the data needed for the calculation of solubility parameters of active salts [23].

1.2. Experimental Approach

1.2.1. Extended Hansen's Approach (EHA)

The extended Hansen approach is a method that uses experimental data in a regression model in order to obtain values of solubility parameters of selected substances. Bustamante *et al.*, complemented the original extended Hansen approach [5,24] and used a regression model that links the solubility parameters of solvents and $\ln_{\chi 2}$, with the solubility parameters of a drug:

$$ln\,\chi_2 = C_0 + C_1\delta_{1d}^2 + C_2\delta_{1d} + C_3\delta_{1p}^2 + C_4\delta_{1p} + C_5\delta_{1h}^2 + C_6\delta_{1h} \quad (8)$$

where the numbers 1 and 2 refer to the solvent and solute, respectively, C_{0-6} are the coefficients of the regression analysis and $\ln_{\chi 2}$ is a mole fraction solubility of a drug in a given solvent. Partial solubility parameters are then calculated using the constants of the regression analysis, C_{0-6}:

$$\delta_{2d} = -\left(\frac{C_2}{2C_1}\right) \quad (9)$$

$$\delta_{2p} = -\left(\frac{C_4}{2C_3}\right) \tag{10}$$

$$\delta_{2h} = -\left(\frac{C_6}{2C_5}\right) \tag{11}$$

1.2.2. Inverse Gas Chromatography (IGC)

With the development of more and more complex technological processes, which enable more sophisticated pharmaceutical dosage forms, a tendency towards alternative methods of evaluation of physiochemical properties of solid pharmaceutical substances has arisen. Thus, in 1967 a modified form of gas chromatography, called inverse gas chromatography (IGC) was developed [25]. Unlike its conventional form, with the IGC method a solid sample is placed in a chromatography column, which represents the stationary phase. The stationary phase is washed with the mobile phase, which is represented as an inert gas, such as nitrogen or helium. Into the flow of inert gas are injected vapor phases of different organic solvents with known characteristics. The extent of interactions between the vapor of a selected solvent and solid sample is reflected in a specific retention volume, V_g. Once the value of V_g is obtained, chromatographic data can be converted into thermodynamic parameters using Equation (12) [26]:

$$\chi_{1,2}^{\infty} = \ln\left(\frac{273.15R}{p_1^0 V_g M_{r,1}}\right) - \frac{p_1^0}{RT}(B_{11} - V_1) + \ln\left(\frac{\rho_1}{\rho_2}\right) - \left(1 - \frac{V_1}{V_2}\right) \tag{12}$$

where the numbers 1 and 2 refer to the solvent and solute, respectively, and $M_{r,1}$, p01, B_{11}, V_g, V_1, V_2, ρ_1, ρ_2, represent the molecular mass, saturated vapor pressure, second virial coefficient, specific retention volume, molar volume and density of the solvent and solute, respectively.

There is a linear correlation between χ∞(1,2)i and δ_{1i}:

$$\frac{\delta_{1i}^2}{RT} - \frac{\chi_{(1,2)i}^{\infty}}{V_1} = \frac{2\delta_2}{RT}\delta_{1i} - \left(\frac{\delta_2^2}{RT} + \frac{\chi_s^{\infty}}{V_1}\right) \tag{13}$$

where $\frac{2\delta_2}{RT}$ represents the inclination of the straight line and is proportional with the solubility parameter of the substance, δ_2. Based on the given equation, δ_2 can also be calculated from the intersection with the abscissa $\left(\frac{\delta_2^2}{RT}+\frac{\chi_s^\infty}{V_1}\right)$.

Voelkel and Janas [27] upgraded the already known Price method [28] for the calculation of Hansen solubility parameters, and thereby contributed to the usefulness of IGC in determination of partial solubility parameters for solid substances [29,30,31,32]. In order to obtain values of partial solubility parameters, tested organic solvents, with known properties, must be classified regarding to their ability for intermolecular interactions. Solvents are divided into three groups, representing dispersion, polar and hydrogen bonding interactions. By using Equation (13) for each group of solvents, one can obtain three straight lines in graphical representation. Partial solubility parameters can then be calculated from the slope of the straight line, using [27]:

$$\delta_d = \frac{m_{n-\text{alkanes}} RT}{2} \quad (14)$$

$$\delta_p = \frac{\left(m_1 - m_{n-\text{alkanes}}\right) RT}{2} \quad (15)$$

$$\delta_h = \frac{\left(m_2 - \mathrm{m}_{n-\text{alkanes}}\right) RT}{2} \quad (16)$$

where $m_{n\text{-alkanes}}$ represents the value of the slope for the group of *n*-alkanes, m_1 represents the value of the slope for the group of polar solvents (aromatic hydrocarbons, ketones, 1-nitropropane, acetonitrile, 1,2-dichloroethane) and $m_2$represents the value of the slope for the group of solvents with the ability to form hydrogen bonds (alcohols, pyridine, 1,2-dioxane).

The crucial step in partial solubility determination, by using IGC, is selection of the solvents. In the case of a group of solvents that can form only dispersive interactions, the choice is rather simple. The chosen *n*-alkanes only show this type of interaction. However, this is not the case with the selection of the solvents in the other two groups. The criteria are not always precise, as in general, organic solvents can interact through both polar interactions and hydrogen bonding. One of the most commonly used criteria is the ratio between δh and δp. The higher the ratio, δp/δh, the more appropriate is the solvent to be classified in the polar solvents group. The same applies to the ratio δh/δp and the classification of solvents in the group which is able of forming hydrogen bonds [31].

The following work considers a classical and universally accepted concept of physical chemistry, the solubility parameter. The basis of solubility parameter theory lies within similarity or “like dissolves like”. However, the reader should be aware that the principle of similarity is often not sufficient enough to evaluate the solubility of a solute in a solvent. Other principles, such as complementarity, should also be taken into consideration. Similar observations have been previously stated on the interactions between the surfaces of liquids, solids-liquids and solids-solids as well.

Fowkes [33] noted that surface interactions could only occur between surfaces of similar type; for example no interaction due to the permanent dipoles can take place across an interface between polar and non-polar materials. This is the similarity principle. However, this concept failed to predict interfacial interactions between certain liquids. Van Oss *et al.*[34] described difficulties in considering polar interactions as all being of a similar type, and he pointed out that there are materials which could be described as polar, which, however, were dipolar, hydrogen bonding, Lewis acids, or Lewis bases. Considering the polar materials as either electron donor or electron acceptors explains why two polar materials can repel each other, when they are of a monopolar type. The same principle might be used in solubility observations as well.

There is therefore a continuing effort towards new definitions and understanding, which account for both similarity and complementarity components and thus can overcome some of the restrictions of the original definition. However, in this work we are still only using the principle of similarity approach and therefore the complementarity principle is not a part of further discussion.

The issue of further discussion is predicting the solubility of a given drug and its salt in polymers and non-aqueous solvents. By taking knowledge about the solubility of a drug into account, one can overcome limits, which arise during manufacturing process, as well as other stressful situations.

In this work, solubility parameters were chosen as potential parameters for predicting, estimating and evaluating the solubility of ibuprofen lysinate and the main goal was their determination.

2. RESULTS

2.1. Calculation Methods

2.1.1. Ibuprofen

Solubility parameters of ibuprofen were calculated, using six different group contribution methods. The results are summarized in Table 1.

Table 1. Values of solubility parameters for ibuprofen by using six different group contribution methods.

Method						
Hansen and Beerbower	Fedors	Hoftyzer and Van Krevelen	Hoy	Stefanis and Panayiotou	Just and Breitkreutz	Median δ_t ($MPa^{0.5}$)
$\delta_t = 19.89$	$\delta_t = 20.91$	$\delta_t = 19.36$ $\delta_d = 17.85$ $\delta_p = 2.22$ $\delta_h = 7.15$	$\delta_t = 19.71$	$\delta_t = 19.7$ $\delta_d = 17.56$ $\delta_p = 3.22$ $\delta_h = 8.31$	$\delta_t = 19.20$ $\delta_d = 16.59$ $\delta_p = 4.27$ $\delta_h = 8.67$	$\delta_t = 19.71$

Total solubility parameter values were consistent throughout the calculations, regardless to the method chosen. The lowest value of the total solubility parameter, $\delta_t = 19.20$ $MPa^{0.5}$, was obtained by using the group contribution method of Just and Breitkreutz. The highest value, $\delta_t = 20.91$ $MPa^{0.5}$, was obtained when using group contribution method of Fedors. The partial solubility parameters values also did not differ much between methods. The largest value in the case of the dispersion solubility parameter was obtained by using the group contribution method of Hoftyzer and Van Krevelen and the smallest value using the method of Just and Breitkreutz, where the values were $\delta_d = 17.85$ $MPa^{0.5}$ and $\delta_d = 16.59$ $MPa^{0.5}$, respectively. The opposite trend had been noticed in the case of polar and hydrogen bonding solubility parameters, where the lowest values were obtained by using the group contribution method of Hoftyzer and Van Krevelen, $\delta_p = 2.22$ $MPa^{0.5}$ in $\delta_h = 7.15$ $MPa^{0.5}$, and the highest values using the method of Just and Breitkreutz, $\delta_p = 4.27$ $MPa^{0.5}$ in $\delta_h = 8.67$ $MPa^{0.5}$. In the following, the median of all six group contribution methods was considered as the value of the total solubility parameter of ibuprofen.

2.1.2. Polymers and Excipients

Solubility parameters of some frequently used polymers and excipients were calculated using the same group contribution methods as in the case of ibuprofen. The results are summarized in Table 2.

Table 2. Values of solubility parameters for polymers and excipients by using six different group contribution methods.

Excipient	Method						
	Hansen and Beerbower	Fedors	Hoftyzer and Van Krevelen	Hoy	Stefanis and Panayiotou	Just and Breitkreutz	Median δ_t ($MPa^{0.5}$)
Copovidone	$\delta_t = 19.01$	$\delta_t = 22.89$	$\delta_t = 24.37$ $\delta_d = 19.23$ $\delta_p = 11.15$ $\delta_h = 9.67$	$\delta_t = 20.25$	$\delta_t = 22.39$ $\delta_d = 18.18$ $\delta_p = 10.15$ $\delta_h = 7.82$	$\delta_t = 21.24$ $\delta_d = 15.45$ $\delta_p = 8.61$ $\delta_h = 11.42$	$\delta_t = 21.82$
Povidone K30	$\delta_t = 18.96$	$\delta_t = 23.75$	$\delta_t = 26.28$ $\delta_d = 20.44$ $\delta_p = 13.67$ $\delta_h = 9.28$	$\delta_t = 20.05$	$\delta_t = 23.60$ $\delta_d = 19.05$ $\delta_p = 12.09$ $\delta_h = 6.86$	$\delta_t = 21.75$ $\delta_d = 16.67$ $\delta_p = 9.85$ $\delta_h = 9.9$	$\delta_t = 22.68$
PEG 6000	$\delta_t = 18.38$	$\delta_t = 19.17$	$\delta_t = 22.86$ $\delta_d = 17.78$ $\delta_p = 11.1$ $\delta_h = 9.13$	$\delta_t = 21.44$	$\delta_t = 20.45$ $\delta_d = 17.33$ $\delta_p = 7.93$ $\delta_h = 7.42$	$\delta_t = 17.76$ $\delta_d = 13.6$ $\delta_p = 11.09$ $\delta_h = 2.75$	$\delta_t = 19.81$
Eudragit® E PO	$\delta_t = 17.34$	$\delta_t = 19.59$	$\delta_t = 19.70$ $\delta_d = 17.35$ $\delta_p = 3.06$ $\delta_h = 8.81$	$\delta_t = 18.28$	$\delta_t = 23.51$ $\delta_d = 16.90$ $\delta_p = 15.45$ $\delta_h = 5.31$	$\delta_t = 16.04$ $\delta_d = 12.19$ $\delta_p = 1.89$ $\delta_h = 10.26$	$\delta_t = 19.00$
Eudragit® S100	$\delta_t = 20.02$	$\delta_t = 21.86$	$\delta_t = 21.1$ $\delta_d = 18.01$ $\delta_p = 3.64$ $\delta_h = 10.38$	$\delta_t = 19.29$	$\delta_t = 22.37$ $\delta_d = 16.31$ $\delta_p = 12.3$ $\delta_h = 9.11$	$\delta_t = 18.63$ $\delta_d = 12.18$ $\delta_p = 4.54$ $\delta_h = 13.34$	$\delta_t = 20.56$
Poloxamer 188	$\delta_t = 18.30$	$\delta_t = 18.81$	$\delta_t = 22.10$ $\delta_d = 17.49$ $\delta_p = 10.23$ $\delta_h = 8.74$	$\delta_t = 20.9$	$\delta_t = 20.33$ $\delta_d = 17.30$ $\delta_p = 7.72$ $\delta_h = 7.37$	$\delta_t = 17.31$ $\delta_d = 13.66$ $\delta_p = 10.24$ $\delta_h = 2.64$	$\delta_t = 19.57$
Poloxamer 407	$\delta_t = 18.26$	$\delta_t = 18.64$	$\delta_t = 21.75$ $\delta_d = 17.36$ $\delta_p = 9.85$ $\delta_h = 8.57$	$\delta_t = 20.66$	$\delta_t = 20.28$ $\delta_d = 17.29$ $\delta_p = 7.63$ $\delta_h = 7.35$	$\delta_t = 17.12$ $\delta_d = 13.68$ $\delta_p = 9.89$ $\delta_h = 2.58$	$\delta_t = 19.46$
Methylcellulose (DS = 1)	$\delta_t = 27.05$	$\delta_t = 29.78$	$\delta_t = 32.29$ $\delta_d = 20.73$ $\delta_p = 10.26$ $\delta_h = 22.54$	$\delta_t = 27.23$	$\delta_t = 24.82$ $\delta_d = 18.38$ $\delta_p = 7.15$ $\delta_h = 14.82$	$\delta_t = 27.45$ $\delta_d = 17.91$ $\delta_p = 17.74$ $\delta_h = 10.66$	$\delta_t = 27.34$
Ethylcellulose (DS = 3)	$\delta_t = 15.86$	$\delta_t = 18.79$	$\delta_t = 20.18$ $\delta_d = 17.59$ $\delta_p = 4.59$ $\delta_h = 8.77$	$\delta_t = 20.1$	$\delta_t = 19.58$ $\delta_d = 17.79$ $\delta_p = 6.44$ $\delta_h = 5.05$	$\delta_t = 21.42$ $\delta_d = 16.09$ $\delta_p = 14.05$ $\delta_h = 1.61$	$\delta_t = 19.84$
Xylitol	$\delta_t = 36.86$	$\delta_t = 40.60$	$\delta_t = 41.53$ $\delta_d = 20.75$ $\delta_p = 12.68$ $\delta_h = 33.67$	$\delta_t = 36.60$	$\delta_t = 30.56$ $\delta_d = 18.62$ $\delta_p = 11.87$ $\delta_h = 21.13$	$\delta_t = 31.3$ $\delta_d = 16.07$ $\delta_p = 21.15$ $\delta_h = 16.54$	$\delta_t = 36.73$
Sucrose	$\delta_t = 35.81$	$\delta_t = 39.95$	$\delta_t = 40.38$ $\delta_d = 21.76$ $\delta_p = 9.87$ $\delta_h = 32.55$	$\delta_t = 33.53$	$\delta_t = 40.83$ $\delta_d = 21.35$ $\delta_p = 18.44$ $\delta_h = 29.51$	$\delta_t = 31.15$ $\delta_d = 18.55$ $\delta_p = 19.63$ $\delta_h = 15.53$	$\delta_t = 37.88$
Lactose	$\delta_t = 34.07$	$\delta_t = 38.86$	$\delta_t = 40.46$ $\delta_d = 22.62$ $\delta_p = 9.63$ $\delta_h = 32.14$	$\delta_t = 31.48$	$\delta_t = 37.45$ $\delta_d = 21.78$ $\delta_p = 19.71$ $\delta_h = 23.14$	$\delta_t = 38.94$ $\delta_d = 20.07$ $\delta_p = 27.90$ $\delta_h = 18.30$	$\delta_t = 38.16$

The values of the total solubility parameter for each polymer and excipient did not differ significantly between different methods. The highest value was obtained using the group contribution method of Hoftyzer and Van Krevelen, in eight out of twelve cases. The lowest value showed up most often (in six out of twelve cases) when the method of Just and Breitkreutz was used.

In Table 2 it is clearly seen that the use of different methods gives different partial solubility parameter trends. In the case of dispersion solubility parameter the trend is similar to that of total solubility parameter. It doesn't vary significantly between methods. The highest values are obtained using the group contribution method by Hoftzyer and Van Krevelen, and the lowest values are obtained using the method by Just and Breitkreutz, in all cases. The highest values of the polar solubility parameter were obtained using the method by Just and Breitkreutz in most cases and the lowest values using the method by Hoftyzer and Van Krevelen. The opposite trend was noticed in the case of calculating the values of the hydrogen bonding solubility parameter. The main reason for the deviation between the results is the fact that different group contribution methods evaluate the same increments differently. For example, the increment –O– has a lower value for the polar solubility parameter contribution, Fpi, and a higher value for the hydrogen bonding solubility parameter contribution, Ehi, when comparing the group contribution method authored by Hoftyzer and Van Krevelen to the one by Just and Breitkreutz.

In what follows, the median of all six group contribution methods was considered as a value of the total solubility parameter of selected excipients and polymers. In the case of partial solubility parameters averaging was not taken into consideration, because of the significant deviation among methods. For this reason, further processing of partial solubility parameters considers the results from all calculations.

As mentioned in the introduction, values of the solubility parameters of ibuprofen lysinate cannot be determined using group contribution methods. Therefore, the use of experimental approaches was needed.

2.2. Experimental Approach

2.2.1. Extended Hansen's Approach (EHA)

The Extended Hansen approach was already applied to ibuprofen and its salt, sodium ibuprofenate, in order to obtain solubility parameter values [35]. However, it has not been used for the determination of solubility parameters of a complex salt, like ibuprofen lysinate. The aim of this work was therefore to obtain the solubility parameters of ibuprofen lysinate, by using EHA.

The solubility results, converted into mole fraction units, are listed in Table 3, together with the mole fraction solubility for ibuprofen and sodium ibuprofen, taken from [35]. The difference in solubility between all processed drug substances can be seen in Figure 1. The graphs can serve us as criterion of polarity and also for qualitative estimation of maximum solubility of a substance in the tested solvents. Since "like dissolves like" is a basis of solubility parameter theory, one can roughly estimate solubility parameters of given substance by finding the area of maximal solubility (Figure 1).

For further evaluation of solubility parameters, the experimental results, $\ln_{\chi 2}$, from the aforementioned procedure were also used in the regression model. By using simple linear regression for the solubility of a given drug, in all solvents, it was not possible to obtain an appropriate calculation procedure for determination of solubility parameters. Also, several types of robust regression, carried out in the statistical program R, gave us results that differ significantly. For this reason, an entirely different approach for processing the experimental data was chosen.

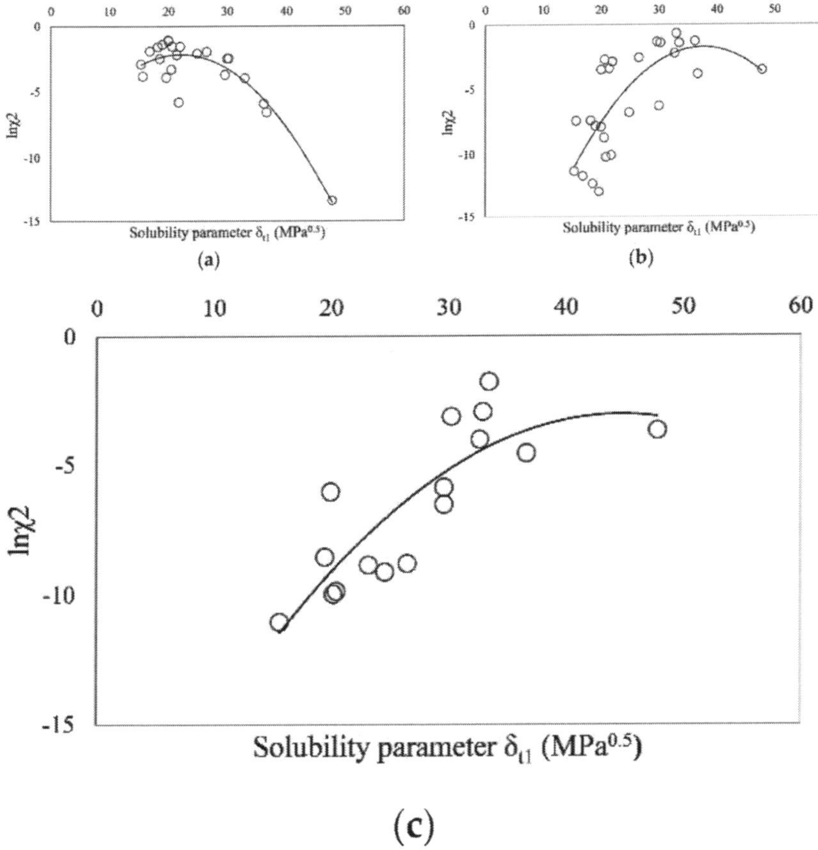

Figure 1. $\ln_{\chi 2}$ as a function of total solubility parameter for (**a**) ibuprofen; (**b**) sodium ibuprofen and (**c**) ibuprofen lysinate.

Table 3. Experimental determination of solubility, $\ln_{x2}$, for ibuprofen and sodium ibuprofen (data taken from [35]) and ibuprofen lysinate.

Solvent	δ_t ($MPa^{0.5}$)	Ibuprofen	Sodium Ibuprofen	Ibuprofen Lysinate
			$\ln x_2$	
Ethanol	26.52	−1.9503	−2.5573	−8.8013
Chloroform	18.95	−1.3712	−7.7910	/
Methanol	29.61	−3.7017	−1.3116	−5.8776
Benzene	18.51	−2.4684	−12.3017	/
Dioxane	20.47	−3.2918	−8.7174	−9.8535
Acetic acid	21.37	−2.1746	−3.3999	−3.0117
1-Pentanol	21.93	−1.5436	−2.8702	/
1-Butanol	23.20	/	/	−8.8584
1-Propanol	24.60	/	/	−9.1191
Cyclohexane	16.80	−1.8834	−11.6705	/
1,2-Propanediol	30.22	−2.4547	−1.4146	−3.1307
Formamide	36.65	−6.5485	−3.8359	−4.5675
Ethylene glycol	32.95	−3.9612	−0.6762	−2.9535
Glycerol	36.16	−5.8932	−1.2651	/
Ethyl acetate	18.15	−1.0942	−7.3852	−9.6817
Propionic acid	19.95	−1.5611	−3.4886	/
1-Octanol	20.56	−1.5134	−2.7018	/
Heptane	15.30	−2.8827	−11.2964	−9.1165
Chlorobenzene	19.58	−3.9057	−12.9644	/
Diethyl ether	15.64	−3.8065	−7.4103	−11.039
Acetone	19.94	−1.0474	−7.8829	−6.0326
Acetophenone	21.72	−5.7807	−10.0583	/
N,N-Dimethylformamide	24.86	−2.058	−6.7721	/
Dichloromethane	20.20	/	/	−9.9672
Water	47.81	−13.398	−3.5326	−3.6882
N-Methylformamide	29.98	/	−6.2656	/
1,4-Butanediol	33.44	/	−1.4053	/
1,3-Propanediol	32.65	/	−2.2121	/
Ethylene dichloride	20.80	−2.4549	−10.1933	/
Tetrahydrofuran	23.20	/	/	8.5397

Solubility parameters of ibuprofen lysinate were calculated using linear regression, for all possible combinations of eight different solvents among all tested solvents, in Wolfram Mathematica. The mathematical procedure of linear regression and combination of eight different solvents did not always deliver reasonable results. Therefore, unreasonable results such as negative values of solubility parameters were defined as failed combinations or errors. In Table 4 data on the number of combinations processed for each individual drug and the number of failed combinations/errors, are shown. Once all of the values of solubility parameters for all of combinations were obtained, the median value for each solubility parameter of given drug was calculated. The results are summarized in Table 4.

Table 4. The values of solubility parameters for ibuprofen, sodium ibuprofen and ibuprofen lysinat obtained by using EHA.

Drug	Solubility Parameter ($MPa^{0.5}$)				Analysis Parameters			
	δ_t	δ_d	δ_p	δ_h	No. of Solvents	No. of Combinations	No. of Errors	R^2
Ibuprofen	20.56	16.60	6.91	9.97	23	490,314	109,416	0.9964
Sodium ibuprofenate	34.12	16.38	12.52	27.19	26	1,562,275	220,884	0.9964
Ibuprofen lysinate	31.15	16.97	22.75	12.83	16	12,870	1573	0.7594

In addition, histograms of all three drugs were drawn by means of the results of regression analysis and are presented inFigure 2. By processing the histograms, we evaluated the most common value for individual solubility parameter and compared it with the calculated median value. Also, histograms were used in comparison of solubility parameters between ibuprofen and it salts, sodium ibuprofen and ibuprofen lysinate. By comparing all three histograms, it can be seen that dispersive solubility parameter takes nearly the same values in all three cases. The values of δ_d are highly condensed in a particular area, resulting in greater reliability of estimation of the dispersion solubility parameter. When comparing the histogram of ibuprofen to that of its salts, one can observe an expected shift of values of the polar and hydrogen bonding solubility parameters to higher values. Movements are in line with the increasing hydrophilicity of the ibuprofen molecule after adding lysine and sodium ions.

The choice of solvents can affect the histogram outcome. Due to the smaller number of solvents used in the case of ibuprofen lysinate, the data in the histogram seems to be more scattered in comparison with the histogram of ibuprofen and sodium ibuprofenate. The broader the distribution, the less reliable the reading on the histogram.

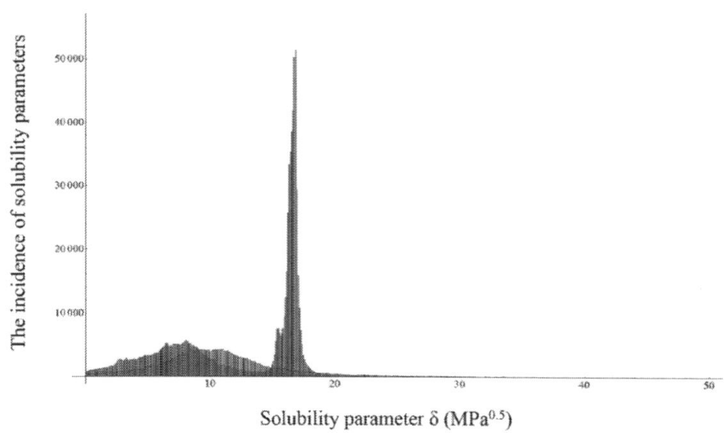

(a)

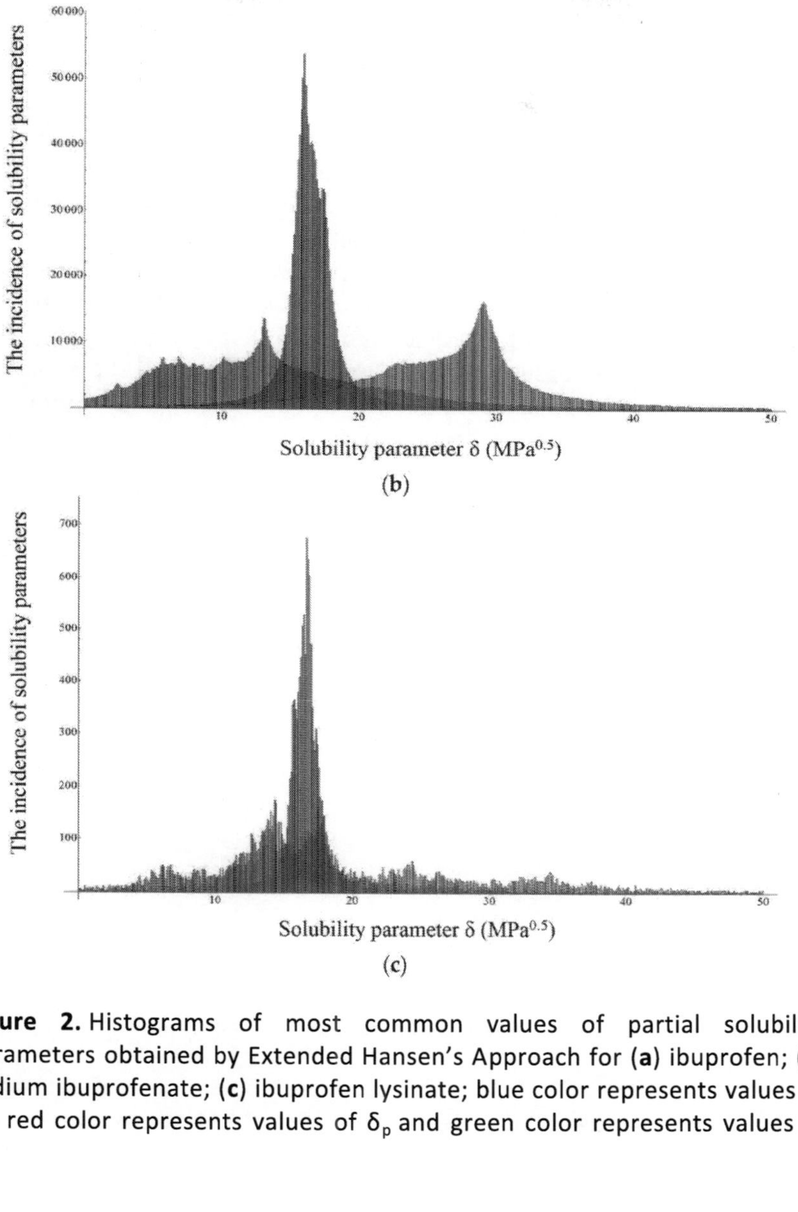

Figure 2. Histograms of most common values of partial solubility parameters obtained by Extended Hansen's Approach for (**a**) ibuprofen; (**b**) sodium ibuprofenate; (**c**) ibuprofen lysinate; blue color represents values of δ_d, red color represents values of δ_p and green color represents values of δ_h.

2.2.2. Inverse Gas Chromatography (IGC)

Inverse gas chromatography is considered to be one of the most reliable methods for the determination of solubility parameters. There are many published research papers where the solubility parameters of excipients and drugs were successfully determined [29,30,31,32,36]. However, there are none where the method of inverse gas chromatography was used in the determination of solubility parameters of active salts. Therefore, our task was to examine the possibility of using IGC in order to obtain the solubility parameters of an active salt.

As mentioned in the method introduction, the solubility parameter of given substance can be calculated using Equation (13) by two approaches: as a slope of a straight line or by the intersection of the straight line with the abscissa. Both approaches can only be used in case where linear fitting can be performed for the data points. On that occasion, the contributions of polar and hydrogen bonding solubility parameters are small. It is clear from the very structure of ibuprofen lysinate molecule that this is not going to be the case and one can easily predict a larger contribution to the polar and hydrogen bonding solubility parameter. We have shown this in practice, since the results couldn't be presented as a single straight line. Due to inability of such presentation, straight lines for each group of solvents were made. Solvents were divided into three groups, as described in the Experimental section, and gave three straight lines, as shown in Figure 3.

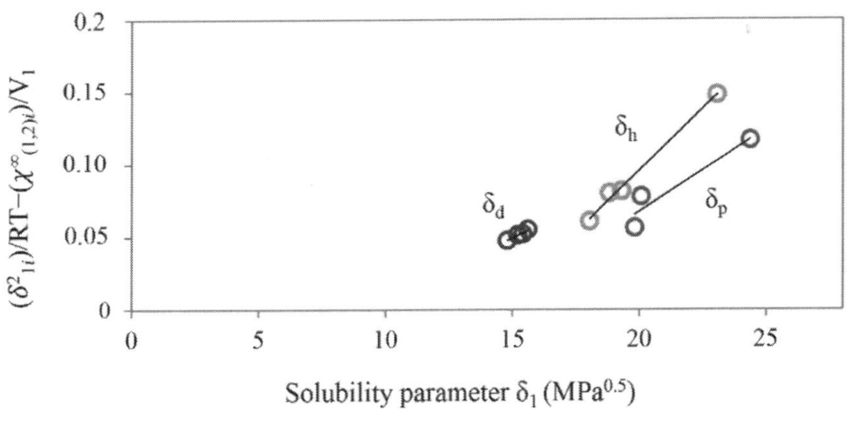

Figure 3. Graph $(\delta^2_{1i})/RT - (\chi^{\infty}_{(1,2)i})/V_1$ for ibuprofen lysinate as a function of total solubility parameter of solvents, δ_1, at T = 303.15 K; the contribution of δ_d is marked in blue color, δ_p in red color and δ_h in green color.

The first group of solvents, where *n*–alkanes were classified, represents the contribution of the dispersion solubility parameter and is marked in blue color. The second group of solvents is the group with predominance of the contribution of polar solubility parameter and is marked in red color. The third group, where the contribution of hydrogen bonding solubility parameter is predominant, is marked in green color. Given the fact that obtained straight lines have different slopes, calculation of solubility parameters of ibuprofen lysinate was made using the approach of intersection with the abscissa for each straight line separately. Partial solubility parameters were calculated for all temperatures, at which the IGC was implemented. Results were extrapolated and solubility parameters were determined at T = 25 °C. Results are shown in Table 5.

Table 5. Solubility parameters of ibuprofen, taken from [36], and ibuprofen lysinate obtained by using inverse gas chromatography.

Drug	Solubility Parameters			
	δ_d	δ_p	δ_h	δ_t
Ibuprofen	12.90	6.30	12.30	18.90
Ibuprofen lysinate	15.74	19.18	24.92	35.17

3. DISCUSSION OF RESULTS

The solubility parameters for ibuprofen and ibuprofen lysinate obtained from all used approaches are summarized inTable 6. For further discussion of solubility parameters, the median value of the group contribution methods for ibuprofen was chosen, as the results don't differ significantly among methods. In the case of ibuprofen lysinate, both experimental methods were taken into consideration, since one of the main goals was also the comparison between them as tools for the accurate determination of solubility parameters.

Table 6. The values of solubility parameters of ibuprofen and ibuprofen lysinate obtained by the calculation approach and by using experimental methods.

Drug	Method	Solubility Parameters ($MPa^{0.5}$)			
		δ_d	δ_p	δ_h	δ_t
Ibuprofen	*Group contribution method*				
	o Median	17.56	3.22	8.31	19.71
	Experimental				
	o Extended Hansen's approach	16.60	6.91	9.97	20.56
	o Inverse gas chromatography	12.90	6.30	12.30	18.90
Ibuprofen lysinate	*Experimental*				
	o Extended Hansen's approach	16.97	22.75	12.83	31.15
	o Inverse gas chromatography	15.74	19.18	24.92	35.17

3.1. Three-Dimensional Evaluation

For three-dimensional evaluation (3D) of solubility parameters, the approach of Hoftyzer and Van Krevelen was used. The position of substances and distance between them in three-dimensional space were determined using Equation (4). According to the authors, a distance smaller than 5 $MPa^{0.5}$, indicates potential solubility of a given drug in the tested excipient or polymer.

Figure 4 shows 3D spaces, in which drugs, polymers and excipients were placed, according to their partial solubility parameters. Polymers and excipients are marked as grey points, while ibuprofen and ibuprofen lysinate are in the form of spheres with a radius of 5 $MPa^{0.5}$. The sphere represents the area of solubility of a given drug, where red represents ibuprofen, green ibuprofen lysinate, whose values of solubility parameters were obtained using EHA and blue ibuprofen lysinate, whose solubility parameters value were obtained with the IGC method. In the case that the distance between drug and polymer or excipient was lower than 5 $MPa^{0.5}$, the interaction in three-dimensional space is specifically colored.

In the case of ibuprofen, at least one of the tested polymers or excipients is within the range of solubility or in the immediate vicinity, irrespective of the group contribution method used for determination of partial solubility parameters of polymers and excipients (Figure 4). The biggest amount of polymers and excipients within the distance of ≤5 $MPa^{0.5}$ is obtained by using method developed by Stefanis and Panayiotou (Figure 4b). In this case the solubilization of ibuprofen in PEG 6000, Poloxamer 188, Poloxamer 407 and ethylcellulose can be expected. In general, one can observe that different group contribution methods gave slightly different solubility predictions. For example, when using method of Hoftyzer and Van Krevelen, solubility of ibuprofen in Eudragit E PO, Eudragit S 100 and ethylcellulose is estimated (Figure 4a), and when using the method authored by Just and Breitkreutz, the estimation of solubility is limited only to Eudragit E PO (Figure 4c).

As it can be seen in Figure 4, the positions of ibuprofen lysinate spheres are different due to differences in the values of the partial solubility parameters. The difference in estimating the solubility of the drug in the 3D space is expected as well. In the case of ibuprofen lysinate, marked as a blue sphere, none of the excipients or polymers got close enough to be evaluated in regards of solubility. By using method authored by Stefanis and Panayiotou, only lactose and sucrose can be observed in the immediate vicinity to the sphere (Figure 4b). In the case of ibuprofen lysinate, marked as a green sphere, solubility in xylitol and sucrose can be estimated by using the method of Just and Breitkreutz (Figure 4c). Methylcellulose is also located near the ibuprofen lysinate sphere.

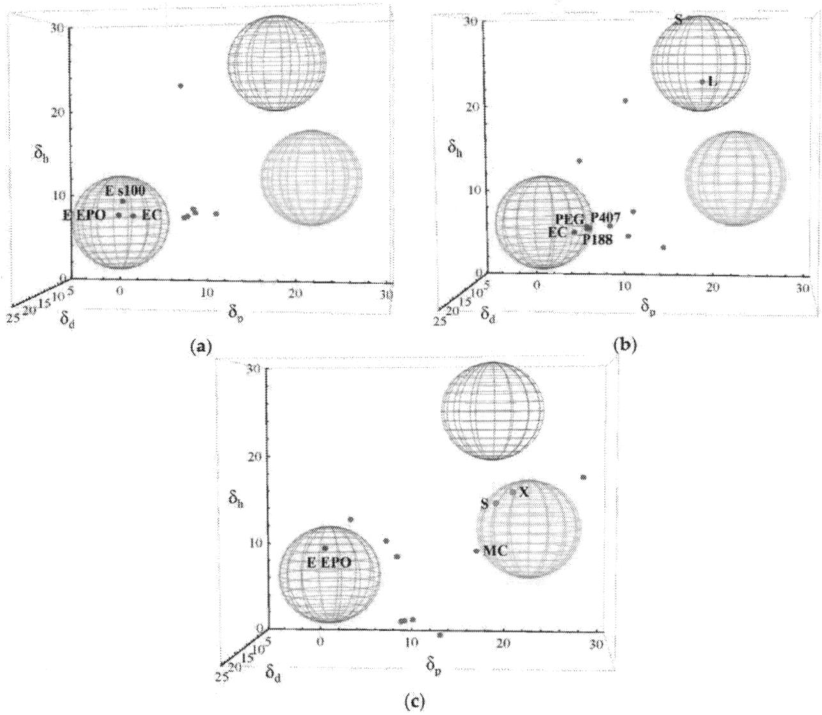

Figure 4. The three-dimensional evaluation of solubility parameters taken from Hoftyzer and Van Krevelen; values of partial solubility parameters for excipients and polymers determined using method authored by (**a**) Hoftyzer and Van Krevelen; (**b**) Stefanis and Panayiotou and (**c**) Just and Breitkreutz; The spheres represent the area of solubility of a given drug, where red represents ibuprofen, green ibuprofen lysinate, whose values of solubility parameters were obtained using EHA and blue ibuprofen lysinate, whose values of solubility parameters were obtained with the IGC method. EC—ethylcellulose; L—lactose; MC—methylcellulose; S—sucrose; X—xylitol; Eudragit E PO, P (188, 407) Poloxamers.

3.2. Two-Dimensional Evaluation

The solubility of given drugs in selected excipients and polymers can also be presented on a two-dimensional (2D) level. For such an evaluation, Bagley's plot was chosen. Given the Equation (6), an ellipse form is obtained in the graphical presentation (Figure 5). The center of the ellipse represents the given drug and the area within a radius of 5.6 $MPa^{0.5}$represents the range of solubility. Bagley's diagrams in Figure 5 differ according to the group contribution method choice used in determination of solubility parameters of polymers and excipients. Polymers and excipients are marked as grey circles, while ibuprofen is presented as a red ellipse, ibuprofen lysinate, whose values of solubility parameters were obtained using EHA, is presented as a green ellipse and ibuprofen lysinate, whose values of solubility parameters were obtained by using the IGC method, is presented as a blue ellipse.

In Figure 5 it is clearly shown that a considerable number of polymers and excipients are in the area of the ibuprofen ellipse in all three Bagley plots. This indicates a potentially good solubility of ibuprofen in these materials. All three Bagley plots predict the solubility of ibuprofen in Poloxamer 188 and Poloxamer 407. In addition, two out of three diagrams also estimate its solubility in ethylcellulose, Eudragit S 100 and PEG 6000.

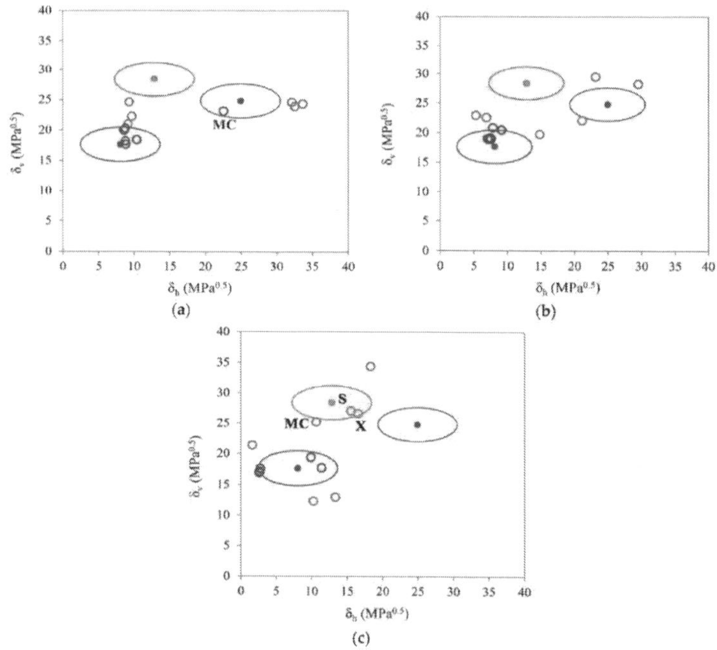

Figure 5. Bagley's plot; values of solubility parameters of polymers and excipients according to (**a**) Hoftyzer and Van Krevelen; (**b**) Stefanis and Panayiotou and (**c**) Just and Breitkreutz; polymers and excipients are marked as grey circles, while ibuprofen is presented as a red ellipse, ibuprofen lysinate, whose values of solubility parameters were obtained using EHA, is presented as a green ellipse and ibuprofen lysinate, whose values of solubility parameters were obtained by using the IGC method, is presented as a blue ellipse. MC—methylcellulose; S—sucrose; X—xylitol.

As it can be seen in Figure 5, the positions of the ibuprofen lysinate ellipses are different due to the differences in the values of the solubility parameters. In the case of the blue ellipse, when using the method authored by Hoftyzer and Van Krevelen, methylcellulose appears in the solubility area (Figure 5a). In the immediate vicinity of the blue ellipse, when using method authored by Stefanis and Panayiotou, xylitol also appears. In the case of the green ellipse of ibuprofen lysinate, when using method of Just and Breitkreutz a distance ≤5.6 $MPa^{0.5}$ is calculated only for xylitol, sucrose and methylcellulose (Figure 5c).

3.3. One-Dimensional Evaluation

The correlation between the solubility parameters of drugs and polymers and excipients can also be presented in one-dimensional (1D) form. For the 1D approach to the solubility, Greenhalgh's approach was chosen (Equation (7)). The one-dimensional approach only takes into consideration the difference between total solubility parameters. The difference should not exceed 7 $MPa^{0.5}$ in regards of solubility. Figure 6 illustrates a bar graph of the absolute difference between solubility parameters of ibuprofen and selected polymers and excipients. According to Greenhalgh's approach, solubility of ibuprofen was predicted in the majority of selected substances, with the exception of lactose, sucrose, xylitol and methylcellulose.

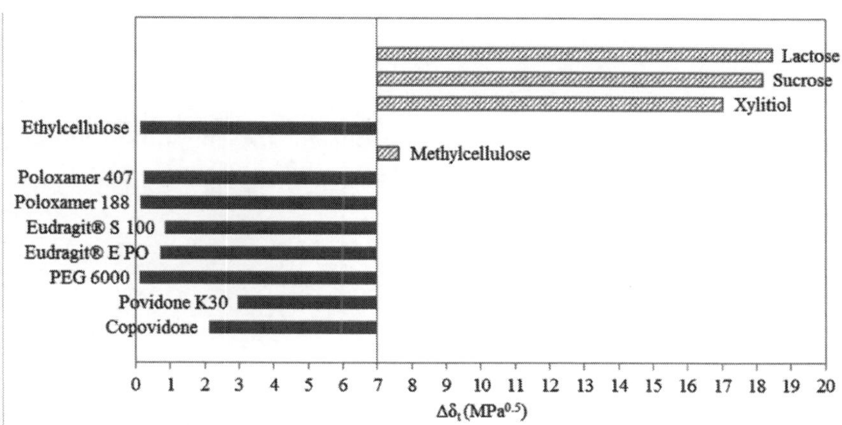

Figure 6. Bar graph for ibuprofen solubility according to Greenhalgh.

Bar graphs of the solubility evaluation of ibuprofen lysinate in selected polymers and excipients are shown in Figure 7and Figure 8. One can observe a common trend around the upper limit of solubility in both cases, where ibuprofen lysinate is estimated to be soluble in lactose, sucrose, xylitol and methylcellulose. Therefore, one-dimensional solubility prediction, according to Greenhalgh, gives comparable results between the two methods for determination of solubility parameters of ibuprofen lysinate–EHA and IGC. In the case, when EHA was used as a method for determination of solubility parameters (Figure 7), the lowest difference in the total solubility parameter, $\Delta\delta_t$, was calculated with methylcellulose. The value is $\Delta\delta_t$= 3.81 $MPa^{0.5}$. In the case of using IGC method for determination of solubility parameters (Figure 8), the lowest difference in the total solubility parameter, $\Delta\delta_t$, was calculated with xylitol. The value is $\Delta\delta_t$ = 1.56 $MPa^{0.5}$.

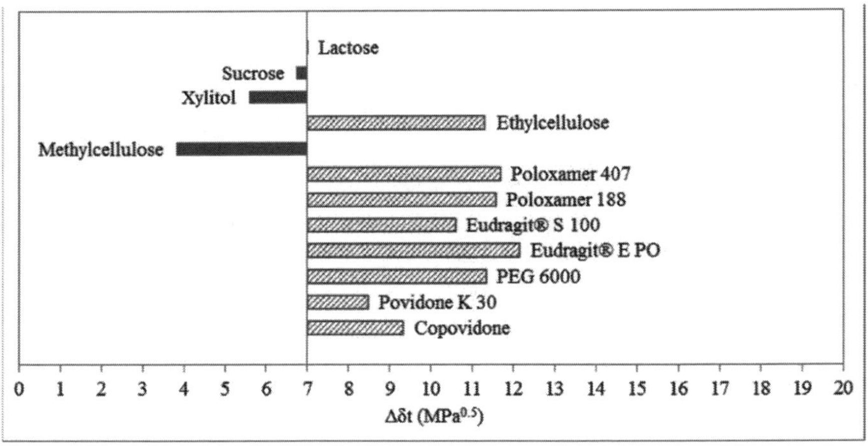

Figure 7. Bar graph according to Greenhalgh for ibuprofen lysinate; solubility parameters of ibuprofen lysinate were obtained using EHA.

4. MATERIALS AND METHODS

4.1. Materials

For the purpose of this work, white crystalline powders of ibuprofen and ibuprofen lysinate were used. The chemical structures of the two substances are shown in Figure 9. Substances were supplied by Zentiva, a part of the Sanofi group (Prague, Czech Republic). The following solvents of analytical grade were chosen for implementation of the extended Hansen approach: 1,2-propanediol, 1-butanol, 1-propanol, absolute ethanol, acetic

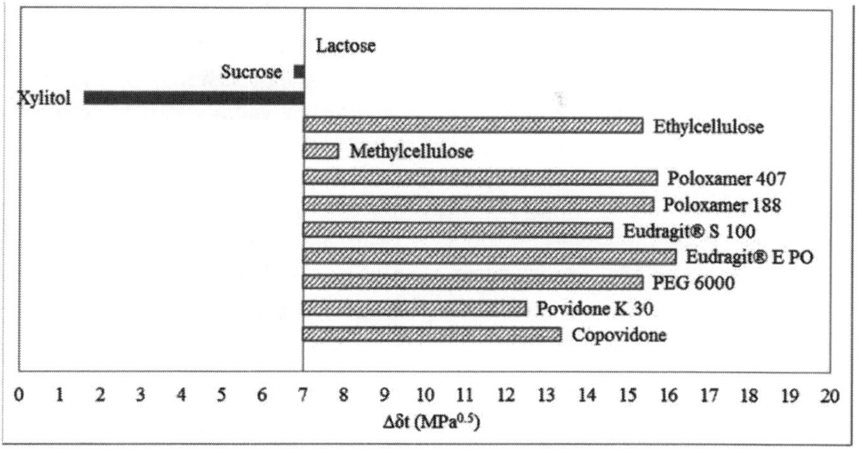

Figure 8. Bar graph according to Greenhalgh for ibuprofen lysinate; solubility parameters of ibuprofen lysinate were obtained using IGC.

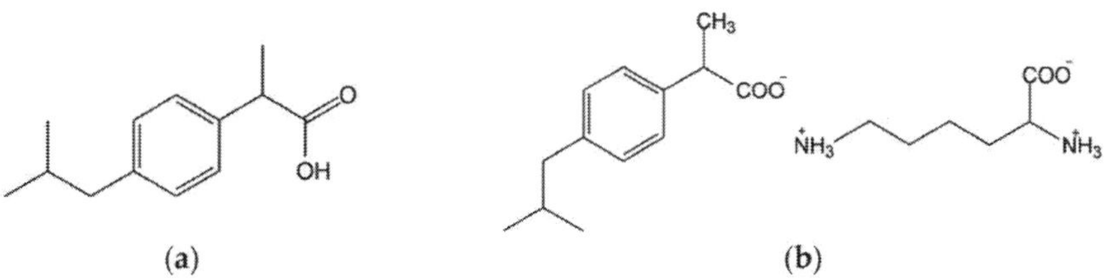

Figure 9. (**a**) Ibuprofen; (**b**) ibuprofen lysinate.

acid, acetone, acetonitrile, dichloromethane, diethyl ether, dioxane, ethyl acetate, ethylene glycol, formamide, heptane, methanol, tetrahydrofuran and distilled water. The following solvents were chosen for implementation of inverse gas chromatography: 1-butanol, acetone, acetonitrile, ethyl acetate, hexane, chloroform, methane, nonane, octane, tetrahydrofuran and toluene.

4.2. Methods

4.2.1. Calculation Methods

Group Contribution Methods (GCM)

Solubility parameters for the drug ibuprofen and several excipients typically used in hot melt processes, were calculated using six different group contribution methods, authored by: Hansen and Beerbower [37], Fedors [38], Hoftyzer and Van Krevelen [3], Hoy [39], Stefanis and Panayiotou [40], and Just and Breitkreutz [41]. Group contribution methods, authored by Hansen and Beerbower, Fedors and Hoy, gave a value of total solubility parameter, δ_t, directly, whilst other three group contribution methods, authored by Hoftyzer and Van Krevelen, Stefanis and Panayiotou and Just and Breitkreutz, gave the values of partial solubility parameters. The total solubility parameter, δ_t, is calculated using Equation (3), if needed. An example of calculation of solubility parameters for a drug ibuprofen and a polymer polyvinylpyrrolidone, using one of the group contribution methods, is shown in Table 7 and Table 8, respectively.

Table 7. Calculation of solubility parameters for the drug ibuprofen, using the group contribution method of Hoftyzer and Van Krevelen.

Individual Functional Group	Frequency	F_{di} $(MJ/m^3)^{0.5}\cdot mol^{-1}$	F_{pi}^2 $(MJ/m^3)^{0.5}\cdot mol^{-1}$	E_{hi} (J/mol)	V_m (cm^3/mol)
$-CH_3$	3	420 × 3	0	0	33.5 × 3
>CH–	2	80 × 2	0	0	−1.0 × 2
$-CH_2-$	1	270	0	0	16.1
phenylene (p)	1	1270	110^2	0	52.4
–COOH	1	530	420^2	10,000	28.5
Sum	-	3490	188,500	10,000	195.5
Calculations and results		$\delta_d = \frac{\sum F_{di}}{V_m}$		$\delta_d = 17.85$	
		$\delta_p = \frac{\sqrt{\sum F_{pi}^2}}{V_m}$		$\delta_p = 2.22$	
		$\delta_h = \sqrt{\frac{\sum E_{hi}}{V_m}}$		$\delta_h = 7.15$	
		$\delta_t^2 = \delta_d^2 + \delta_p^2 + \delta_h^2$		$\delta_t = 19.36$	

4.2.2. Experimental Approach

Extended Hansen Approach (EHA)

The solubility of ibuprofen lysinate was determined in 17 different solvents, chosen to represent various chemical classes and to cover a wide range of the solubility parameter scale. An excess of the drug was added to the solvent in a conical flask equipped with a ground-glass joint. The stoppered conical flasks were allowed to stir on a magnetic stirplate for 48 h, at T = 25 °C, until the equilibrium solubility had been attained. The suspensions were

Table 8. Calculation of solubility parameters for the polymer polyvinylpyrrolidone, using the group contribution method of Hoftyzer and Van Krevelen.

Individual Functional Group	Frequency	F_{di} $(MJ/m^3)^{0.5} \cdot mol^{-1}$	F_{pi}^2 $(MJ/m^3)^{0.5} \cdot mol^{-1}$	E_{hi} (J/mol)	V_m (cm^3/mol)
$-CH_2-$	4	270 × 4	0	0	16.1 × 4
>CH–	1	80	0	0	−1.0
–N<	1	20	800^2	5000	−9.0
>C=O	1	290	770^2	2000	10.8
ring (5)	1	190	-	-	16.0
Sum	-	1660	1,232,900	7000	81.2
Calculations and results		$\delta_d = \frac{\sum F_{di}}{V_m}$		$\delta_d = 20.44$	
		$\delta_p = \frac{\sqrt{\sum F_{pi}^2}}{V_m}$		$\delta_p = 13.67$	
		$\delta_h = \sqrt{\frac{\sum E_{hi}}{V_m}}$		$\delta_h = 9.28$	
		$\delta_t^2 = \delta_d^2 + \delta_p^2 + \delta_h^2$		$\delta_t = 26.28$	

then filtered through PTFE or RC filters, depending on the solvent used, with pore size of 0.45 μm. 2 mL of saturated solution was pipetted into three different glass flask and rotary evaporated. Distilled water (5–10 mL) was added to the dry residue and absorbance analyzed at 273 nm by using a UV-VIS spectrophotometer (Agilent HP UV-VIS 8453, HP Agilent Technologies Deutschland, Germany). In the case of non-volatile solvents, such as 1,2-propandiol, ethylene glycol, formamide and acetic acid, individual calibration curves were plotted. The densities of saturated solutions were determined using a 10 mL pycnometer, at T = 25 °C, three times for each saturated solution All the data from solubility measurements were used in further calculation of partial solubility parameters in Wolfram Mathematica, by using Equation (8) and a specific regression model. The applied regression model was a linear regression for all possible combinations of eight different solvents among all tested solvents.

Inverse Gas Chromatography (IGC)

Inverse gas chromatography was implemented on GC SMS (Agilent Technologies 6890N, Agilent Technologies, Santa Clara, CA, USA), using the following conditions: helium, with a flow rate of 5 mL/min was used as a carrier gas, methane was used as a non-interacting marker in order to correct the dead time retention, the measurements were carried out at four different temperatures, 303.15, 313.15, 323.15 and 333.15 K, with 13 different solvents. The solvents were chosen and classified into three groups, according to the ability to form different intermolecular interactions. Solvents in the first group (*n*-alkanes) were only able to interact with dispersion forces. In the second group were placed solvents with predominant ability to form polar interactions: acetone, acetonitrile and toluene. The third group included solvents which were capable of forming hydrogen bonds like tetrahydrofuran, ethyl acetate, chloroform and 1-butanol. Three injections of the vapor of each solvent were made for each probe at each temperature and the retention time was determined from the maximum of the symmetric peak. By knowing the retention time values of each solvent, we were able to evaluate partial solubility parameters for our sample, ibuprofen lysinate.

5. CONCLUSIONS

The determination of solubility parameters of a drug in a salt form is still in its beginnings. When dealing with such salts, determination is limited to experimental approaches only, such as the use of the extended Hansen approach (EHA) and inverse gas chromatography (IGC). Within the context of solubility parameter determination, the main focus was on experimental determinations for ibuprofen lysinate and critical evaluation of the results. The values of the total solubility parameter, δ_t, did not differ significantly among methods, whereas, the values of partial solubility parameters deviated. The largest deviation was estimated in the case of the polar, δ_p, and hydrogen bonding, δ_h, solubility parameters. Since the line between both of the aforementioned solubility parameters is rather thin, the better selection and greater number of solvents, used in EHA and IGC, could give us smaller deviations and more reliable data. Due to the simpler nature and structure of the ibuprofen molecule, we were able to use a calculation approach in order to obtain solubility parameter values. The calculation results were in good agreement when compared to experimental data, regarding all three solubility parameters. The interpretation of these results was nevertheless rather challenging, due to the complex behavior of salts in the presence of solvents. The existence of reliable models for salts, on which one could rely, is negligible so far.

ACKNOWLEDGMENTS

This research was carried out in R & D Labs of Zenitva k.s., Czech Republic, and on Faculty of Pharmacy, University of Ljubljana, Slovenia. The financial support for T. Kitak from both institutions is highly appreciated. The authors are expressing their special thanks to Saša Leskovšek, past Head of R & D Zentiva.

AUTHOR CONTRIBUTIONS

Conceived and designed the experiments: A.D., T.K., O.P., R.Š., S.S.; Experiments performed and analyzed: T.K., R.Š., O.P., S.S.; Paper prepared: T.K., R.Š., A.D., S.S.

CONFLICTS OF INTEREST

The authors declare no conflict of interest.

REFERENCES

1. Hildebrand, J.H.; Scott, R.L. The Solubility of Nonelectrolytes; Reinhold Pub. Corp.: New York, NY, USA, 1950; p. 514.

2. Hildebrand, J.H.; Scott, R.L. Regular Solutions; Prentice-Hall: Saddle River, NJ, USA, 1962; p. 202.

3. Van Krevelen, D.W.; te Nijenhuis, K. Properties of Polymers: Their Correlation with Chemical Structure; Their Numerical Estimation and Prediction from Additive Group Contributions; Elsevier: Amsterdam, The Netherland; Tokyo, Japan, 2009; p. 189.

4. Hancock, B.C.; York, P.; Rowe, R.C. The use of solubility parameters in pharmaceutical dosage form design. Int. J. Pharm. 1997, 148, 1–21.

5. Hansen, C.M. The Three Dimensional Solubility Parameter—Key to Paint Component Affinities I.—Solvents, Plasticizers, Polymers, and Resins. J. Paint Technol. 1967, 39, 104–117.

6. Hansen, C.M. The Three Dimensional Solubility Parameter—Key to Paint Component Affinities II.—Dyes, Emulsifiers, Mutual Solubility and Compatibility, and Pigments. J. Paint Technol. 1967, 39, 511–514.

7. Hansen, C.M. The Universality of the Solubility Parameter. Ind. Eng. Chem. Prod. Res. Dev. 1969, 8, 2–11.

8. Hansen, C.M. Hansen Solubility Parameters: A User's Handbook, 2nd ed.; CRC Press, Taylor Francis Group: Boca Raton, FL, USA, 2007; p. 544.

9. Güner, A. The algorithmic calculations of solubility parameter for the determination of interactions in dextran/certain polar solvent systems. Eur. Polym. J. 2004, 40, 1587–1594.

10. Swarbrick, J. Encyclopedia of Pharmaceutical Technology: Stat-Zeta; Informa Healthcare: New York, NY, USA; London, UK, 2007; p. 888.

11. Mohammad, M.A.; Alhalaweh, A.; Velaga, S.P. Hansen solubility parameter as a tool to predict cocrystal formation.Int. J. Pharm. 2011, 407, 63–71.

12. Bagley, E.B.; Nelson, T.P.; Scigliano, J.M. Three-dimensional solubility parameters and their relationship to internal pressure measurements in polar and hydrogen bonding solvents. J. Paint Technol. 1971, 43, 35–42.

13. Albers, J. Hot-Melt Extrusion with Poorly Soluble Drugs; Cuvillier Verlag: Goettingen, Germany, 2008; p. 145.

14. Greenhalgh, D.J.; Williams, A.C.; Timmins, P.; York, P. Solubility parameters as predictors of miscibility in solid dispersions. J. Pharm. Sci. 1999, 88, 1182–1190.

15. Ghebremeskel, A.N.; Vemavarapu, C.; Lodaya, M. Use of surfactants as plasticizers in preparing solid dispersions of poorly soluble API: Selection of polymer-surfactant combinations using solubility parameters and testing the processability. Int. J. Pharm. 2007, 328, 119–129.

16. He, H.; Yang, R.; Tang, X. In vitro and in vivo evaluation of fenofibrate solid dispersion prepared by hot-melt extrusion. Drug Dev. Ind. Pharm. 2010, 36, 681–687.

17. Chokshi, R.J.; Sandhu, H.K.; Iyer, R.M.; Shah, N.H.; Malick, A.W.; Zia, H. Characterization of physico-mechanical properties of indomethacin and polymers to assess their suitability for hot-melt extrusion processs as a means to manufacture solid dispersion/solution. J. Pharm. Sci. 2005, 94, 2463–2474.

18. Nair, R.; Nyamweya, N.; Gönen, S.; Martínez-Miranda, L.J.; Hoag, S.W. Influence of various drugs on the glass transition temperature of poly(vinylpyrrolidone): A thermodynamic and spectroscopic investigation. Int. J. Pharm.2001, 225, 83–96.

19. Zheng, X.; Yang, R.; Tang, X.; Zheng, L. Part I: Characterization of solid dispersions of nimodipine prepared by hot-melt extrusion. Drug Dev. Ind. Pharm. 2007, 33, 791–802.

20. Babu, P.R.S.; Subrahmanyam, C.V.S.; Thimmasetty, J.; Manavalan, R.; Valliappan, K. Extended Hansen's solubility approach: Meloxicam in individual solvents. Pak. J. Pharm. Sci. 2007, 20, 311–316. [PubMed]

21. Liu, J.; Xiao, Y.; Allen, C. Polymer-drug compatibility: A guide to the development of delivery systems for the anticancer agent, ellipticine. J. Pharm. Sci. 2004, 93, 132–143.

22. Forster, A.; Hempenstall, J.; Tucker, I.; Rades, T. Selection of excipients for melt extrusion with two poorly water-soluble drugs by solubility parameter calculation and thermal analysis. Int. J. Pharm. 2001, 226, 147–161.

23. Barra, J.; Peña, M.A.; Bustamante, P. Proposition of group molar constants for sodium to calculate the partial solubility parameters of sodium salts using the van Krevelen group contribution method. Eur. J. Pharm. Sci. Off. J. Eur. Fed. Pharm. Sci. 2000, 10, 153–161.

24. Beerbower, A.; Wu, P.L.; Martin, A. Expanded solubility parameter approach. I: Naphthalene and benzoic acid in individual solvents. J. Pharm. Sci. 1984, 73, 179–188.

25. Kiselev, A.V.; I͡Ashin, I.I. Gas-Adsorption Chromatography; Plenum Press: New York, NY, USA, 1969; p. 272.

26. Voelkel, A.; Milczewska, K.; Jęczalik, J. Characterization of the interactions in polymer/silica systems by inverse gas chromatography. Macromol. Symp. 2001, 169, 45–55.

27. Voelkel, A.; Janas, J. Solubility parameters of broad and narrow distributed oxyethylates of fatty alcohols. J. Chromatogr. A 1993, 645, 141–151.

28. Price, G.J.; Lloyd, D.R.; Ward, T.C.; Schreiber, H.P. Calculation of Solubility Parameters by Inverse Gas Chromatography. In ACS Symposium Series; Oxford University Press: Oxford, UK, 1989; Volume 391, pp. 48–58.

29. Voelkel, A.; Kopczyński, T. Inverse gas chromatography in the examination of organic compounds: Polarity and solubility parameters of isoquinoline derivatives. J. Chromatogr. A 1998, 795, 349–357.

30. Adamska, K.; Voelkel, A.; Héberger, K. Selection of solubility parameters for characterization of pharmaceutical excipients. J. Chromatogr. A 2007, 1171, 90–97.

31. Adamska, K.; Bellinghausen, R.; Voelkel, A. New procedure for the determination of Hansen solubility parameters by means of inverse gas chromatography. J. Chromatogr. A 2008, 1195, 146–149.

32. Adamska, K.; Voelkel, A. Inverse gas chromatographic determination of solubility parameters of excipients. Int. J. Pharm. 2005, 304, 11–17.

33. Fowkes, F.M. Dispersion force contribution to surface and interfacial tension, contact angles and heats of immersions.Adv. Chem. Ser. 1964, 43, 99–111.

34. Van Oss, C.J.; Chaudhury, M.K.; Good, R.J. Monopolar surfaces. Adv. Colloid Interface Sci. 1987, 28, 35–64.

35. Bustamante, P.; Pena, M.A.; Barra, J. The modified extended Hansen method to determine partial solubility parameters of drugs containing a single hydrogen bondi ng group and their sodium derivatives: Benzoic acid/Na and ibuprofen/Na. Int. J. Pharm. 2000, 194, 117–124.

36. Voelkel, A.; Adamska, K.; Strzemiecka, B.; Batko, K. Determination of Hansen solubility parameters of solid materials by inverse gas-solid chromatography. Acta Chromatogr. 2008, 20, 1–14.

37. Standen, A.; Kirk, R.E.; Othmer, D.F.; Hansen, C.M.; Beerbower, A. "Solubility Parameters" v Kirk-Othmer Encyclopedia of Chemical Technology, 2nd Completely Rev., Ed ed; Adamantane to Units. Interscience: New York, NY, USA, 1971; pp. 889–910.

38. Fedors, R.F. A method for estimating both the solubility parameters and molar volumes of liquids. Polym. Eng. Sci.1974, 14, 147–154.

39. Hoy, K.L. The Hoy: Tables of Solubility Parameters; Union Carbide Corp., Solvents & Coatings Materials, Research & Development Dept.: South Charleston, WV, USA, 1985.

40. Stefanis, E.; Panayiotou, C. Prediction of Hansen Solubility Parameters with a New Group-Contribution Method. Int. J. Thermophys. 2008, 29, 568–585.

41. Just, S.; Sievert, F.; Thommes, M.; Breitkreutz, J. Improved group contribution parameter set for the application of solubility parameters to melt extrusion. Eur. J. Pharm. Biopharm. 2013, 85 (3, Part B), 1191–1199.

Chapter 6

Role of Polymeric Excipients in the Stabilization of Olanzapine when Exposed to Aqueous Environments

Maria Paisana [1], Martin Wahl [2] and João Pinto [1,*]

[1]*iMed.ULisboa, Departamento de Farmácia Galénica e Tecnologia Farmacêutica, Faculdade de Farmácia, Universidade de Lisboa, Av. Prof. Gama Pinto, P-1649-003 Lisboa, Portugal*

[2]*Institut für Pharmazeutische Technologie und Biopharmazie, Pharmazeutisches Institut, Eberhard Karls Universität Tübingen, Auf der Morgenstelle 8, D-72076 Tübingen, Germany*

ABSTRACT

Hydrate formation is a phase transition which can occur during manufacturing processes involving water. This work considers the prevention of hydration of anhydrous olanzapine and hydrate conversions in the presence of water and polymers (polyethyleneglycol; hydroxypropylcellulose; polyvinylpyrrolidone) in forming pellets by wet extrusion and spheronisation. Anhydrous olanzapine was added to water with or without those polymers prior to extrusion with microcrystalline cellulose. Assessment of olanzapine conversion was made by XRP-Diffraction; FTIR spectroscopy; calorimetry (DSC) and microscopy (SEM for crystal size and shape). The addition of water converted the anhydrous form into dihydrate B and higher hydrate; whereas polyethyleneglycol promoted a selective hydrate conversion into the higher hydrate olanzapine form. Both polyvinylpyrrolidone and hydroxypropylcellulose prevented the hydrate transformations of the anhydrous drug; the latter even in the presence of hydrate seeds. This may be explained by the higher H-bond ability; higher network association and higher hydrophobicity of hydroxypropylcellulose by comparison with polyethyleneglycol and polyvinylpyrrolidone; which could contribute to its higher affinity to the crystal surfaces of the hydrate nuclei/initial crystals and promoting steric hindrance to the incorporation of other drug molecules into the crystal lattice; thus, preventing the crystal growth. The addition of microcrystalline cellulose needed for the pellets production (final product) did not eliminate the protector effect of both hydroxypropylcellulose and polyvinylpyrrolidone during pellets' processing and dissolution evaluation.

Keywords: anhydrous; extrusion; hydrate; hydroxypropylcellulose; olanzapine; pellet; polyethyleneglycol; polymorphism; polyvinylpyrrolidone; spheronisation

1. INTRODUCTION

Active pharmaceutical ingredients (APIs) are exposed to water in many pharmaceutical processes, such as wet granulation or wet extrusion, which can lead to the conversion of the anhydrate form into the hydrate form during processing. The subsequent drying processes may induce the dehydration of the API, possibly resulting in a solid-state form different from the original [1]. Phase transformation of an API is of pharmaceutical interest because such transformations may change the physical and chemical properties of the drug with an impact on its dissolution and bioavailability [2]. The hydrate formation in wet masses takes place via solvent-mediated transformations (SMT) involving three different steps: (1) dissolution of the anhydrate phase in the solvent used for the wet massing of the powders; (2) nucleation of the hydrate species and (3) growth of the hydrate phase [3,4]. Inhibiting one of these stages could be advantageous to maintaining the original crystal form of the API and avoiding a product containing a mixture of solid-state forms. This inhibition can be accomplished by the use of selected excipients, thus avoiding hydrate and hydrate conversions of the API during processing.

The process of extrusion-spheronisation is a technique which includes various processing steps, such as blending of the dry mass, wet massing, extrusion of the moist mass, spheronisation of the extrudate and drying. Depending on the API and excipients used, solvent-mediated polymorphic transformations (SMT) may take place during the wetting step [5], which may or may not be reversible after the drying step. For example, it has been observed that both theophylline and nitrofurantoin can hydrate during the wet extrusion process, transforming into theophylline and nitrofurantoin monohydrates, respectively. However, anhydrous theophylline was shown to have better wetting properties and, thus, formed hydrates more easily. This demonstrates that each drug has different stabilities in aqueous environments requiring different conditions for hydration to occur [5,6]. During drying, theophylline monohydrate was shown to dehydrate easier and at relatively lower temperatures (60 °C) than nitrofurantoin monohydrate (135 °C). Therefore, inhibiting hydrate formation could be advantageous to maintaining the original crystal form by avoiding a product containing a mixture of solid-state forms, which may have an impact on the properties and stability of the API.

Microcrystalline cellulose (MCC) is extensively used in wet extrusion processes [7]. The effect of MCC on the reduction of the hydrate formation of the drugs during the wet massing step was shown to be originating from its ability to hold large amounts of water in its internal structure, which can retard the hydrate formation [2]. However, the use of high contents of water (e.g., 40%–60% *w/w*) may reduce the protection of MCC on retarding the hydrate formation. The addition of other excipients to the formulation (e.g., polymers) may reduce the amount of water needed for the powder to agglomerate. Furthermore, polymers are known to inhibit the nucleation and/or growth of hydrated pharmaceutical crystals in aqueous slurries [3,4]. For instance, polyvinylpyrrolidone (PVP) has been demonstrated to inhibit the growth rate of sulfathiazole and indomethacin in solutions [3]. The cross-linked poly(acrylic acid) (PAA) has been shown to completely inhibit the caffeine anhydrate to hydrate transformation in aqueous environments [4]. Cellulose derivatives (e.g., HPC and HPMC) have been shown to be successful in inhibiting the hydrate conversion of carbamazepine but exhibit no effect on the transformation kinetics of anhydrous sulfaguanidine [4]. The different inhibitory behaviors of different polymers, and the same polymer with different APIs, reveal that there is no single polymer which promotes the inhibition of hydrate conversions in all APIs. This may be a consequence of different specific interactions that the polymer may establish with individual crystal surfaces [8,9].

Olanzapine, an antipsychotic drug, is an example of an API which can hydrate into three different crystalline dihydrates (B, C and E) and a higher hydrate form [10]. This ability to form multiple hydrates makes it difficult to control the anhydrate to hydrate transformation during the process of wet extrusion. To the best of our knowledge, studies on the monitoring of the polymorphic transitions of olanzapine during the entire extrusion-spheronisation process have not been reported previously. Furthermore, there is a lack of information about how olanzapine behaves in moist environments and how excipients may control the transformations that are likely to take place during the pharmaceutical processes involving moisture.

This study aims (1) to access the time needed for olanzapine to undergo hydrate conversion when olanzapine comes in contact with moisture (which occurs during the second step of wet extrusion process); (2) to study whether excipients (e.g., PEG, HPC, PVP or MCC) can influence the hydrate conversions of olanzapine both during and after the wet massing step; (3) to characterize different hydrate crystal forms of olanzapine that may be generated during processing and (4) to find a formulation which allows the maintenance of the original crystal form of the API throughout the entire extrusion-spheronisation process. In particular, the understanding of the mechanisms underpinning the effect of polymeric excipients on the polymorphic conversion of olanzapine in aqueous environments was analyzed.

2. RESULTS

The different polymers used in this work may be classified into three categories according to their molecular structure. In group I polyethyleneglycol with different molecular weights (PEG 6000 and 40,000) were considered; this polymer has both hydrogen bond donor and acceptor groups at the two ends of the polymer chain (Figure 1). Group II is constituted by a cellulose derivative polymer, hydroxypropylcellulose (HPC) with both hydrogen bond donor and acceptor groups in their ring structure (Figure 1). Group III includes polyvinylpyrrolidone (PVP) which present only one acceptor hydrogen bonding group in their ring structure. Furthermore, these polymeric materials also present different surface free energies which affect their spreading over OLZ particles.

2.1. Polymer Screening Method to Identify Polymers that can Stabilize OLZ FI during Wet Massing

The X-ray powder diffraction (XRPD) patterns for OLZ, OLZ:Polymer physical mixtures and the respective wet/dried masses are represented in Figure 2. In order to characterize the changes in the crystal structure of OLZ in each formulation, the XRPD patterns were compared to the XRPD patterns of different crystal structures of OLZ, such as, anhydrous form I, dihydrate B, dihydrate E and higher hydrate.

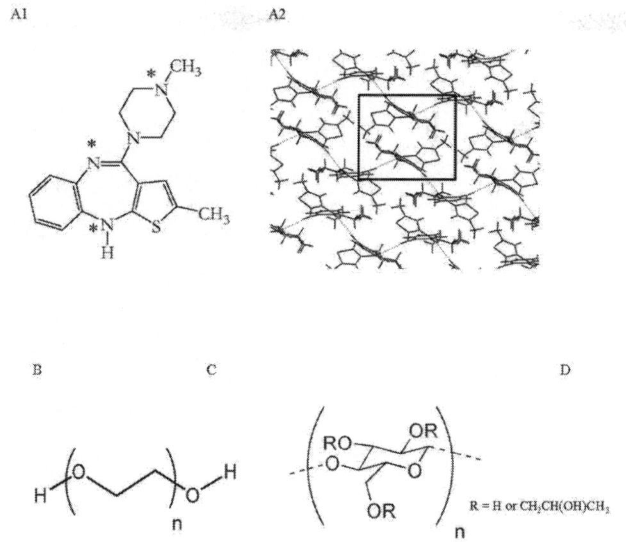

Figure 1. Olanzapine (OLZ) molecular structure (**A1**) and molecular packing (**A2**) of anhydrous OLZ FI and structures of PEGs (**B**); HPC (**C**) and PVPs (**D**). (*) Hydrogen bond acceptor or donor groups.

Figure 2A shows the XRPD pattern of OLZ and recovered OLZ after 360 min of being in contact with water. The XRPD diffractograms show that OLZ was able to hydrate into the dihydrate-B form. However, the peaks at 8.58°, 19.20°, and 24.14° 2θ, which are characteristic of the higher hydrate form, suggest a concomitant crystallization of both hydrates, although in different proportions. The subsequent resting of the wet mass at 25 °C/55% RH for 24 h was followed by the elimination of the higher hydrate form. The higher hydrate is a metastable crystal form that contains 2–2.5 mol of water per mol of OLZ and has only been observed in wet cakes of OLZ [10], which explains its elimination after removal of the adsorbed water.

OLZ which was physically mixed with PEG 6000/PEG 40,000 was shown to hydrate selectively into the higher hydrate form (Figure 2B). The product from the dehydration of the higher hydrate (2.5 mol water per mol of OLZ) was the dihydrate E. The XRPD study of the wet material enabled the observation of the intermediate higher hydrate form which is generated and is only stable in a wet environment. Similar diffraction patterns of OLZ before and after wetting suggest a complete inhibition of hydrate transformations of OLZ during the time of the experiment. However, a slight widening of the OLZ peaks was observed after drying the OLZ: HPC wet mass (Figure 2C). Similarly, OLZ: PVP mixtures also failed to show transformations of OLZ. No peaks due to excipients were observed after the manufacturing of physical mixtures with PVP k12, PVP k30 and HPC-LF reflecting the amorphous nature of these polymers.

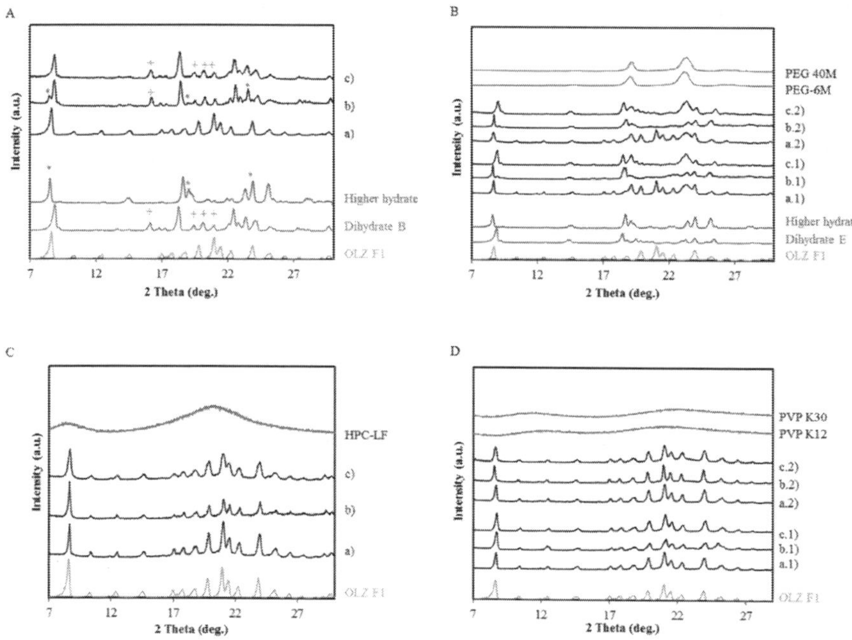

Figure 2. Diffractograms of six formulations before wetting, after wetting and after drying. The wet samples were analyzed after storage in a sealed container for 360 min. (**A**) OLZ before (a) and after wetting (b) and drying (c); (+) represents the characteristic peaks of the OLZ dihydrate B and (*) represents the characteristic peaks of the OLZ higher hydrate; (**B**) Formulations B1 and B2 physical mixture (a.1 and a.2), after wetting (b.1 and b.2) and after drying (c.1 and c.2); (**C**) Formulation C physical mixture (a), after wetting (b) and after drying (c); (**D**) Formulation D1 and D2 physical mixture (a.1 and a.2) after wetting (b.1 and b.2) and after drying (c.1 and c.2).

FT-Infrared spectroscopy was also used to monitor the anhydrate to hydrate conversion of OLZ since the addition of water (wet massing, t = 0) until the end of the study (360 min) and the respective spectra are shown in Figure 3. The changes in the region 950–990 cm^{-1} confirmed the hydrate conversion independently of the type of hydrate formed, due to a shift on the bands at 965 cm^{-1}. This shift was related to the deformation of the piperazinyl group coupled to the azepine and thiophene moieties (965 cm^{-1}) in OLZ molecules [11]. Therefore, Figure 3 shows that hydrate formation of OLZ started at an early stage when polymers were absent from the formulation. This was visible by the shift of the bands at 965 cm^{-1} to 971 cm^{-1} for the samples taken immediately after the completion of the wet massing. For formulation B1 (OLZ:PEG 6000), the shift started after 30 min, showing a higher induction period for hydration of OLZ in the presence of this polymer. By increasing the polymer weight (B2), the induction time increased (Figure 3B2). In both PEGs' wet masses, anhydrous OLZ was not visible after 360 min (Figure 3B1,B2). These results have shown that by using PEG as a binder it was possible to prevent hydration, provided the time allowed for OLZ to contact the aqueous solution was less than 30 min for PEG 6000 or less than 180 min for PEG 40,000. When PVPs and HPC replaced PEGs in the formulation, no signs of hydrate formation were observed during the period of the test (Figure 3C,D1,D2, 360 min).

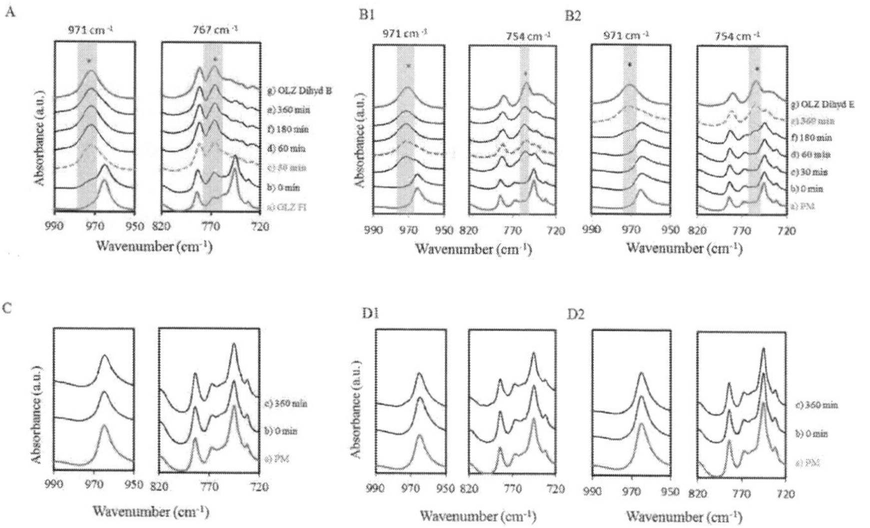

Figure 3. Spectra (FTIR) regions (950–990 cm^{-1} and 720–820 cm^{-1}) of OLZ wet masses (1.5 OLZ:1 polymer) after storage in a sealed container. Formulations A (**A**); B_1 (**B1**) and B_2 (**B2**) show complete hydration after 30, 60 and 360 min (dashed lines); Formulations C (**C**); D_1 (**D1**) and D_2 (**D2**) show no hydration during the time of storage (360 min). (*) Represents the characteristic peaks of the OLZ dihydrate E.

The region 720–780 cm^{-1} was selected to distinguish the different hydrate conversion occurrences in OLZ wet masses containing no polymer (A) or the PEGs (B1 and B2) as excipients. The out of plane deformation of the CH bond of the benzene group at 745 cm^{-1} was no longer present in both wet masses after 360 min. Wet mass A revealed a new absorption maximum at 767 cm^{-1} specific for the dihydrate B whereas wet masses B1 and B2 presented a new band at 754 cm^{-1}, characteristic of the dihydrate E.

Figure 4A displays the spectra within the range of 950–990 cm^{-1} of the wet masses (L1, L2 and L3, respectively 50, 75 and 100 g of water) collected 360 min after their preparation. Both PVP and HPC were shown to prevent hydrate formation regardless of the amount of water used. In Figure 4B, the fraction of the polymer was reduced by two orders of magnitude, in order to differentiate the ability of PVP and HPC to prevent hydrate conversion. The ratio of OLZ:Polymer was then fixed at 1.5:0.01. Figure 4B displays the spectra of the wet masses collected at three different times (t = 60, 180, 360 min) after their preparation. The reduction on the polymers fraction in formulations D was reflected by a reduction of the induction time for the hydrate conversion (especially for the formulation containing PVP k12), which started to hydrate after 360 min in a sealed container. Conversely, no hydrate transformation was observed in formulation C containing a low percentage of HPC in the wet mass.

Seeding experiments were carried out to determine if the HPC and PVP could still prevent the transformation in the presence of nuclei, and to better understand the inhibitory mechanism of these polymers. Five percent of the anhydrous API was thus replaced by OLZ dihydrate E before the wetting step. Shifts in the spectra of samples based on formulations D1 and D2 were observed and assumed to be related to hydration of anhydrous OLZ, whereas samples of formulation C did not produce shifts in the characteristic bands (Figure 4C).

Figure 5 A1 shows that immediately after wet massing (t = 0), OLZ particles were irregularly shaped and agglomerated. OLZ particles, alone (A) or in the presence of PEG (B1), after 360 min in direct contact with water, had their morphology changed to plate-like shaped crystals (Figure 5 A2) characteristic of the hydrate and the size of the latter larger than OLZ particles alone (Figure 5 B2). Conversely, formulations C and D1 showed no transformation of OLZ particles over the same period (Figure 5 C,D).

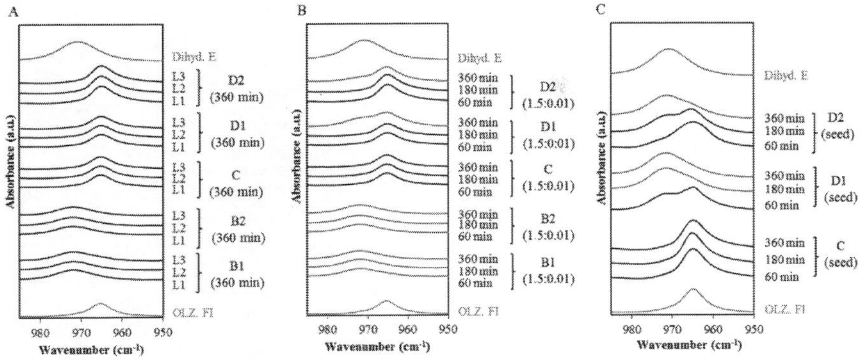

Figure 4. Spectra (FTIR) in the region (990–950 cm^{-1}) for the different OLZ:polymer formulations (B1, B2, C, D1, and D2). (**A**) Formulations were wet massed with 3 levels of water L1 (50 g), L2 (75 g), L3 (100 g). The spectra were collected after 360 min mass preparation; (**B**) The polymer content had a two order of magnitude reduction (1.50 OLZ:0.01 Polymer); (**C**) Wet masses C, D1 and D2 with seeds (5%). The spectra were collected at 60, 180 and 360 min after the wet mass preparation.

The thermal behavior of anhydrous OLZ and physical mixtures of formulations B1, B2, C, D1 and D2 are shown in Figure 6 (line 1, A to D) in which thermograms 2 and 5 correspond to the samples recovered immediately after the wet massing (t = 0 min) and the curves 3 and 6 correspond to the samples which were kept in a sealed container (t = 360 min).

Figure 6A, curve 1 (OLZ raw-material) shows only one endotherm event at 195.29 °C due to the melting of OLZ. Figure 6A, curves 2 and 3 reveal a new endotherm at lower temperatures corresponding to the release of water in the molecular structure, as a result of the presence of OLZ hydrate in the samples. The small fraction of the hydrate content in OLZ samples which were produced immediately after the wet massing (a mixture of anhydrous and hydrous content is present,Figure 6) resulted in a water loss endotherm with a low enthalpy.

Figure 5. Micrographs of wet masses for formulations A, B_1, C and D_1. Formulation A (t = 0 min (**A1**) and t = 360 min (**A2**); Formulation B_1 (t = 0 min (**B1**) and t = 360 min (**B2**)); Formulation C (t = 360 min (**C**)); Formulation D_1 (t = 360 min (**D**)).

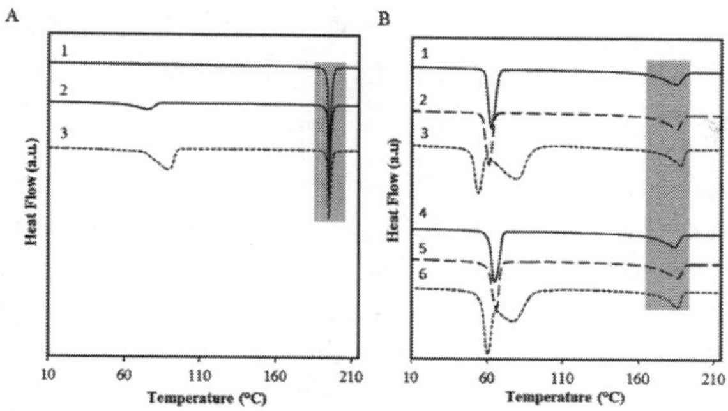

curve	Tm (°C)	ΔH (J/g)
A 1)	195.29 ± 0.08	132.05 ± 1.01
A 2)	194.95 ± 0.27	130.70 ± 1.23
A 3)	194.95 ± 0.05	113.02 ± 1.31
B 1)	185.64 ± 1.69	55.58 ± 3.70
B 2)	186.89 ± 1.85	53.29 ± 5.01
B 3)	187.42 ± 1.83	55.776 ± 4.84
B 4)	184.66 ± 1.22	55.80 ± 2.54
B 5)	186.22 ± 0.10	50.46 ± 2.589
B 6)	186.06 ± 1.52	50.48 ± 2.77

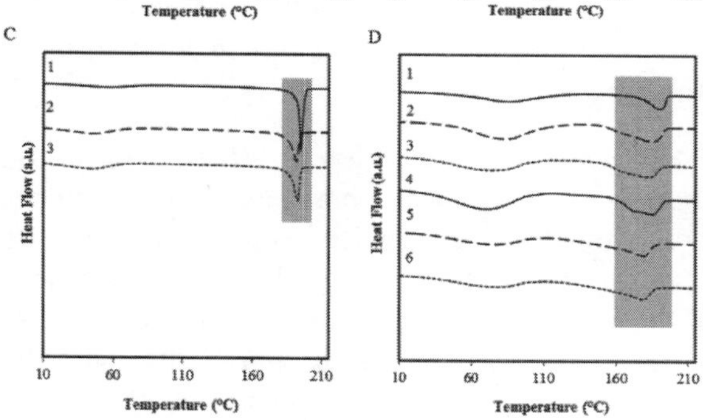

curve	Tm (°C)	ΔH (J/g)
C 1)	194.79 ± 0.06	76.20 ± 2.51
C 2)	192.28 ± 0.89	62.62 ± 3.52
C 3)	192.45 ± 0.13	66.04 ± 2.63
D 1)	190.72 ± 1.14	60.01 ± 0.63
D 2)	188.22 ± 0.90	52.77 ± 2.17
D 3)	189.83 ± 1.95	52.64 ± 1.83
D 4)	186.95 ± 1.67	59.77 ± 1.44
D 5)	182.09 ± 1.54	52.07 ±2.34
D 6)	181.01 ± 1.88	51.74 ± 1.25

Figure 6. Thermograms of formulations A (**A**); B_1 (**B**, curves 1–3; B_2 (**B**, 4–6); C (**C**); D_1(**D**, 1–3) and D_2 (**D**, 4–6). Curve 1 shows the physical mixture of raw materials, curves 2 and 5 show wet masses at t = 0 min, and curves 3 and 6 show the wet masses at t = 360 min. Tables present the melting point of OLZ and the respective enthalpy of fusion (Tmax and ΔH) for each formulation (grey shadow in figures).

Figure 6B, curves 1 and 4 show that the physical mixtures (PM) obtained from formulations B_1 and B_2 present two endothermic peaks due to the melting of both PEG and OLZ. The depression on OLZ melting temperature and enthalpy results from the solubilization of the drug within the molten PEG during heating. Figure 6B, curves 2 and 5 show no traces of water evaporation allowing the assumption that no hydrate was present in the sample. However, curves 3 and 6 (Figure 6B) reveal an endothermic peak due to dehydration of the dihydrate E. Furthermore, a reduction in the melting temperature of the PEG polymers was also observed in the same thermograms, suggesting conformational changes of the polymer chains with an impact on their melting temperature. This is in line with the ability of the PEG chains to adapt their conformation when in contact with water to establish extensive hydrogen bonds [12]. In a different experiment, OLZ was heated up to 120 °C with dehydration, cooled and heated again (Figure 6B, curves 3 and 6). These samples revealed an increase of the PEG melting temperature in line with its initial value (data not shown). This finding suggests reversibility of the conformational changes of the polymer chains occurred after removal of water in the sample.

There was no apparent conversion to the OLZ dihydrate for the samples which were formulated in the presence of both PVP and HPC, however, some adsorbed water created a broad endotherm (Figure 6C,D). In thermograms C (curves C2 and C3) and D (curves D2, D3, D5 and D6), the OLZ endotherms due to melting became broader, with a lower enthalpy and temperature than those observed in the physical mixture (curve 1D). This shift towards lower temperatures indicates a higher miscibility of drug in the polymer after the wet massing process.

Table 1 shows the surface energy and the polarity for each polymer in the formulation. The results show the ranking for the measured hydrophobicity as HPC > PVP > PEG. Generally, for the binder (polymer) to spread over the substrate (API) in a wet massing process, the work of cohesion of the binder must be lower than that of the substrate, allowing the spreading of the first on the latter. This only occurred for HPC. These results were in line with the results obtained for the contact angles (Table 2) which have shown the wettability of the powders, as measured by the contact angle (°) of the substances in water. While the contact angle of the particles in formulations B1, B2, D1 and D2 after wet massing showed a contribution of the contact angles of both polymers and OLZ, the contact angle of particles in formulation C1 (OLZ with HPC) seemed to account only for the contribution of HPC (Table 2). Furthermore, a slightly lower contact angle was observed for these particles than for the isolated HPC suggesting that this polymer was not only completely covering the crystalline particles of OLZ, but also a decreasing number of hydrophobic groups were exposed on the surface of the particles, as a result of polymer chain conformation.

Table 1. Surface energy of olanzapine and polymers.

	γ^d [1] (mN/m)	γ^p [1] (mN/m)	γ [1] (mN/m)	Polarity (%)	W_c [2] (mJ/m^2)	W_a [2] (mJ/m^2)	$W_a - W_c$ [3] (mJ/m^2)
OLZ	41.31	1.19	42.50	3	85.00		
PEG 6000	35.71	14.62	50.33	29	100.66	81.01	−19.65
PEG 40,000	34.78	12.93	47.71	27	95.42	79.89	−15.53
PVP k12	43.22	7.48	50.71	15	101.42	88.59	−12.83
PVP k30	40.46	6.73	47.19	14	94.38	85.81	−8.57
HPC LF	31.90	4.80	36.70	13	73.40	75.81	2.41

[1] Surface free energy (γ) and its dispersive (γ^d) and polar (γ^p) components; [2] W_c and W_a are, respectively, the work cohesion and the work of adhesion (to OLZ); [3] $W_a - W_c$ reflect the spreading coefficient (λ).

Table 2. Contact angles of the raw materials and different formulations [1].

Raw-Material	Contact Angle (θ/°)	Formulations after Massing	Contact angle (θ/°)
OLZ	104.65 ± 0.64	A	104.87 ± 0.74
PEG 6000	63.60 ± 2.05	B1	81.24 ± 2.98
PEG 40,000	64.30 ± 1.95	B2	79.52 ± 1.07
HPC-LF	91.84 ± 0.47	C	87.41 ± 1.72
PVP k12	56.46 ± 1.36	D1	79.73 ± 1.54
PVP K30	75.77 ± 2.17	D2	89.14 ± 3.40

[1] Contact angle measured in water ($n = 6$).

2.2. Characterization of OLZ Wet Masses and Pellets Containing Microcrystalline Cellulose

For the formulations containing MCC, the ratio OLZ:polymer was maintained, as well as the amount of liquid used for the wet massing. The examination of the diffractograms acquired off-line after the wet massing (t = 360 min) (Figure 7) revealed that the contact with the liquid binder triggered the hydrate formation in formulations A_M, $B1_M$ and $B2_M$ (Figure 7A,B). In all these formulations, both OLZ dihydrate B and OLZ higher hydrate were observed in the wet mass. For formulations $B1_M$ and $B2_M$, the conversion of the anhydrous OLZ into two hydrates, instead of only one hydrate (B1 and B2), suggests that PEG was not interacting with the drug at the same extent as without the MCC. Drying the pellets (Figure 7Ac,Bc1,c2) at 45 °C for 60 min was not sufficient to remove the dihydrate B, which was visible in both OLZ formulations (Figure 7Ac,Bc1,c2).

Figure 8 shows a cross section of the pellets $B1_M$ and C_M. The appearance of plate-like crystals of OLZ in the inner part of the pellet has shown that the changes in shape of OLZ particles due to hydration were also occurring in the formulations containing MCC.

Figure 9a shows the dissolution curves of OLZ released from pellets prepared immediately before the test. For each formulation, dissolution profiles were obtained from A_M, $B1_M$ (PEG 6000), C_M (HPC) and $D2_M$ (PVP k30) pellets made from non-stored (#1, *i.e.*, t = 0) and stored (#2, *i.e.*, t = 360 min) wet masses. A significant change on OLZ dissolution rate ($p < 0.05$) was observed between #1 and #2 pellets for A_M and $B1_M$ formulations. On the contrary, no significant differences in the OLZ dissolution rate from pellets #1 and #2 were observed from formulations C_M (HPC) and $D2_M$ (PVP), although a slight increase in OLZ dissolution rate was verified for pellets #2.

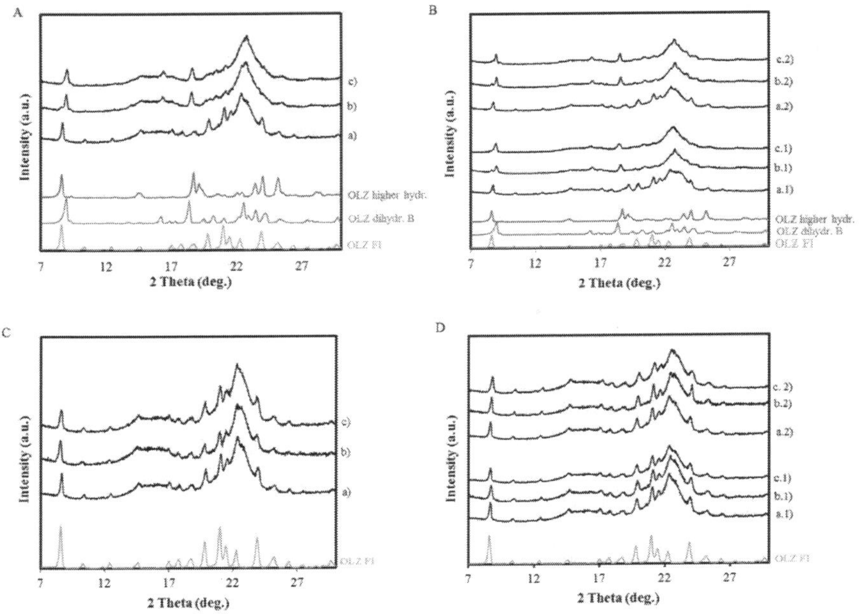

Figure 7. Diffractograms of formulations as physical mixtures (a), as wet masses (t = 360 min), (b) and dried pellets (c). (**A**) Formulation A_M; (**B**) Formulation $B1_M$ (a.1–c.1) and $B2_M$ (a.2–c.2); (**C**) Formulation C_M; (**D**) Formulation $D1_M$ (a.1–c.1) and $D2_M$ (a.2–c.2).

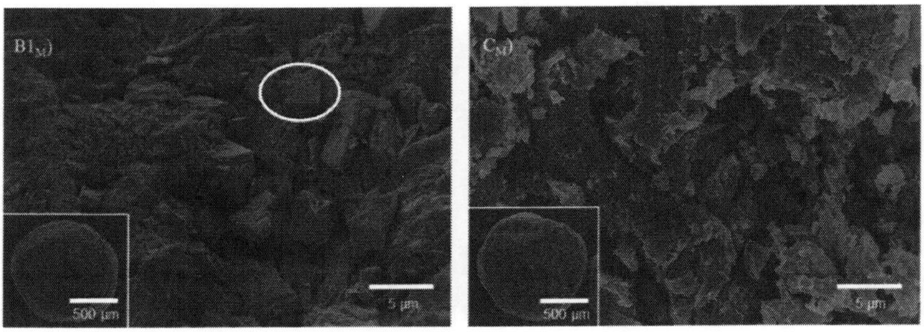

Figure 8. Micrographs (SEM) of a cross section and pellets (inserts) from formulations B_{1M} (**left**) and C_M (**right**). The circle in the left micrographs shows a crystal of OLZ dihydrate.

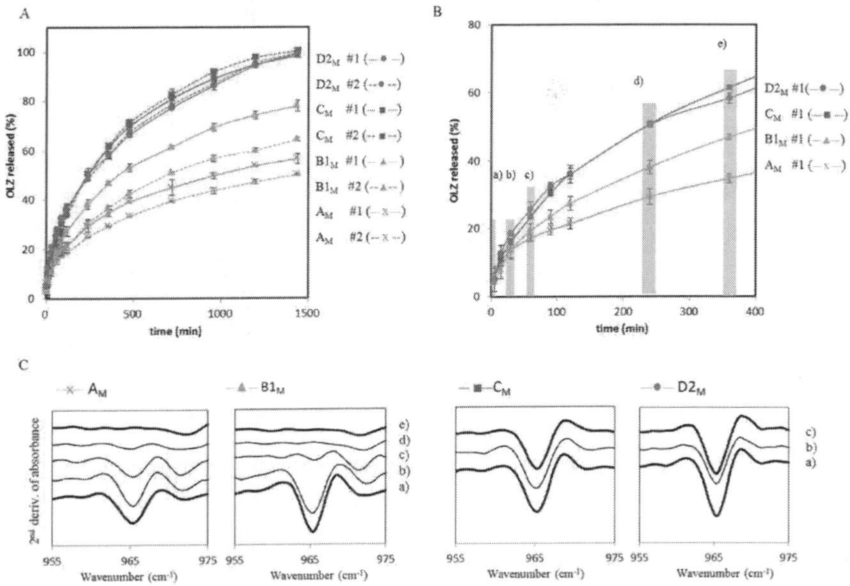

Figure 9. Dissolution profiles of OLZ from pellets (top) and Spectra (FTIR) analysis made from different formulations (A_M, $B1_M$, C_M and $D2_M$). (**A**) Dissolution profiles of OLZ from pellets #1 (t = 0 min) and #2 (t = 360 min); (**B**) Formulations and the time periods (a–e) in which pellets were collected for FTIR characterization; (**C**) Second derivative of FTIR absorbance spectra from collected samples (times a–e for formulations A_M and $B1_M$ and times c–d for formulations C_M and $D2_M$).

During dissolution, it was possible to characterize the OLZ solid-state form remaining in the pellets. The second derivative from FTIR absorbance spectra of the pellets taken from the dissolution media has shown that PVP and HPC can also have a suppressing effect on the OLZ hydrate conversion in the dissolution media, by opposition to pellets A_M (no polymer) and $B1_M$ (PEG) in which OLZ suffered hydration. This observation strongly suggests that HPC and PVP, which had shown a strong protection of OLZ over the production of wet masses, also stabilize the anhydrous OLZ during their release from pellets into the dissolution media.

Table 3 summarizes the results of OLZ release modelling for the pellets under investigation. The goodness of fit for the various models were ranked in the order: Korsmeyer-Peppas > Higuchi > First-order for the formulations A_M, $B1_M$ and C_M. For formulation $D2_M$, the goodness of fit for the first-order model was better than for the Higuchi model. The fact that drug release from pellets follows Korsmeyer-Peppas model and the values of the exponent *n* are lower than 0.5 (0.383–0.441) indicates that the OLZ, independently of the polymorphic form present in the pellets, follows a diffusion controlled release.

Table 3. Fitting of dissolution models to OLZ release data for pellets.

Formulation		First-Order Model			Higuchi Model			Korsmeyer-Peppas Model			
		Adj R^2	MSE Root	K_1	Adj R^2	MSE Root	k_H	Adj R^2	MSE Root	k_{KP}	n
A_M	#1	0.669	9.946	0.001	0.952	3.797	1.648	0.997	0.958	3.585	0.383
	#2	0.706	8.624	0.001	0.97	2.728	1.441	0.995	1.127	2.676	0.407
$B1_M$	#1	0.82	10.15	0.002	0.962	4.674	2.218	0.994	1.794	4.442	0.395
	#2	0.839	8.466	0.001	0.986	2.464	1.821	0.995	1.554	2.703	0.441
C_M	#1	0.951	6.910	0.002	0.986	3.81	2.696	0.995	2.226	4.450	0.431
	#2	0.926	8.408	0.003	0.973	5.044	2.755	0.995	2.353	4.966	0.411
D_M	#1	0.972	5.697	0.003	0.958	6.991	2.995	0.980	4.801	5.510	0.408
	#2	0.978	5.115	0.003	0.956	7.321	3.069	0.981	4.658	5.494	0.410

#1 and #2 represent pellets manufactured at $t = 0$ min and $t = 360$ min of wet masses, respectively.

3. DISCUSSION

The different crystalline forms of OLZ are built through the assemblage of (OLZ) centrosymmetric racemic pairs and then connected to each other through hydrogen bonds, either from atoms like in the OLZ structure (e.g., anhydrous form) or by water molecules (e.g., hydrate forms). In form I, the hydrogen bond interactions between the amine and imine (NH-N) are responsible for the connection of two different OLZ pairs [10]. When the hydrate formation occurs, the NH–N interactions are disrupted and replaced by water bridging molecules between the dimmers. The presence of water balances the hydrogen bond donor/acceptor ratio of OLZ, thus allowing two acceptors to participate in hydrogen-bonding, which was not possible in the anhydrates. Therefore, in OLZ hydrates, a single hydrogen-bond donor (NH) and two good acceptors (imine and piperidine) may participate in hydrogen bonding with water.

In manufacturing processes in which water is added (e.g., wet extrusion), OLZ is exposed to conditions in which the hydrate forms are expected to be the most stable due to high water activity [13]. Consequently, starting from the addition of water until the final drying, the API may undergo an anhydrate to hydrate transformation which may or, may not, be completed by the end of the process. The API transformations in an aqueous environment are expected to occur via a solvent-mediated transformation (SMT) [13] which encompasses a three-stage process. Firstly, the metastable phase in this work, the anhydrous OLZ, starts to dissolve. Since the dissolution of the metastable phase is a prerequisite for the transformation to take place, the solubility of the metastable form will initially provide the thermodynamic driving force for the phase transformation. Dissolution under non-sink conditions will lead to a supersaturation with respect to the stable hydrate phase and provide a thermodynamic driving force for nucleation of the stable phase (hydrate). The next stage involves nucleation of the stable hydrate once supersaturation has been reached. During wet massing of powders, which can be regarded as a suspension, there is an abundance of particles in the form of the anhydrate phase, which might aid heterogeneous nucleation [14]. By opposition to homogeneous nucleation, heterogeneous nucleation lowers the thermodynamic free energy barrier. Thus, heterogeneous nucleation can occur under lower driving forces [15]. Finally, once nuclei of the stable form are present in the solution, growth of that form continues until all the metastable phase has dissolved. Therefore, the growth of the stable phase (hydrate) is maintained by the supersaturation created by the dissolution of the metastable phase [14,16]. The kinetics of solvent-mediated transformations is therefore governed by dissolution of the unstable phase and nucleation and growth of the stable phase. It follows that, if nucleation of the hydrate phase could be inhibited by polymers, then crystal growth could not occur, *i.e.*, inhibition of crystal growth only became important if seeding crystals were present. When observing the XRPD data obtained from the masses without the presence of excipients, one could observe that a concomitant crystallization of both dihydrate B and higher hydrate took place on the OLZ wet mass A. This phenomenon may be associated with the nucleation rates, favoring the concomitant crystallization of both hydrates [15]. Also, it was noted that the hydration of all of the material was not finished within the time scale of the wet massing step. The kinetics of this process was controlled by the dissolution rate of the unstable form, the rate of crystallization of the more stable form and their relative solubilities. It means that in wet massing, the extent of hydrate formation is proportional to the solubility of the anhydrate and the volume of the wetting liquid used [17]. Furthermore, since wetting of powders is a factor governing hydrate formation kinetics [18], it is clear that the APIs with poor wettability will require an excessive amount of moisture and/or prolonged contact with wet massing liquid for the hydrate to form. Thus, the short wet massing period together with the low wettability of OLZ ($\theta \approx 104°$) led to the incomplete hydrate conversion of OLZ during wet massing. Furthermore, its low solubility (0.043 mg/mL [19], *i.e.*, a small quantity of OLZ can be dissolved in the liquid at any time) resulting in a slower hydrate rate conversion, which contrasts with other drugs such as theophylline (solubility in water of 12 mg/mL) [20] or sulfaguanidine (solubility in water of 1.38 mg/mL) that was shown to hydrate in less than 5 min during wet granulation processes [13].

It has been reported that the inhibition of polymorphic transitions by the polymers is related to their hydrophobicity, viscosity and steric hindrance. Polymers with higher hydrophobicity (lower polarity) are less likely to be present in the aqueous phase rather than being adsorbed on the API particles' surface, inhibiting therefore the API interaction with the water molecules [13]. The addition of PEG 6000 and 40,000 to the formulations, respectively B1 and B2, has retarded the hydrate conversion without concomitant crystallization into two different hydrates. In these wet masses, the OLZ higher hydrate crystallized as a single metastable form that was shown to dehydrate easily into the dihydrate E form when the samples were dried. The addition of PEG 40,000 (higher viscosity) was shown to increase the lag time for hydration in comparison with PEG 6000. The rate by which the new crystals are formed in supersaturated solutions is related to the diffusion rate of the drug molecules [14]. The addition of PEG might increase the viscosity of the media affecting the drug's diffusivity, which in turn affects the dissolution, nucleation and crystal growth rates [21]. The change of the nucleation kinetics caused by PEG favored the appearance of a single form—higher hydrate—which is never obtained as an isolated form in water environments without any additive. On the other hand, a complete inhibition of OLZ hydration was achieved by addition of both HPC and PVP. According to the literature, there are two major factors that lead to an efficient hindrance of the hydrate formation by the polymeric excipients: the capacity to form hydrogen bonds with the drug and a sufficient hydrophobicity. Both these features prevent the drug from binding with water and force it to bind with the polymer instead [9,13,22], which are important factors affecting the nucleation and growth rates of crystals.

For nucleation to occur, drug-polymer hydrogen bonds have to be broken so that drug molecules may diffuse and form a nucleus. In the wet masses with HPC (C) and PVP (D), OLZ molecules may interact with the polymer molecules to form OLZ-polymer hydrogen bonds in the aqueous environment. These interactions may then inhibit or retard the formation of nuclei due to collisions. The strength of the drug-polymer association determines the time required for the nucleation to take place [13].

When the content of HPC, PVP k30 and PVP k12 was reduced by two orders of magnitude in formulations C, D1 and D2, respectively, only HPC was shown to continue to inhibit hydrate conversion. This can easily be explained by the hydrogen's bond ability of the involved polymer and its spreading over OLZ particles. While HPC has both hydrogen bond donor and acceptor groups in its ring structure, PVP has just one hydrogen bond acceptor group per monomer unit. In a different experiment, where hydrate seeds were added to the wet masses, transformation times decreased for PVPs. When hydrate seeds were added to the HPC formulation, the hydrate transformation continued to be inhibited during the 6 h of the test, indicating that HPC can effectively inhibit the crystal growth. The stronger inhibition of hydration by HPC may be justified by the higher H-bond ability and higher network association of this polymer compared to PVP, which could have contributed to a stronger bond with the crystal surfaces of the hydrate nuclei/initial crystals and promote steric hindrance to the incorporation of other drug molecules into the crystal lattice, thus preventing the crystal growth. Although PVP was not successful in preventing transformations, polymer molecular weight had an impact (formulations D1 and D2). In both cases, the nucleation step was no longer a limiting factor but nuclei grew at different rates: faster for PVP k12 than for PVP k30. This was also observed by other authors [9] for which the inhibition effect of polymers on carbamazepine was also related to their hydrophobicity, since polymers with a higher hydrophobicity had a stronger interaction with the carbamazepine crystal surface. Data has shown that HPC had a lower polarity (higher hydrophobicity) and a lower cohesion ability than its ability to adhere to the OLZ molecules. This showed the ability of this polymer to spread over the OLZ crystal surfaces.

Moreover, the OLZ powders, which were subjected to wet massing with HPC, showed that this polymer can reduce the contact angle of OLZ to the same values than the polymer alone, revealing that the polymer could effectively spread on and bind to the crystal surface of OLZ particles. The other more hydrophilic polymers were not so successful in spreading on OLZ surfaces. These results indicate affinity of this polymer to OLZ surfaces, which when absorbed at relevant growth sites, may have led to a blockage of the growth of the nuclei.

The addition of cellulose to the wet masses as an excipient did not have an effect on the inhibition of hydrate conversion, and the respective formulations A_M, $B1_M$ and $B2_M$, OLZ were shown to hydrate in the same manner as had occurred in formulations without HPC or PVP. However, while in formulation B1 and B2 only one hydrate was obtained (higher hydrate), in formulations with cellulose ($B1_M$ and $B2_M$), two hydrates were visible in the wet mass (dihydrate B and higher hydrate). This loss of selectively of hydrate conversion indicates a reduction of the close contact between the polymer (PEG) and the drug.

A better understanding of the dissolution process can be accomplished by analyzing not only the drug in the liquid dissolution medium but also in the solid state of the dissolving sample, which provides a more complete understanding of the dissolution behavior.

The fast and similar dissolution rate of OLZ present in all the different pellets formulations at the very beginning can be partly explained by the dissolution of OLZ anhydrate before the hydrates began to crystallize.

Moreover, the dissolution of OLZ particles on the surface of the pellet can promote the burst effect during the first few minutes.

The mechanism by which OLZ hydrates in the dissolution vessel is expected to be the same as the one that occurs when OLZ is subjected to the wet massing process, since in this situation the drug is also exposed to a solvent. Aaltonen *et al.* [23] reported hydration by solvent-mediated phase transformation of theophylline and nitrofurantoin during dissolution in purified water. The FTIR absorbance spectra of OLZ present in the pellets after being for different periods in the dissolution medium (not dissolved OLZ fraction) showed that both PVP and HPC had a suppressing effect on the OLZ hydrate formation during the dissolution test, which coincides with the best performance of OLZ.

During the dissolution process of the pellets, several processes competing with each other may take place simultaneously, such as: dissolution of OLZ, recrystallization of anhydrous OLZ to OLZ hydrated, swelling of the polymer and dissolution of the polymer. The polymers that were shown to strongly interact with the drug during the wet massing of OLZ may also stabilize the anhydrous OLZ during the dissolution by inhibiting aggregation due to a stronger interaction between the drug and polymer in the dissolution media. Therefore, the interaction with the polymers prevents the formation of the hydrate crystals, thus rendering the drug dispersed in the matrix of the polymer easier to dissolve.

4. EXPERIMENTAL SECTION

4.1. Materials

Olanzapine anhydrous form I [OLZ FI, molecular weight (Mw) = 312.43 g/mol, density (d) = 1.30 gcm^{-3}] was purchased from Pharmorgana, India. The OLZ molecule and representation of its three-dimensional structure of anhydrous OLZ form I is represented in Figure 1.

The polymers considered in the study were polyethyleneglycol with different molecular weights (PEG, Mw = 6000 and 40,000, Sigma-Aldrich, Munich, Germany), hydroxypropylcellulose (HPC LF Pharm grade, Mw = 95,000, Klucel™, Ashland, Düsseldorf, Germany) and polyvinylpyrrolidone (PVP, Kollidon 12, 2600 < Mw < 5500 and Kollidon 30, 44,000 < Mw < 54,000, BASF Chemicals, Ludwigshafen, Germany). Other chemical were supplied by Sigma-Aldrich, Munich, Germany.

The following OLZ forms were produced in order to identify the transformations that OLZ FI may undergo during processing:

4.1.1. OLZ Dihydrate B

(CSD ref. code: AQOMAU01, [10]): OLZ Form I (2.5 g) was suspended in ethyl acetate (25 mL) and toluene (2 mL) under stirring before heating (80 °C) for complete dissolution. Then, the solution was cooled down to 60 °C prior to the addition of water (30 mL), and further cooled down to room temperature, producing a crystalline slurry. The crystals of the dihydrate B were collected by filtration under vacuum;

4.1.2. OLZ Dihydrate E

(CSD ref. code: AQOMAU02, [10]): OLZ Form I (3 g) was suspended in ethyl acetate (60 mL) and toluene (3.6 mL). The suspension was stirred and heated to 80 °C for complete dissolution. The solution was then cooled to ambient temperature to produce a slurry. The suspension was filtered, washed with water and dried at room temperature.

4.1.3. OLZ Higher Hydrate

(CSD ref. code: AQOMEY, [10]): OLZ (2 g) was suspended in CH_2Cl_2 (12 mL). The suspension was stirred and heated to reflux to dissolve the solids. Water (1.5 mL) was added as the solution was cooled to ambient temperature, at which time the crystal slurry formed. The crystal slurry was cooled to 0 °C, and a wet cake (2.7 g) was isolated by vacuum filtration and washed with CH_2Cl_2.

4.2. Methods

4.2.1. Conversion of OLZ FI during Wet Massing in the Presence of Polymers

Four wet masses were prepared using anhydrous OLZ form I (OLZ FI, 15 g, *"formulation A"*) or physical mixtures of OLZ with each one of the different polymers commonly used in extrusion processes [24,25], belonging to the three groups described above, namely PEG-6000/ PEG 40,000 (group I, *"formulation B1 and B2"*, 10 g), HPC-LF (group II, *"formulation C"*, 10 g) and PVP k12/ PVP k30 (group III, *"formulation D1 and D2"*, 10 g) (Table 1). OLZ or physical mixtures of OLZ and polymer (1.5:1) were blended for 5 min in a planetary mixer (Kenwood, Hampshire, UK). Demineralized water (75 g, 5 times the OLZ mass) was added for 1 min to the blends. The wet masses were mixed thoroughly for 10 min and then placed into sealed glass vials filled to the top. Samples were taken at different time periods (t = 0, 30, 60, 180 and 360 min) after production. A fraction of each sample was analyzed immediately by XRPD and the remaining portion was dried in Petri dishes (24 h at 25 °C/55% RH, until constant weight) and analyzed by infrared spectroscopy (FTIR), thermal analysis (calorimetry, DSC) and microscopy (SEM). These formulations (B1, B2, C, D1 and D2) and processing conditions took into consideration other quantities of water, namely 50 and 100 g and, fractions of polymer, ranging from 1.5:1 to 1.5:0.01.

In a different set of experiments, 5% of anhydrous OLZ was replaced by OLZ dihydrate, to promote crystal growth within the anhydrous OLZ containing formulations.

4.2.2. Manufacture of Pellets Containing Polymers

Taking into consideration the same fractions of both API and excipients, 75 g MCC was added to the blends described previously (Table 4), to turn the manufacture of pellets possible: OLZ:MCC (A_M), OLZ:MCC:PEG-6000/40,000 ($B1_M$ and $B2_M$, respectively), OLZ:MCC:HPC (C_M) or PVP k12/k30 ($D1_M$ and $D2_M$, respectively) were dry-mixed in a planetary mixer (Kenwood, UK) for 5 min prior to the addition of 75 g (5 times the olanzapine mass) of demineralized water. Each wet mass (175 g) was further mixed for another 10 min. Half of the wet mass was extruded immediately (#1) whereas the other half was allowed to equilibrate for 360 min (#2) in a sealed polyethylene bag before extrusion. The storage of wet masses to equilibrate for several hours is commonly considered in daily practice to ensure a uniform distribution of water [26,27] and in the present case to promote hydration of OLZ. All measurements were performed off-line at defined time points after wet massing (t = 0, 30, 60, 180, and 360) by FTIR and XRPD.

Extrudates were manufactured from 2 wet masses (#1 and #2) in a ram extruder (Lurga, Portugal) adjusted to an universal testing machine (LR 50K, Lloyds Instruments, UK) fitted with a load cell to allow the collection of data for the applied force to the ram and its displacement (200 mm/min). The wet mass was forced to pass through a die (Length/Diameter = 6) and, after extrusion, the extrudates were immediately placed in a spheroniser (radial friction plate rotating at 500 rpm, Caleva, UK) for 5 min. The wet pellets were dried in a fluid bed drier (Aeromatic-Fielder AG, Bubendorf, Switzerland) at 40 °C for 30 min. Dried pellets were evaluated for powder diffraction (XRPD) infrared spectroscopy (FTIR) and OLZ dissolution studies into an aqueous solution.

Table 4. Formulations considered in the manufacture of pellets [1].

Formulation	Components (in Parts)		
	OLZ	Polymer	MCC
A_M	1.5	-	8.5
B_{1M}	1.5	1.0 (PEG 6000)	7.5
B_{2M}	1.5	1.0 (PEG 40,000)	7.5
C_M	1.5	1.0 (HPC-LF)	7.5
D_{1M}	1.5	1.0 (PVP k12)	7.5
D_{2M}	1.5	1.0 (PVP k30)	7.5

[1] These formulations were compared to the respective ones without MCC (e.g., C (1.5 OLZ:1 HPC-LF) *vs.* C_M (1.5 OLZ:1HPC-LF:7.5 MCC)).

4.2.3. Characterization of Blends and Pellets

X-ray Powder Diffraction (XRPD)

The structural characterization of OLZ formulations were carried out by X-ray powder diffraction (XRPD) on an Analytical X'Pert PRO apparatus set with a vertical PW 3050/60 goniometer (θ–2θ mode) equipped with X'Celerator detector with automatic data acquisition (X'Pert Data Collector (v2.0b) software), using a monochromatized CuKα radiation as incident beam, 40 kV–30 mA. Diffractograms were obtained by continuous scanning in a 2θ range of 7–35° with a step size of 0.017° and scan step times of 20 s.

X-ray diffraction characterization was performed to the different formulations (A, B1, B2, C, D1, D2) before and after the end of the study (360 min) in order to observe the hydrate conversions of OLZ in each sample.

Differential Scanning Calorimetry (DSC)

Thermograms of different OLZ forms were obtained with a Differential Scanning Calorimeter (TA instruments, Q200, New Castle, DE, USA) after calibration with indium (TA instruments; T_{fus} = 156.55 °C, $\Delta_{fus}H$ = 28. 51 J/g). Dry N_2 was used as the purge gas (Air Liquide, 50 mL/min). Dried samples (2–3 mg) were placed in pinhole crucibles and heated at 10 °C/min within 0–210 °C temperature range.

FT-Infrared Spectroscopy (FTIR)

Each sample was mixed with KBr in a 1:1 proportion and the mixture (200 mg) compacted into compacts (12 mm diameter). The infrared spectra were collected in the absorption mode (IR Affinity-1 Shimadzu spectrophotometer, Kyoto, Japan) based on 64 scans with a resolution of 2 cm^{-1}.

Scanning Electron Microscopy (SEM)

OLZ wet masses and pellets were analyzed by scanning electron microscopy. The samples were mounted on aluminum stubs and coated with gold by ion sputtering (JEOL JFC-1200 Fine Coater, Tokyo, Japan) and observed under (JSM-5200 LV, Tokyo, Japan) scanning electron microscope.

Wettability and Surface Energy Measurements

In order to measure the wettability and the surface energy of the powders before and after being exposed to hydrating conditions, the contact angles in two liquids (bi-distilled water, Ström and diiodometane) of the samples were measured (Wilhelmy plate technique, Krüss Tensiometer K12, Hamburg, Germany). Samples of powders were made to adhere on a rectangle shape substrate (20 × 20 mm) uniformly coating its surface. All measurements were performed at a controlled temperature 25 ± 0.5 °C, by flowing water from a circulator (Krüss). The surface energy of the polymers and OLZ was measured according to the Wu equation [28].

Dissolution Studies

Dissolution studies were performed for OLZ (5 mg) from dried pellets (60 min/45 °C, size of 1–1.4 mm). Tests were performed in a dissolution apparatus (paddle method, 50 rpm, 900 mL of phosphate buffer at 6.8, 37 ± 0.5 °C, n = 6, AT7, SOTAX, Aesch, Switzerland) with collection of samples at time intervals (t = 5 to 1440 min). The residual water present in the pellets was taken into consideration (Sartorius Moisture Analyser, Goettingen, Germany). OLZ was quantified by UV spectrophotometry (λ = 254 nm, Hitachi U-2000, Tokyo, Japan) and the analysis of OLZ release and dissolution profiles was possible by comparison of dissolution models. Simultaneously, dissolution tests were performed with pellets A_M, $B1_M$, $C_M$and $D2_M$ for off-line FT-Infrared evaluation of the anhydrate to hydrate transformations occurring in non-dissolved fraction of OLZ present in the pellets (t = 0 min, 30 min, 60 min, 180 min and 360 min). Each sample collected for FT-IR analysis was recovered from a different vessel.

5. CONCLUSIONS

In an attempt to clarify the mechanisms underlying the effect of polymeric excipients on the hydrate conversion of anhydrous OLZ in aqueous environments, the influence of various polymers on the hydration of OLZ in wet masses was investigated. The polymer screening methodology used to identify polymers that could stabilize OLZ during wet massing proved to be useful to identify which polymers may have the strongest effect of inhibition of OLZ. For the polymers selected in this study, it was found that both hydrogen bond ability and hydrophobicity (with impact on the spreading ability of the polymer) had an impact on the inhibition ability of OLZ hydration. In general all polymers promoted the inhibition of the transformation with different degrees of success, provided the contact time of OLZ anhydrous with water was low (e.g., less than 30 min). PEGs delayed the conversion of OLZ, but after 360 min of storage in a moist environment, conversion was observed. The HPC had the strongest effect on protecting the hydrate conversion of OLZ. The major reduction of the wettability of the OLZ powders in a blend with HPC revealed that this polymer can adsorb to OLZ's surface inhibiting the water attack and, therefore, causing a hydrate formation. The presence of microcrystalline cellulose in formulations, needed for the pellets production, neither improves the protection against water nor did it eliminate the protector effect of both PVP and HPC. Conversion of the anhydrous form into hydrate was partially reversible and occurred during the wetting, drying and dissolution test in which the anhydrous form was converted into a hydrate. The combination of data retrieved from the different analytical techniques (FTIR spectroscopy, XRPD, calorimetry and microscopy) was successfully used to investigate the hydrate conversion of OLZ in various excipients allowing to visualize the different hydrate conversions that OLZ may undergo in the presence of excipients.

ACKNOWLEDGMENTS

The Authors acknowledge *"Fundação para a Ciência e a Tecnologia"*, Lisbon, Portugal for providing the financial support to this work (PTDC/CTM/098688/2008 and SFRH/BD/90118/2012).

AUTHOR CONTRIBUTIONS

M.P. was involved in the conception, design, production and analysis of data; M.W. and J.P. were involved in the conception, design and analysis of the data. All authors were involved in writing the manuscript.

CONFLICTS OF INTEREST

The authors declare no conflict of interest.

REFERENCES

1. Phadnis, N.V.; Suryanarayanan, R. Polymorphism in anhydrous theophylline—Implications on the dissolution rate of theophylline tablets. J. Pharm. Sci. 1997, 86, 1256–1263.

2. Airaksinen, S.; Luukkonen, P.; Jorgensen, A.; Karjalainen, M.; Rantanen, J.; Yliruusi, J. Effects of excipients on hydrate formation in wet masses containing theophylline. J. Pharm. Sci. 2003, 92, 516–528.

3. Patel, D.D.; Anderson, B.D. Adsorption of Polyvinylpyrrolidone and its Impact on Maintenance of Aqueous Supersaturation of Indomethacin via Crystal Growth Inhibition. J. Pharm. Sci. 2015, 104, 2923–2933.

4. Gift, A.D.; Luner, P.E.; Luedeman, L.; Taylor, L.S. Influence of polymeric excipients on crystal hydrate formation kinetics in aqueous slurries. J. Pharm. Sci. 2008, 97, 5198–5211.

5. Sandler, N.; Rantanen, J.; Heinamaki, J.; Romer, M.; Marvola, M.; Yliruusi, J. Pellet manufacturing by extrusion-spheronisation using process analytical technology. AAPS PharmSciTech 2005, 6, E174–E183.

6. Jorgensen, A.; Rantanen, J.; Karjalainen, M.; Khriachtchev, L.; Rasanen, E.; Yliruusi, J. Hydrate formation during wet granulation studied by spectroscopic methods and multivariate analysis. Pharm. Res. 2002, 19, 1285–1291.

7. Dukić-Ott, A.; Thommes, M.; Remon, J.P.; Kleinebudde, P.; Vervaet, C. Production of pellets via extrusion-spheronisation without the incorporation of microcrystalline cellulose: A critical review. Eur. J. Pharm. Biopharm.2009, 71, 38–46.

8. Katzhendler, I.; Azoury, R.; Friedman, M. Crystalline properties of carbamazepine in sustained release hydrophilic matrix tablets based on hydroxypropyl methylcellulose. J. Control. Release 1998, 54, 69–85.

9. Tian, F.; Saville, D.J.; Gordon, K.C.; Strachan, C.J.; Zeitler, J.A.; Sandler, N.; Rades, T. The influence of various excipients on the conversion kinetics of carbamazepine polymorphs in aqueous suspension. J. Pharm. Pharmacol.2007, 59, 193–201.

10. Reutzel-Edens, S.M.; Bush, J.K.; Magee, P.A.; Stephenson, G.A.; Byrn, S.R. Anhydrates and Hydrates of Olanzapine: Crystallization, Solid-State Characterization, and Structural Relationships. Cryst. Growth Des. 2003, 3, 897–907.

11. Ayala, A.P.; Siesler, H.W.; Boese, R.; Hoffmann, G.G.; Polla, G.I.; Vega, D.R. Solid state characterization of olanzapine polymorphs using vibrational spectroscopy. Int. J. Pharm. 2006, 326, 69–79.

12. Thijs, H.M.L.; Becer, C.R.; Guerrero-Sanchez, C.; Fournier, D.; Hoogenboom, R.; Schubert, U.S. Water uptake of hydrophilic polymers determined by a thermal gravimetric analyzer with a controlled humidity chamber. J. Mater. Chem. 2007, 17, 4864–4871.

13. Gift, A.D.; Luner, P.E.; Luedeman, L.; Taylor, L.S. Manipulating hydrate formation during high shear wet granulation using polymeric excipients. J. Pharm. Sci. 2009, 98, 4670–4683.

14. Wikström, H.; Rantanen, J.; Gift, A.D.; Taylor, L.S. Toward an Understanding of the Factors Influencing Anhydrate-to-Hydrate Transformation Kinetics in Aqueous Environments. Cryst. Growth Des. 2008, 8, 2684–2693.

15. Rodriguez-Hornedo, N.; Murphy, D. Significance of controlling crystallization mechanisms and kinetics in pharmaceutical systems. J. Pharm. Sci. 1999, 88, 651–660.

16. Mangin, D.; Puel, F.; Veesler, S. Polymorphism in Processes of Crystallization in Solution: A Practical Review. Org. Process Res. Dev. 2009, 13, 1241–1253.

17. Davis, T.D.; Peck, G.E.; Stowell, J.G.; Morris, K.R.; Byrn, S.R. Modeling and monitoring of polymorphic transformations during the drying phase of wet granulation. Pharm. Res. 2004, 21, 860–866.

18. Otsuka, M.; Ishii, M.; Matsuda, Y. Effect of surface-modification on hydration kinetics of nitrofurantoin anhydrate.Colloids Surf. B 2002, 23, 73–82.

19. Thakuria, R.; Nangia, A. Polymorphic form IV of olanzapine. Acta Crystallogr. Sect. C Cryst. Struct. Commun. 2011, 67, o461–o463.

20. Wikstrom, H.; Carroll, W.J.; Taylor, L.S. Manipulating theophylline monohydrate formation during high-shear wet granulation through improved understanding of the role of pharmaceutical excipients. Pharm. Res. 2008, 25, 923–935.

21. Otsuka, M.; Ohfusa, T.; Matsuda, Y. Effect of the binders on polymorphic transformation kietics of carbamazepine in aqueous solution. Colloids Surf. B Biointerfaces 2000, 17, 145–152.

22. Raghavan, S.L.; Trividic, A.; Davis, A.F.; Hadgraft, J. Crystallization of hydrocortisone acetate: Influence of polymers.Int. J. Pharm. 2001, 212, 213–221.

23. Aaltonen, J.; Heinanen, P.; Peltonen, L.; Kortejarvi, H.; Tanninen, V.P.; Christiansen, L.; Hirvonen, J.; Yliruusi, J.; Rantanen, J. In situ measurement of solvent-mediated phase transformations during dissolution testing. J. Pharm. Sci.2006, 95, 2730–2737.

24. Karim, S.; Baie, S.H.; Hay, Y.K.; Bukhari, N.I. Development and evaluation of omeprazole pellets fabricated by sieving-spheronisation and extrusion—Spheronisation process. Pak. J. Pharm. Sci. 2014, 27, 425–438.

25. Kleinebudde, P. Application of low substituted hydroxypropylcellulose (L-HPC) in the production of pellets using extrusion/spheronisation. Int. J. Pharm. 1993, 96, 119–128.

26. Majidi, S.; Motlagh, G.H.; Bahramian, B.; Kaffashi, B.; Nojoumi, S.A.; Haririan, I. Rheological evaluation of wet masses for the preparation of pharmaceutical pellets by capillary and rotational rheometers. Pharm. Dev. Technol.2013, 18, 112–120.

27. Sousa, J.J.; Sousa, A.; Podczeck, F.; Newton, J.M. Factors influencing the physical characteristics of pellets obtained by extrusion-spheronisation. Int. J. Pharm. 2002, 232, 91–106.

28. Wu, S. Calculation of interfacial tension in polymer systems. J. Polym. Sci. C Polym.Symp. 1971, 34, 19–30.

Chapter 7

Hydrogel-Based Drug Delivery Systems for Poorly Water-Soluble Drugs

Matthew McKenzie, David Betts, Amy Suh, Kathryn Bui, London Doyoung Kim and Hyunah Cho *

Pharmaceutical and Administrative Sciences, St. Louis College of Pharmacy, 4588 Parkview Place, St. Louis, MO 63110, USA

ABSTRACT

Hydrogels are three-dimensional materials that can withstand a great amount of water incorporation while maintaining integrity. This allows hydrogels to be very unique biomedical materials, especially for drug delivery. Much effort has been made to incorporate hydrophilic molecules in hydrogels in the field of drug delivery, while loading of hydrophobic drugs has not been vastly studied. However, in recent years, research has also been conducted on incorporating hydrophobic molecules within hydrogel matrices for achieving a steady release of drugs to treat various ailments. Here, we summarize the types of hydrogels used as drug delivery vehicles, various methods to incorporate hydrophobic molecules in hydrogel matrices, and the potential therapeutic applications of hydrogels in cancer.

Keywords: hydrogels; thermosensitive; block copolymers; hydrophobic drugs

1. INTRODUCTION

Hydrogels are three-dimensional networks of chemically or physically cross-linked polymers that swell in water. Since the first synthetic poly(2-hydroxyethymethacrylate) hydrogels introduced by Wichterle and Lim in the late 1950s, natural and synthetic polymers with different structural features and physicochemical properties have been widely adopted to prepare “smart” or “intelligent” hydrogels as a depot-based drug delivery system to treat various diseases [1,2]. Hydrogels can be tailored to be sensitive to different environmental conditions, such as temperature, pH, and enzymatic activities at the diseased sites [3]. pH is one of the common stimulants for a successful sol-gel transformation. Hydrogels can also be temperature-sensitive: the majority of temperature-sensitive hydrogels are in a sol-state at the room temperature and undergo transformation to become a gel at a higher temperature, commonly the body temperature [4]. Hydrogels are generally biocompatible as reflected in their applications as contact lenses and as an exogenous barrier in the peritoneum [5,6]. Due to their unique physicochemical properties, hydrogels have emerged as a safe and effective depot-based drug delivery system in cancer therapy [4,7]. However, hydrogels, as a drug delivery system, have been limited to carrying hydrophilic drugs rather than hydrophobic drugs due to the limited quantity/homogeneity of loaded hydrophobic drugs in hydrogel matrices. To improve drug loading capacity, in recent studies, hydrogels have been confined into networks composed of small micelles with the average particle size of ≤200 nm [8]. Due to their micellar structure (hydrophobic core-hydrophilic shell), in matrices, hydrogels can encapsulate both hydrophilic and hydrophobic compounds to loco-regionally deliver multi-drugs in a single dose [4]. Being able to deliver hydrophobic drugs via hydrogel-based delivery systems is important, in particular, for local chemotherapy [9], noting that a large number of chemotherapeutic drug candidates possess poor water solubility and have failed to reach sufficient concentrations in injectable solutions required to exhibit therapeutic potency in preclinical studies and clinical trials [10]. Incorporation of poorly water-soluble drugs into hydrogels can enhance aqueous solubility of drugs and achieve extended release of drugs thereby increasing chances of intratumoral uptake of drugs than free drugs. In this review, we focus on hydrogels for hydrophobic drug delivery, covering the types of hydrogels used as hydrophobic drug delivery vehicles, methods to load hydrophobic drugs in hydrogel matrices, and the potential applications of hydrogels in cancer therapy.

2. TYPES OF HYDROGELS

2.1. Thermosensitive Hydrogels

The majority of thermosensitive hydrogel-forming polymers are block copolymers containing poly(ethylene glycol) (PEG) coupled to other hydrophobic polymer blocks. These block copolymers are composed of A-blocks and B-blocks arranged as ABA or BAB. A-blocks, PEG, provide biocompatibility, low immunogenicity, and water solubility whereas B-blocks (Figure 1) including poly(propylene oxide) (PPO), polyesters such as (poly(D,L-lactide) (PLA), poly(D,L-lactide-co-glycolide) (PLGA), poly(ε-caprolactone) (PCL) and poly[(*R*)-3-hydroxybutyrate] (PHB)), polyphosphazenes, and even polypeptides, impart hydrophobicity and provide drug loading capacity for hydrophobic drugs and properties of micellization and gelation [11].

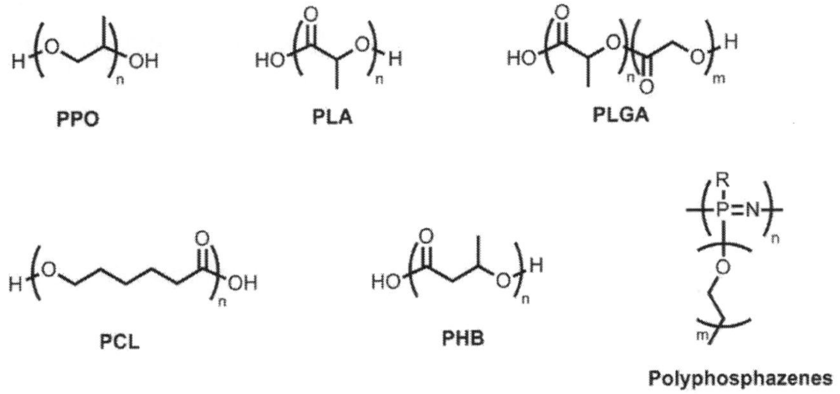

Figure 1. Hydrophobic polymer blocks in thermosensitive gel-forming triblock copolymers. PPO: poly(propylene oxide), PLA: (poly(D,L-lactide), PLGA: poly(D,L-lactide-co-glycolide), PCL: poly(ε-caprolactone), PHB: poly[(*R*)-3-hydroxybutyrate].

2.1.1. Hydrogels with Hydrophobic PPO Blocks

Poloxamers (also known as Pluronics) are an entity widely used to create hydrogels. Poloxamers are composed of one PPO block sandwiched between two poly(ethylene oxide) (PEO) blocks. In fact, PEO-PPO-PEO triblock copolymers exist in different physical forms and vary in their properties based on molecular weights and PPO/PEO ratios of copolymers [11,12]. Poloxamers have the ability to morph into self-assembling micelles from single triblock copolymer molecules. Below the critical micelle concentration (CMC), unimeric copolymers are dispersed in water. Above the CMC, aggregation of copolymers occurs causing micellization. These triblock copolymers at a given concentration are also identifiable by their critical micelle temperature (CMT): a specific temperature at which micelles form. At the CMT, triblock copolymers form micelles. At even higher temperatures, aggregations of micelles occur due to the dehydration of both PEO and PPO blocks, leading to a phase separation [13]. Due to the unique micelle-forming structure of triblock copolymers, poloxamers have excellent thermogelling behavior; a number of reports demonstrated that dispersions of poloxamer copolymers in water are in a sol-state at lower temperatures and in a gel-state at higher temperatures [14,15,16]. The sol-gel transition has been correlated to internal changes in micellar properties, to entropic variation in the water molecules closest to the PPO segments, or to formation of a physically cross-linked, three-dimensional structure able to hold water in its network [15,17,18,19]. Overall, both micellization and gelation depend on different factors, especially temperature, polymer concentration, and PEO block length [20]. The various applications of poloxamers as a surfactant, a stabilizing agent, a suppository base, a solubilizer, a gelling agent, and an efflux pump inhibitor have been extensively studied [21]. In particular, poloxamers 407, 188, 127, and 388 have been used to deliver various compounds, such as vancomycin, clotrimazole, ciprofloxacin, 5-aminolevulinic acid, ibuprofen and even protein drugs [21]. As two hydrophilic blocks sandwiched one PPO hydrophobic block in poloxamers, the loading of hydrophobic drugs into the poloxamer-based hydrogels is limited. To improve the solubility capacity for hydrophobic compounds, alpha-cyclodextrin was used to include poloxamer-based micelles carrying payloads to form crystalline inclusion complexes called polypseudorotaxanes [21]. These inclusion complexes enable incorporation of hydrophobic compounds in polypseudorotaxanes and further modulate release properties of payloads.

In addition to the limited loading capacity of poloxamers for hydrophobic drugs, the high CMC of poloxamers has restricted the application of poloxamers in the broad range of biomedical fields. Poloxamers have higher CMC values than many other triblock copolymers because there is a weaker hydrophobic interaction among PPO blocks requiring a greater quantity of copolymers to yield micelles and hydrogels. Hydrogels consisting of poloxamers with the high CMC values may be instable in patients. Upon the administration, a rapid physical dissociation of hydrogels (the burst effect) occurs due to the dilution of hydrogels with a substantial quantity of physiological liquids. The issues regarding the loading capacity for hydrophobic drugs, non-biodegradability, higher CMC values, and rapid dissociation of poloxamer-based hydrogels can be tackled by substituting PPO blocks with more hydrophobic and biodegradable polyesters.

2.1.2. Hydrogels with Hydrophobic PLA Blocks

Two polyesters, PLA and PLGA, as hydrophobic B-blocks of triblock copolymers have been shown to provide outstanding biodegradability, a higher gelation temperature, and extended release profiles for hydrophobic drugs [22,23,24,25,26]. Lee *et al.* synthesized PLA-PEG-PLA with acrylated groups at the ends of the PLA blocks which turned the self-assembled micelles into photo-crosslinked nanogels upon ultraviolet irradiation [22]. Photo-crosslinked PLA-PEG-PLA nanogels were 150 to 250 nm in size which can be easily manipulated by changing the concentration of crosslinkers and ultraviolet irradiation time [22]. Photo-crosslinked PLA-PEG-PLA nanogels served as a promising carrier for a hydrophobic drug, camptothecin, permitting a steady and extended release of camptothecin [22].

Asadi *et al.* prepared PLA-PEG-PLA hydrogels carrying a hydrophobic narcotic antagonist, naltrexone, in the size range of 128–200 nm using a different gelation technique without ultraviolet irradiation [23]. Briefly, PLA-PEG-PLA diacrylate copolymer micelles carrying naltrexone were first prepared by the nanoprecipitation method and then thermally crosslinked by heating the micelles up to 70 °C under nitrogen for 24 h. The resulting hydrogel (called nanogel in the article) suspension was further lyophilized for 48 h to remove solvents. A higher ratio of PLA/PEG blocks caused a greater naltrexone loading efficiency in PLA-PEG-PLA nanogels which theoretically provides a greater core loading capacity for hydrophobic drugs. Naltrexone incorporated in PLA-PEG-PLA nanogels using 50% of ethylene glycol dimethacrylate as a crosslinker reached <40% drug release within 35 days whereas naltrexone incorporated in PLA-PEG-PLA nanogels using 10% of ethylene glycol dimethacrylate reached 96% drug release within 15 days. This result demonstrated that PLA-PEG-PLA nanogels encapsulated a hydrophobic drug, naltrexone, in their hydrophobic core of micellar networks, and the drug release profile relied on the dissociation of crosslinked polymers.

2.1.3. Hydrogels with Hydrophobic PLGA Blocks

Given that PLA is hydrophobic, easy to crystallize, and readily forming precipitates, designing water-soluble thermogelling copolymers using PLA as hydrophobic blocks remains challenging [24]. To modify the hydrophobicity of hydrophobic blocks, lactide was copolymerized with glycolide which is more hydrophilic than lactide to yield PLGA [24]. Hydrogels with PEG and PLGA segments have been extensively investigated since 1999. One of the most popular thermosensitive products is ReGel.

ReGel is a triblock copolymer comprised of PLGA and PEG arranged into PLGA-PEG-PLGA. A PLGA-PEG-PLGA triblock copolymer is a free-flowing water soluble solution (sol-state) at lower temperatures (2–15 °C) and converted from sol to gel (semi-solid gel-state) at body temperature, 37 °C [11,25,26]. At temperatures below the critical gelation temperature (CGT), as shown in Figure 2, PLGA-PEG-PLGA copolymers create loops ("flower petals") sharing PLGA segments at the center and form "flower-like micelles" [11,25,26]. Triblock copolymers which do not contribute to the loop formation create bridges between flower-like micelles. As the temperature increases toward the CGT, hydrophobic interactions among PLGA segments increase, leading to strong micellar aggregation, loss of flowability, and inevitably gelation [11,25,26]. The gelation mechanism allows compounds with different hydrophilicity to be incorporated into the micellar matrix forming the basic structure of hydrogels. Hydrophobic drug molecules can be entrapped within the hydrophobic (PLGA-dense) regions, and hydrophilic drug molecules can be encased near the PEG regions (micellar bridging network).

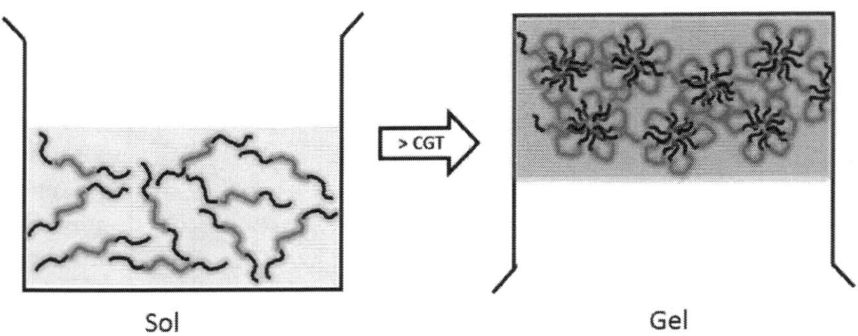

Figure 2. A schematic illustration of a sol-to-gel transition of PLGA-PEG-PLGA triblock copolymers. CGT: critical gelation temperature.

Nguyen *et al.* found that PEG-PLGA-PEG hydrogels had similar properties to ReGel [11]. Similarly, PEG-PLGA-PEG hydrogels were soluble in water at lower temperatures and gelled at higher temperatures [11]. Gelation occurred based on micellar growth/tight-packing with increasing temperature. With further increase in temperature, a sol-gel phase transition occurred due to micellar structure breakdown and dehydration of PEG and PLGA blocks.

Jeong *et al.* proved that release profiles of drugs differ regarding the hydrophobicity of drugs incorporated in PEG-PLGA-PEG hydrogels [27]. For example, upon *in situ* injections of PEG-PLGA-PEG hydrogels carrying ketoprofen or spironolactone into a 37 °C aqueous environment, a relatively hydrophilic drug, ketoprofen (log *p* 0.97), was released in two weeks with a first-order release profile whereas spironolactone (log *p* 2.78) was released in two months with a sigmoidal release profile. This suggests that payloads incorporated in hydrogels are released based on the different drug release mechanisms: The diffusion mechanism is responsible for the release of hydrophilic drugs whereas the physical degradation/erosion of hydrogels permits release of hydrophobic drugs.

2.1.4. Hydrogels with Hydrophobic PCL, PHB, or Polyphosphazene Blocks

PCL is a semi-crystalline biodegradable polyester, which yields hydrogels with stronger integrity of gels and extended release profiles of hydrophobic drugs. Gong *et al.* synthesized PEG-PCL-PEG block copolymers and successfully prepared PEG-PCL-PEG thermosensitive hydrogels carrying a hydrophobic drug, honokiol [28]. Within 14 days, less than 50% of honokiol was gradually released from PEG-PCL-PEG hydrogels. Upon a subcutaneous injection (30% *w/w*, 0.5 mL) of PEG-PCL-PEG hydrogels in mice, PEG-PCL-PEG hydrogels remained at the injection site at least 14 days whereas Poloxamer 407 (Pluronic F-127) hydrogels were no longer visible at the injection site in mice on the 14th day post injection.

PHB is a polyester naturally produced by bacteria. PHB has a higher crystallinity and hydrophobicity than PLA or PCL. Due to the great hydrophobicity, copolymers carrying PHB shows a sol-gel transition with temperature change at very low concentrations (2%–5% *w/w*) [11]. The mechanism of gelation by micellar packing is similar to the other polyesters.

Biodegradable polyphosphazenes are a type of common thermosensitive hydrogels used today. They consist of a hydrophilic PEG block and hydrophobic amino acids, such as L-isoleucine ethyl ester (IleOEt), D,L-leucineethyl ester (LeuOEt), L-valine ethyl ester (ValOEt), or di-, tri-, and oligo-peptides in the side groups [11]. A few different studies have been performed in regards to these copolymers. One involved incorporating IleOEt with PEG which exhibited a sol-gel transition as a function of temperature change [29]. Gelation properties were modified by changing the composition of the substituents and also the molecular weight of PEG. A similar result was observed when polyphosphazenes with oligopeptides (tri- or tetra peptides) and methoxy PEG 350 as side groups exhibited a phase transition [30].

2.2. Thermo- and pH-Sensitive Hydrogels

Block copolymers with a pH-sensitive moiety have been added to the existing thermosensitive copolymers to overcome the following obstacles [11,31,32,33,34,35]. First, injecting thermosensitive hydrogels into the physiological environment at 37 °C tends to cause rapid gelation and blockage in the delivery syringe/needle. In addition, the lack of side chain functional groups on the thermosensitive copolymers diminishes the ability to deliver ionic species and proteins. Block copolymers containing sulfamethazine oligomers (OSMs), poly(*b*-amino ester) (PAE), and poly(amidoamine) (PAA) were able to form thermo- and pH-sensitive hydrogels, providing improved sensitivity of gelation upon the environmental changes (Figure 3).

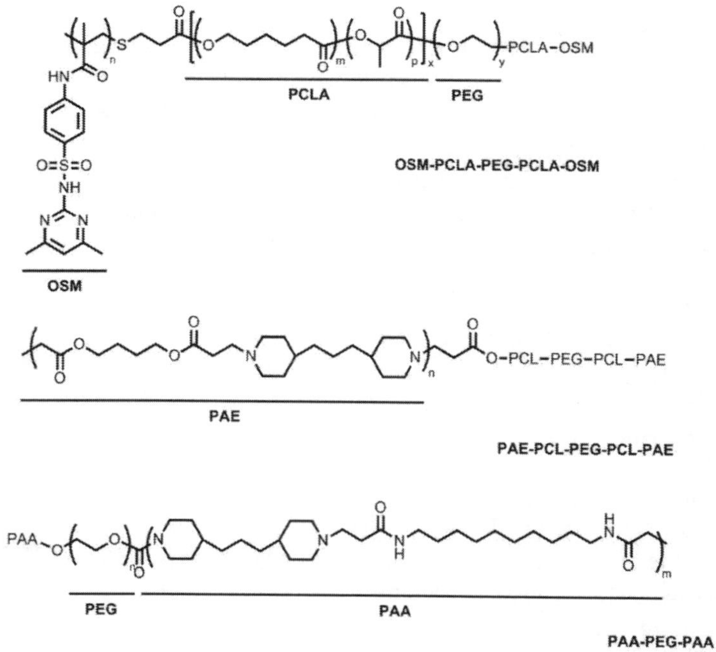

Figure 3. Thermo- and pH-sensitive hydrogel-forming block copolymers. OSM-PCLA-PEG-PCLA-OSM: sulfamethazine oligomers-poly(ε-caprolactone-co-lactide)-poly(ethyleneglycol)-poly(ε-caprolactone-co-lactide)-sulfamethazine oligomers, PAE-PCL-PEG-PCL-PAE: poly(*b*-aminoester)-poly(ε-caprolactone)-poly(ethylene glycol)-poly(ε-caprolactone)-poly(*b*-amino ester), PAA-PEG-PAA: poly(amidoamine)-poly(ethylene glycol)-poly(amidoamine).

OSMs have been used to create thermo- and pH-sensitive hydrogels. Thermosensitive poly(ε-caprolactone-co-lactide)-PEG-poly-(ε-caprolactone-co-lactide) (PCLA-PEG-PCLA) triblock copolymers were combined with OSMs to create thermo- and pH-sensitive OSM-PCLA-PEG-PCLA-OSM copolymers [31]. A thermosensitive PCLA-PEG-PCLA triblock copolymer was used as a parent molecule. PCLA-PEG-PCLA showed a sol-gel phase transition in an aqueous solution (15% *w/w*) as a function of temperature, however, it was not pH-sensitive [31]. Upon addition of OSMs at the endcaps, OSM-PCLA-PEG-PCLA-OSM copolymers (15% *w/w*) showed a sol-gel phase transition in response to both temperature and pH changes. Similarly, the gelation window grew wider with increasing pH as well as a greater PEG/PCLA ratio. It was proposed that an association of bridged micelles was responsible for the gelation of the pentablock copolymers [32]. When tested at pH 8.0 and between the temperatures of 10–70 °C, the sol-state was detected in the pentablock copolymer solution due to the ionized OSMs. However, at physiological pH of 7.4 and 15 °C, the OSMs became unionized and hence, more hydrophobic, but the pentablock copolymer still existed as in the sol-phase because of weak interactions among PCLA blocks at this lower temperature. This suggests that OSM-PCLA-PEG-PCLA-OSM hydrogels will stay in the sol-phase at pH 8.0 in the wide temperature ranges which will permit easier delivery of formulations using a syringe. In contrast, at physiologic pH of 7.4 and temperature of 37 °C, the PCLA blocks started to become hydrophobic, causing stronger hydrophobic interactions between the PCLA-OSM blocks. This led to mass micelle bridging and a gelation process ensued [31].

Basic PAE has also been used in studies to create block copolymers [33,34,35]. PAE-PCL-PEG-PCL-PAE is thermo- and pH-sensitive, forming hydrogels at pH above 6.0 with increasing temperatures [33]. Gelation was theorized to be due to micellar packing upon heating. PAE-PCL-PEG-PCL-PAE pentablock copolymers were synthesized using PCL-PEG-PCL as a parent copolymer. PCL-PEG-PCL, a parent copolymer, is sensitive to temperature change but not to pH changes. In contrast, the PAE-PCL-PEG-PCL-PAE pentablock copolymers at pH above 6.0 underwent a sol-gel phase transition in response to both temperature and pH changes [34]. The gelation window could be manipulated by changing PEG molecular weight, PAE block length, PCL/PEG ratio, and concentration of copolymers [35].

PAA has been shown to be both thermo- and pH-sensitive [11]. PAA-PEG-PAA triblock copolymers were in a sol-state at pH below 3.0; however, PAA-PEG-PAA triblock copolymers earned hydrophobic properties at pH above 3.0. When subjected to higher temperatures, pK_a of PAA-PEG-PAA copolymers decreased. PAA-PEG-PAA aqueous solutions subsequently underwent a sol-gel-to-condensed gel transition with a very high viscosity at physiological temperature and pH. The sol–gel window could be tailored by varying the polymer compositions and concentrations [11].

3. PREPARATION OF HYDROGELS CARRYING HYDROPHOBIC DRUGS

Hydrogel systems as drug delivery vehicles have been commonly used to deliver hydrophilic compounds. More recently, studies focusing on delivery of hydrophobic drugs using hydrogel systems have been reported to improve solubility of hydrophobic drugs in water, achieve a steady release of drugs, and minimize the burst effect of drugs in patients.

3.1. Simple Mixing below the CGT

One of the ways to incorporate hydrophobic drugs within a hydrogel matrix was seen in the study conducted by Almonen *et al.*, in which progesterone (P4) was incorporated into amphiphilic glycol chitin-based hydrogels [36]. Progesterone is a lipophilic female hormone, making it difficult to treat a patient without the proper vehicle. Almonen *et al.*incorporated P4 in a glycol chitin hydrogel matrix to safely and effectively administer P4 to patients via the vaginal route. Glycol chitin-based hydrogels provided an enhanced aqueous solubility of P4, was biodegradable, and made a sol-gel transition at the body temperature [36,37]. Glycol chitin, a linear polysaccharide composed of D-glucosamine and *N*-acetyl-D-glucosamine in addition to partial deacetylation of chitin, has shown to be temperature-sensitive, nontoxic, biocompatible, and biodegradable [38]. In this study, glycol chitin carried P4 in the hydrogel matrix and gradually released P4 in the vaginal environment (37 °C, pH 4.2). The pK_a value of the amine groups on glycol chitin backbone is *ca.* 6.84. Thus, ionization of amines takes place at the vaginal pH and induces the repulsion between polymer chains. The ionization of amines, in turn, leads to an increase in the CGT. Therefore, glycol chitin carrying P4 maintained its gel-like integrity in the vaginal environment, and is in a sol-state at the storage condition of pH 7.2 and 4 °C. Almomen *et al.* was able to successfully incorporate P4 by simply adding P4 above the level of solubility in 5% *w/v* glycol chitin at pH 4.2, 4 °C. Once added, the mixture was stirred for 24 h. The resulting mixture was found to contain 0.1% *w/v* of P4 [36]. This method, making dispersions of a payload at a greater concentration than its intrinsic solubility (*ca.* 7 μg/mL (intrinsic solubility) *vs.*148 μg/mL), aimed to saturate extramicellar channels in hydrogels so that hydrogels release drugs in a sustained manner without altering the structure of micellar aggregates and the CGT.

In a study conducted by Xuan *et al.*, doxorubicin was loaded into a poloxamer (P407 and P188)-based hydrogel in the presence of hydrochloric acid [39]. First, P188 and P407 were completely dissolved in distilled water at 4 °C using gentle stirring and allowed to rest overnight at 4 °C. Doxorubicin and hydrochloric acid were then added the following day with gentle stirring and left to rest overnight at 4 °C to load doxorubicin in poloxamers.

3.2. Caging Drug-Loaded Nanoparticles or Complexes in Hydrogels

Another example of incorporation of hydrophobic drugs in a gel matrix was found in a study conducted by Kulkarni *et al.*, in which κ-carrageenan (KC) hydrogel was used to incorporate nanostructured lipid particles, known as internally self-assembled some-particles (ISAsomes) within its hydrogel matrix [40]. The purpose of this study was to find an efficient vehicle for the drug-loaded ISAsome that would allow slow drug release and maintain the intact ISAsome structure. A hydrophobic drug was first encapsulated in the ISAsome. Once encapsulated, ISAsome was incorporated into KC hydrogels followed by dehydration resulting in a thin film. This preparation decreased porosity of hydrogels and allowed the ISAsome carrying a hydrophobic drug to be released as the hydrogel liquefied. Briefly, nanostructured lipid particles were prepared and formed into emulations. Separately, a 4% KC solution was prepared by stirring KC powder in water, and heated at 60 °C for 20 min. Equal parts of lipid emulsion and KC solution were then combined and allowed to be gently stirred at 60 °C until the mixture is visually homogenized [40].

In a study conducted by Kushwaha *et al.*, camptothecin (CPT) was incorporated in a chitosan polymer matrix that had glycerol-2-phosphate (β-GP) added to the base structure. In addition, cyclodextrin was added as a cosolvent in order to increase solubility and facilitated the delivery of CPT by complexing with the drug using guest-host interactions [41]. By creating a complex, a chitosan polymer could remain a liquid at cool to room temperature and form a gel at 37 °C, remain stable at a physiological pH, and sufficiently hold the CPT-cyclodextrin payload, allowing for a slow release of the drug locally at the injection site. The preparation started with a base chitosan solution prepared by slowly adding chitosan powder to 0.1 M hydrochloric acid at room temperature, stirred for 3 h, and autoclaved to sterilize the solution. β-GP was then added (drop-wise) to the chitosan solution, and the solution was mixed for an additional 10 min at 4 °C. Once the base vehicle was made, β-cyclodextrin was slowly added at room temperature to the chitosan β-GP solution. Once that solution was homogenized, CPT was added to the solution [41].

Another use of poloxamer-based hydrogels was reported by Ju *et al.* Paclitaxel-loaded *N*-octyl-*O*-sulfate chitosan (NOSC) micelles were prepared to increase water solubility of paclitaxel up to 1000-fold. Paclitaxel-loaded NOPSC micelles were dispersed in carboxymethyl chitosan (CMCS)-conjugated poloxamer P407 gels and mixed with glutaraldehyde to create crosslinks with NH_2 groups on CMCS [42]. The CMCS-conjugated P407 carrying paclitaxel-loaded NOSC micelles made a gel depot upon an intratumoral injection and extended retention time at tumor sites, resulting in the improved antitumor efficacy and reduction in hepatic metabolism.

3.3. Solvent Evaporation Method

A study was performed by Gao *et al.*, in which docetaxel (DTX) was incorporated in a hydrogel matrix comprised of PLGA-PEG-PLGA. In order to incorporate docetaxel within the PLGA-PEG-PLGA micelles, both were dissolved in acetone, mixed and dried under vacuum. Once dried, distilled water was added to form DTX-loaded micelles. Upon the intratumoral injection of DTX-loaded PLGA-PEG-PLGA micelles, PLGA-PEG-PLA formed a gel depot at the injection site at 37 °C [43].

3.4. Modified Nanoprecipitation Method

In a study conducted by Pillai *et al.*, folic acid-conjugated PEG acrylic polymer (FA-CLAP) hydrogel was synthesized for curcumin delivery [44]. Curcumin was loaded into the FA-CLAP hydrogels using a lyophilizaiton procedure. FA-CLAP was dissolved in water at 1 g/100 mL, and curcumin was dissolved in chloroform. The curcumin solution was added to the hydrated FA-CLAP, and the mixture was continuously vortexed and sonicated. The mixture was then lyophilized which can be rehydrated with a sterile solvent prior to the injection.

4. HYDROGELS IN CANCER THERAPY

Depending on the dosage form of a medication, the results of drug release can vary by numerous degrees. Over years, researchers have been trying to create a dosage form that would provide better bioavailability and tolerability with a decrease in side effects. Hydrogels are one of the rising drug delivery systems that accommodates for these challenges today, especially with chemotherapeutic agents. First, hydrogels could improve water solubility of hydrophobic drugs enabling concurrent delivery of multiple hydrophobic drugs. OncoGel, a non-Cremophor, ReGel (PLGA-PEG-PLGA)-based formulation of paclitaxel, was designed for locoregional delivery of paclitaxel to solid tumors [45]. OncoGel as monotherapy demonstrated acceptable local tolerability and no systemic toxicity in preclinical efficacy studies. In a CRL-1666 breast cancer spinal metastases model, OncoGel increased the survival rate *vs.* the untreated control group and improved hind limb function with minimal side effect toxicity. In a MDA-MB-231 subcutaneous breast cancer-bearing mouse model, intraperitoneally administered OncoGel at 1/10 of the systemic taxol dose exhibited a similar long-term survival result as a monotherapy. OncoGel as an adjuvant to radiation therapy in an intracranial glioblastoma mouse model demonstrated improved long-term survival rate of 37.5% whereas there was no long-term survivors when treated with the radiation therapy alone. In addition, OncoGel as an adjuvant to surgery in a CRL-1666 breast cancer spinal metastases model (placement of OncoGel into the resection cavity post-surgery) delayed the onset of paresis.

Cho *et al.* prepared PLGA-PEG-PLGA thermosensitive hydrogels carrying three hydrophobic drugs, paclitaxel, 17-AAG and rapamycin (Triogel) at 6, 6, 3 mg/mL, respectively [9]. Drug release profiles for Triogel reached 46% paclitaxel, 46% 17-AAG and 44% rapamycin in 48 h. In contrast, PEG-*b*-PLA micelles carrying paclitaxel, 17-AAG and rapamycin at 6, 6, 3 mg/mL, respectively (Triolimus, micellar solution) released 78% paclitaxel, 91% 17-AAG, and 68% rapamycin in 24 h. Triogel was highly effective in eradicating metastatic peritoneal ovarian tumors in ES-2-luc human ovarian cancer-bearing mice. A single intraperitoneal injection of Triogel carrying paclitaxel, 17-AAG and rapamycin at 60, 60 and 30 mg/kg, respectively, decreased tumor burden from 100% to 6% on the 14th day post treatment. Tumor regression upon a single intraperitoneal injection of Triogel was 70-fold superior than a treatment result of intravenously injected PEG-*b*-PLA micelles carrying paclitaxel, 17-AAG and rapamycin at 60, 60, 30 mg/kg, respectively. This interesting result was presumably due to the different release rates of payloads for hydrogels (slower) and liquid micelles (faster) and duration of drug exposure at the peritoneum of mice (longer residence time for hydrogels than micellar solutions).

Various research has been done on the use of hydrogels during peritoneal cancer surgery. One area in which hydrogels play a huge role is the possible prevention of post-surgical adhesions from intra-abdominal surgeries. Results through various studies on animals have shown that prevention of post-surgical adhesion is possible with the use of hydrogels. These postsurgical adhesions are not inevitable, still occurring in approximately 67% to 93% of humans who undergo abdominal or pelvic surgery [46,47].

Adhesions commonly occur after gynecological procedures and intraperitoneal surgeries leading to problems such as infertility, intestinal obstruction, chronic abdominal pain, and sometimes requiring re-surgery. During the first three days of post-surgery, abnormal attachments between tissues and the adjacent organs may occur [46,48]. The formation of internal scars that develop from post-surgery, infection, trauma, or radiation has caused complications in many patients. Strategies to prevent the abdominal adhesion formation using thermosensitive polymeric hydrogels have been studied in animals. Polymeric hydrogels, such as PEG-PCL-PEG with its sol-gel-sol property, have been proven to be more biocompatible, biodegradable and nontoxic. Hydrogels provide an effective barrier which will allow the peritoneal tissues and organs to separate long enough during the healing process [49]. The solidity of the gel also contributes to its adhesive characteristics that can be beneficial by adhering to internal organs for the prevention of postsurgical adhesions. Noting that both PEG and PCL are FDA-approved biodegradable and biocompatible materials, PEG-PCL-PEG hydrogels were able to adhere to the affected sites in the peritoneum and were eventually absorbed by the body while the wounds were healing [46,47]. In addition, while evaluating the hydrogel on the injured surfaces that were created on the cecum and abdominal wall of rats, it was found that the hydrogel coat disappeared on the fifth day. However, the residual hydrogel that adhered closely to the damaged surface still appeared on day 7 through day 14 [46,47]. The viscous liquid was completely absorbed by the body by day 14. Thermosensitive hydrogels have also been studied as a novel amphiphilic, multi-drug delivery system for intra-abdominal and peritoneal cancers [9]. Intraperitoneal administration of thermosensitive hydrogels carrying multi-drugs will be useful to prevent the post-surgical adhesion formation and treat residual tumor margins and tissues.

5. CONCLUSIONS

In this article, we summarize the types of hydrogels as drug delivery vehicles, various methods to load hydrophobic molecules in hydrogel matrices, and the potential applications of hydrogels in cancer treatment. Hydrogels have shown great potential as a barrier to prevent post-surgical adhesions as well as a support system in the delivery of hydrophilic and hydrophobic drugs to the site of action. Although there have not been many clinical trials over the use of hydrogels for the locoregional treatment of diseases, efforts have been made to tailor hydrogels to impart desirable physicochemical properties, such as biocompatibility and loading capacity for hydrophobic compounds, allowing for optimization of hydrogel-based drug delivery systems.

ACKNOWLEDGMENTS

This work was supported by the Faculty Research Incentive Fund (FRIF), St. Louis College of Pharmacy.

AUTHOR CONTRIBUTIONS

M.M., D.B., A.S., K.B., and L.D.K. wrote the paper; H.C. drafted and revised the paper. All authors read and approved the final manuscript.

CONFLICTS OF INTEREST

The authors declare no conflict of interest. The founding sponsors had no role in the writing of the manuscript and in the decision to publish the results.

REFERENCES

1. Gupta, P.; Vermani, K.; Garg, S. Hydrogels: From controlled release to pH-responsive drug delivery. Drug Discov. Today 2002, 7, 569–579.

2. Kopecek, J. Hydrogels from soft contact lenses and implants to self-assembled nanomaterials. J. Polym. Sci. A Polym. Chem. 2009, 47, 5929–5946.

3. Singh, N.K.; Lee, D.S. In situ gelling pH- and temperature-sensitive biodegradable block copolymer hydrogels for drug delivery. J. Control Release 2014, 193, 214–227.

4. Simoes, S.; Figueiras, A.; Francisco, V. Modular hydrogels for drug delivery. J. Biomater. Nanobiotechnol. 2012, 3, 185–199.

5. Sakai, S.; Ueda, K.; Taya, M. Peritoneal adhesion prevention by a biodegradable hyaluronic acid-based hydrogel formed in situ through a cascade enzyme reaction initiated by contact with body fluid on tissue surfaces. Acta Biomater. 2015, 24, 152–158.

6. Xinming, L.; Yingde, C.; Lloyd, A.W.; Mikhalovsky, S.V.; Sandeman, S.R.; Howel, C.A.; Liewen, L. Polymeric hydrogels for novel contact lens-based ophthalmic drug delivery systems: A review. Contact. Lens Anterior Eye 2008,31, 57–64.

7. Hoare, T.R.; Kohane, D.S. Hydrogels in drug delivery: Progress and challenges. Polymer 2008, 49, 1993–2007.

8. Maya, S.; Sarmento, B.; Nair, A.; Rejinold, N.S.; Nair, S.V.; Jayakumar, R. Smart stimuli sensitive nanogels in cancer drug delivery and imaging: A review. Curr. Pharm. Des. 2013, 19, 7203–7218.

9. Cho, H.; Kwon, G.S. Thermosensitive poly-(D,L-lactide-co-glycolide)-block-poly(ethylene glycol)-block-poly-(D,L-lactide-co-glycolide) hydrogels for multi-drug delivery. J. Drug Target 2014, 22, 669–677.

10. Lipinski, C.A. Drug-like properties and the causes of poor solubility and poor permeability. J. Pharmacol. Toxicol. Methods 2000, 44, 235–249.

11. Nguyen, M.K.; Lee, D.S. Injectable biodegradable hydrogels. Macromol. Biosci. 2010, 10, 563–579.

12. Bonacucina, G.; Cespi, M.; Mencarelli, G.; Giorgioni, G.; Palmieri, G.F. Thermosensitive self-assembling block copolymers as drug delivery systems. Polymers 2011, 3, 779–811.

13. Raymond, E.G. On the theory of lower critical solution points in hydrogen-bonded mixtures. J. Chem. Phys. 1984, 80, 5340–5341.

14. Al-Saden, A.A.; Florence, A.T.; Morrison, H.; Whateley, T.L. Association of poloxamer block copolymers in aqueous solution [proceedings]. J. Pharm. Pharmacol. 1979, 31, 81.

15. Wang, P.; Johnson, T.P. Kinetics of sol-to-gel transition for poloxamer polyols. J. App. Polym. Sci. 1991, 43, 283–292.

16. Brown, W.; Schillen, K.; Almgren, M.; Hvidt, S.; Bahadur, P. Micelle and gel formation in a poly(ethylene oxide)/poly(propylene oxide)/poly(ethylene oxide) triblock copolymer in water solution: Dynamic and static light scattering and oscillatory shear measurements. J. Phys. Chem. 1991, 95, 1850–1858.

17. Wanka, G.; Hoffmann, H.; Ulbricht, W. The aggregation behavior of poly-(oxyethylene)-poly-(oxypropylene)-poly-(oxyethylene)-block-copolymers in aqueous solution. Colloid Polym. Sci. 1990, 268, 101–117.

18. Rassing, J.; Attwood, D. Ultrasonic velocity and light-scattering studies on the polyoxyethylene-polyoxypropylene copolymer pluronic F127 in aqueous solution. Int. J. Pharm. 1982, 13, 47–55.

19. Vadnere, M.; Amidon, G.; Lindenbaum, S.; Haslam, J.L. Thermodynamic studies on the gel-sol transition of some pluronic polyols. Int. J. Pharm. 1984, 22, 207–218.

20. Pandit, N.K.; Wang, D. Salt effects on the diffusion and release rate of propranolol from poloxamer 407 gels. Int. J. Pharm. 1998, 167, 183–189.

21. Devi, D.R.; Sandhya, P.; Hari, B.N.V. Poloxamer: A novel functional molecule for drug delivery and gene therapy. J. Pharm. Sci. Res. 2013, 5, 159–165.

22. Lee, W.C.; Li, Y.C.; Chu, I.M. Amphiphilic poly(D,L-lactic acid)/poly(ethylene glycol)/poly(D,L-lactic acid) nanogels for controlled release of hydrophobic drugs. Macromol. Biosci. 2006, 6, 846–854.

23. Asadi, H.; Rostamizadeh, K.; Salari, D.; Hamidi, M. Preparation and characterization of tri-block poly(lactide)-poly(ethylene glycol)-poly(lactide) nanogels for controlled release of naltrexone. Int. J. Pharm. 2011, 416, 356–364.

24. Loh, X.J.; Li, J. Biodegradable thermosensitive copolymer hydrogels for drug delivery. Expert Opin. Ther. Pat. 2007, 17, 965–977.

25. Cho, H.; Lai, T.C.; Tomoda, K.; Kwon, G.S. Polymeric micelles for multi-drug delivery in cancer. AAPS Pharm. Sci. Technol. 2015, 16, 10–20.

26. Qiao, M.; Chen, D.; Ma, X.; Liu, Y. Injectable biodegradable temperature-responsive PLGA-PEG-PLGA copolymers: Synthesis and effect of copolymer composition on the drug release from the copolymer-based hydrogels. Int. J. Pharm. 2005, 294, 103–112.

27. Jeong, B.; Bae, Y.H.; Kim, S.W. Drug release from biodegradable injectable thermosensitive hydrogel of peg-plga-peg triblock copolymers. J. Control Release 2000, 63, 155–163.

28. Gong, C.; Shi, S.; Dong, P.; Kan, B.; Gou, M.; Wang, X.; Li, X.; Luo, F.; Zhao, X.; Wei, Y.; et al. Synthesis and characterization of PEG-PCL-PEG thermosensitive hydrogel. Int. J. Pharm. 2009, 365, 89–99.

29. Lee, B.H.; Lee, Y.M.; Sohn, Y.S.; Song, S.C. A thermosensitive poly(organophosphazene) gel. Macromolecules 2002, 35, 3876–3879.

30. Seong, J.Y.; Juna, Y.J.; Jeong, B.; Sohn, Y.S. New thermogelling poly(organophosphazenes) with methoxypoly(ethylene glycol) and oligopeptide as side groups. Polymer 2005, 46, 5075–5081.

31. Shim, W.S.; Yoo, J.S.; Bae, Y.H.; Lee, D.S. Novel injectable ph and temperature sensitive block copolymer hydrogel.Biomacromolecules 2005, 6, 2930–2934.

32. Choi, B.G.; Sohn, Y.S.; Jeong, B. Closed-loop sol-gel transition of peg-pec aqueous solution. J. Phys. Chem. B 2007, 111, 7715–7718.

33. Yoo, J.S.; Kim, M.S.; Lee, D.S.; Kim, B.S.; Kim, J.H. Novel pH and temperature-sensitive block copolymers: Poly(ethylene glycol)-b-poly(ε-caprolactone)-b-poly(β-amino ester). Macromol. Res. 2006, 14, 117–120.

34. Huynh, D.P.; Nguyen, M.K.; Pi, B.S.; Kim, M.S.; Chae, S.Y.; Lee, K.C.; Kim, B.S.; Kim, S.W.; Lee, D.S. Functionalized injectable hydrogels for controlled insulin delivery. Biomaterials 2008, 29, 2527–2534.

35. Huynh, D.P.; Nguyen, M.K.; Kim, B.S.; Lee, D.S. Molecular design of novel pH/temperature-sensitive hydrogels.Polymer 2009, 50, 2565–2571.

36. Almomen, A.; Cho, S.; Yang, C.H.; Li, Z.; Jarboe, E.A.; Peterson, C.M.; Huh, K.M.; Janat-Amsbury, M.M. Thermosensitive progesterone hydrogel: A safe and effective new formulation for vaginal application. Pharm. Res.2015, 32, 2266–2279.

37. Li, Z.; Cho, S.; Kwon, I.C.; Janát-Amsbury, M.M.; Huh, K.M. Preparation and characterization of glycol chitin as a new thermogelling polymer for biomedical applications. Carbohydr. Polym. 2013, 92, 2267–2275.

38. De Alvarenga, E.S. Characterization and properties of chitosan. In Biotechnology of Biopolymers; Elnashar, M., Ed.; InTech: Rijeka, Croatia, 2011; pp. 91–108.

39. Xuan, J.J.; Yan, Y.D.; Oh, D.H.; Choi, Y.K.; Yong, C.S.; Choi, H.G. Development of thermo-sensitive injectable hydrogel with sustained release of doxorubicin: Rheological characterization and in vivo evaluation in rats. Drug Deliv. 2011,18, 305–311.

40. Kulkarni, C.V.; Moinuddin, Z.; Patil-Sen, Y.; Littlefield, R.; Hood, M. Lipid-hydrogel films for sustained drug release.Int. J. Pharm. 2015, 479, 416–421.

41. Kushwaha, S.K.S.; Rai, A.K.; Singh, S. Formulation of thermosensitive hydrogel containing cyclodextrin for controlled drug delivery of camptothecin. Trop. J. Pharm. Res. 2014, 13, 1007–1012.

42. Ju, C.; Sun, J.; Zi, P.; Jin, X.; Zhang, C. Thermosensitive micelles-hydrogel hybrid system based on poloxamer 407 for localized delivery of paclitaxel. J. Pharm. Sci. 2013, 102, 2707–2717.

43. Gao, Y.; Ren, F.; Ding, B.; Sun, N.; Liu, X.; Ding, X.; Gao, S. A thermosensitive PLGA-PEG-PLGA hydrogel for sustained release of docetaxel. J. Drug Target 2011, 19, 516–527.

44. Pillai, J.J.; Thulasidasan, A.K.; Anto, R.J.; Chithralekha, D.N.; Narayanan, A.; Kumar, G.S. Folic acid conjugated cross-linked acrylic polymer (fa-clap) hydrogel for site specific delivery of hydrophobic drugs to cancer cells. J. Nanobiotechnol. 2014, 12, 25–33.

45. Elstad, N.L.; Fowers, K.D. Oncogel (regel/paclitaxel)-clinical applications for a novel paclitaxel delivery system. Adv. Drug Deliv. Rev. 2009, 61, 785–794.

46. Pados, G.; Makedos, A.; Tarlatzis, B. Adhesion prevention strategies in laparoscopic surgery. In Endoscopy; Amornyotin, S., Ed.; InTech: Rijeka, Croatia, 2013; pp. 49–72.

47. Yang, B.; Gong, C.; Qian, Z.; Zhao, X.; Li, Z.; Qi, X.; Zhou, S.; Zhong, Q.; Luo, F.; Wei, Y. Prevention of post-surgical abdominal adhesions by a novel biodegradable thermosensitive pece hydrogel. BMC Biotechnol. 2010, 10, 65–73.

48. Di Zerega, G.S. Biochemical events in peritoneal tissue repair. Eur. J. Surg. Suppl. 1997, 577, 10–16.

49. Gao, X.; Deng, X.; Wei, X.; Shi, H.; Wang, F.; Ye, T.; Shao, B.; Nie, W.; Li, Y.; Luo, M.; et al. Novel thermosensitive hydrogel for preventing formation of abdominal adhesions. Int. J. Nanomed. 2013, 8, 2453–2463.

Chapter 8

Polymorph Impact on the Bioavailability and Stability of Poorly Soluble Drugs

Roberta Censi and Piera Di Martino *

School of Pharmacy, University of Camerino, via S. Agostino, 1, Camerino 62032, Italy

ABSTRACT

Drugs with low water solubility are predisposed to poor and variable oral bioavailability and, therefore, to variability in clinical response, that might be overcome through an appropriate formulation of the drug. Polymorphs (anhydrous and solvate/hydrate forms) may resolve these bioavailability problems, but they can be a challenge to ensure physicochemical stability for the entire shelf life of the drug product. Since clinical failures of polymorph drugs have not been uncommon, and some of them have been entirely unexpected, the Food and Drug Administration (FDA) and the International Conference on Harmonization (ICH) has required preliminary and exhaustive screening studies to identify and characterize all the polymorph crystal forms for each drug. In the past, the polymorphism of many drugs was detected fortuitously or through manual time consuming methods; today, drug crystal engineering, in particular, combinatorial chemistry and high-throughput screening, makes it possible to easily and exhaustively identify stable polymorphic and/or hydrate/dehydrate forms of poorly soluble drugs, in order to overcome bioavailability related problems or clinical failures. This review describes the concepts involved, provides examples of drugs characterized by poor solubility for which polymorphism has proven important, outlines the state-of-the-art technologies and discusses the pertinent regulations.

Keywords: polymorphism; poorly soluble drug; polymorphism screening; regulatory issues

1. INTRODUCTION

In the industrial development of a new drug substance and/or product, considerable problems are posed by candidate drugs with poor aqueous solubility, as this characteristic is related to poor bioavailability. Research and Development takes various approaches to enhancing the solubility and/or dissolution rate, and thus oral bioavailability, of poorly water-soluble drugs. One of the most common and effective approaches for increasing the solubility and dissolution rates of acidic and basic drugs is salt formation [1]. More recently, co-crystals, defined as crystalline materials comprised of at least two different components [2], have attracted attention for improving the dissolution rate of poorly water-soluble drugs [3]. Drug particle size reduction, affecting the dissolution rates, has been revealed one of oldest strategies for improving bioavailability of drugs and has been frequently applied in the pharmaceutical industry for routine production [4]. During the last years, the development of nanotechnologies have aroused the interest of researchers who have developed new technologies, easily industrially scalable, to reduce the particle size to nanodimensions [5,6]. Including or dispersing the poorly soluble drug in a carrier such as a cyclodextrin [7,8] or a polymer (solid dispersion) [9] are also common applied approaches. Modifications in the solid state, conversion from one polymorph to another [10], solvation/hydration [11], or amorphization [12,13] have been intentionally or unintentionally considered by the researchers and by the pharmaceutical industry during drug development of poorly-soluble drugs.

When the polymorphic form modification approach is chosen, not only must the effective improvement of drug bioavailability—which is not always obvious—be verified, but problems with the drug substance and product stability can arise. Generally, metastable forms are more soluble than the corresponding stable polymorphic forms, but they transform to the more thermodynamically stable form in a relatively short time [14], and thus it is necessary to monitor the polymorphic transformation during formulation, manufacturing, and storage of dosage forms to ensure reproducible bioavailability after administration [15].

In addition, the change of the polymorphic form has frequently caused clinical failures once it is on the market. This review should provide a useful overview for pharmaceutical industry readers interested in the development of new drug substances and/or products using polymorphic modifications, and offers many examples of such efforts.

Since the US Food and Drug Administration (FDA) and the International Conference on Harmonization (ICH) classify anhydrous, hydrate and solvate forms as polymorphs [16], in this review the term polymorphism will refer to both anhydrous and solvate (hydrate) forms.

2. IMPORTANCE OF SOLUBILITY ON THE BIOAVAILABILITY OF DRUGS

Solubility is the ability of a *solute* to dissolve in a *solvent* to form a homogeneous solution of the solute in the solvent. This property is influenced by temperature and pressure [17]. Typical aqueous solubilities are indicated in several Pharmacopoeia, including the U.S. Pharmacopoeia (Table 1).

Solubility is an essential property of drugs, because they must dissolve in order to be absorbed through membranes and reach the site of action. Consequently, solubility is one of the most critical and important parameters influencing drug bioavailability, that is, the ability of a drug to be available in an appropriate concentration at the site of action, independently of the pharmaceutical dosage form and route of administration.

Table 1. Aqueous solubilities according to the U.S. Pharmacopoeia.

Freely Soluble	100–1000 mg/mL
Soluble	33–100 mg/mL
Sparingly soluble	10–33 mg/mL
Slightly soluble	1–10 mg/mL
Very slightly soluble	0.1–1 mg/mL
Practically insoluble	<0.1 mg/mL

Poor aqueous solubility is the one of the major problems encountered during the development of new drug substances and/or drug products. This aspect becomes even more important if one considers that more than 40%

of marketed immediate release oral drugs are practically insoluble (<100 g/mL) [18,19], and that fully 70% of new drug candidates in the pharmaceutical industry pipelines are practically insoluble in water [20]. Jean-Paul Garnier, CEO of GlaxoSmithKline, said that *"About 50% of drug candidates that enter clinical trials fail due to efficacy and safety concerns, and the remaining 40% fizzle due to patent concerns and issues like solubility and drug interaction"* [21].

To have an idea of the importance of drug solubility and how poor aqueous solubility limits drug bioavailability, one can refer to some examples offered by Amidon *et al.*, [22] concerning the volumes needed to dissolve poorly water soluble drugs according to their dose. Some of the consequences of the inadequate aqueous solubility of a drug are limited and variable absorption, formulation and food effects, and poor tissue distribution and metabolism [23].

The importance of the solubility parameter is confirmed in the Biopharmaceutical Classification System (BCS) [24], a scientific framework for classifying drug substances based on their aqueous solubility and intestinal permeability [22,25]. The BCS takes into account three major factors that govern the rate and extent of drug absorption from immediate release solid oral dosage forms: (1) dissolution; (2) solubility; and (3) intestinal permeability. According to the BCS, drug substances are classified as follows:

- Class 1: High Solubility–High Permeability
- Class 2: Low Solubility–High Permeability
- Class 3: High Solubility–Low Permeability
- Class 4: Low Solubility–Low Permeability

3. THE POLYMORPHISM OF DRUGS: ANHYDROUS AND SOLVATED FORMS

Among the various techniques used to enhance the solubility of poorly soluble drugs are physical and chemical modifications of the drug, and methods such as particle size reduction, salt formation, solid dispersion, use of surfactant, and complexation [23]. Selection of a solubility improving method depends on drug property, site of absorption, and required dosage form characteristics [26].

Crystalline polymorphs have the same chemical composition, but different internal crystal structures, and therefore possess different physicochemical properties [27] because of their different lattice structures and/or different molecular conformations [28]. The phenomenon of polymorphism is quite common among organic molecules, and many drugs can crystallize into different polymorphic forms [29,30,31,32].

Polymorphic forms of drugs can prove interesting for drug developers because their thermodynamic and physicochemical properties, such as energy, melting point, density, stability, and in particular solubility, may offer an improvement on the original form.

Generally, the solubility of metastable polymorphs is kinetically higher than that of a thermodynamically more stable polymorph [33], offering, at least in theory, a solution to bioavailability problems.

Actually, it has been demonstrated that differences between the solubility of one polymorph and another are typically lower than a factor of 2 [34] or more rarely of 5 [35]. Thus, while a polymorph may offer a slight improvement in solubility compared to the original compound, this benefit may be offset the fact that it is also less stable than the original, and thus there may be no advantage in choosing this polymorph over the original compound. Actually, metastable and more soluble forms tend to convert into the more thermodynamic stable form in a relatively short time. The presence of specific excipients, or particular chemical and/or technological processes may accelerate the transition to the solid state [36,37]. This transition may proceed according to the relative thermodynamic stability of metastable forms, or be accelerated by the presence of seeds of one polymorph in another, with important repercussions on clinical practice, as it was the case of ritonavir (refer to the specific paragraph).

Solvates, also inappropriately termed pseudopolymorphs [38], are crystalline solids containing within the crystal structure either stoichiometric or nonstoichiometric proportions of solvent. When the incorporated solvent is water, the solvate is called a hydrate [27]. In general, it is undesirable to use solvates for drugs and pharmaceuticals, as the presence of organic solvent residues may be toxic; regulations for all the organic solvents in products for human use establish specific limits to how much daily exposure to residual solvent in the formulated preparation is allowed.

The solubility and dissolution rate of a drug can significantly differ for different solvates, and in particular hydrates. Important reviews concerning pharmaceutical solvates and hydrates are those of Morris [39] and Khankari and Grant [11].

Hydrates may have a faster or slower dissolution rate than the corresponding anhydrous form, though more frequently, the former are slower than the latter [40], perhaps because there are fewer sites of the drug molecule available for interaction with water during dissolution. A classic example is theophylline anhydrate, which dissolves faster than its hydrate form [41,42].

In other cases, the hydrate form exhibits a more rapid dissolution rate than its anhydrous form: for example, erythromycin dihydrate was found to exhibit a significantly faster dissolution rate than that of monohydrate and anhydrous forms [43,44].

Glibenclamide has been isolated as pentanol and toluene solvates, and these solvates exhibited higher solubility and dissolution rate than two non-solvated polymorphs [45].

The physical stability of hydrates and anhydrous forms strongly depends upon the relative humidity and/or temperature of the environment [46,47,48], and transitions from one form to the other occur as a consequence of variations in storage conditions and/or technological treatments [37,49].

In particular, anhydrous to hydrate transitions can occur during dissolution at the drug/medium interface and can affect dissolution rate and perhaps bioavailability [46].

4. POLYMORPH SCREENING

The polymorph screening process seeks to determine whether a given compound exists in polymorphic forms [50]. In recent decades, several techniques have been developed to improve the polymorph screening of drugs.

The concept of crystal engineering was introduced by Pepinsky in 1955 [51] and first applied by Schmidt in the context of covalent bond formation in the solid state [52]. It is traditionally defined as the deliberate design and control of molecular packing within a crystal structure with the intention of generating a solid that shows a particular desirable characteristic [53,54,55]. Combinatorial chemistry and high-throughput screening used in drug discovery have resulted in an increase of poorly water soluble drug candidates [56,57].

Among traditional methods to generate polymorphs (as well as hydrates and solvates), manual techniques [58] are time and material consuming, and sometimes fail to identify all possible polymorphs for a compound.

The development of computer software tools that consider the arrangement of atoms within a compound to predict the possible crystal structures has been a boon to the pharmaceutical industryenabling savings of time and materials in the process of identifying the most thermodynamically stable polymorph, and making it possible to tailor the manufacturing process for production of the active ingredient [59].

High-throughput polymorphism screening has been developed with the aim of accelerating the identification of potential polymorphs for a drug, and thus avoid problems during drug development [60,61]. The efficiency of screening in HT mode is estimated to be about two orders of magnitude greater than that of traditional bench-scale approaches [62], and it has been applied to numerous drugs.

A high-throughput (HT) crystallization study of an experimental angiotensin II antagonist and sertraline hydrochloride identified new forms, improved understanding of the transitions among different forms, and demonstrated that an HT strategy coupled with critical analysis can be used to rank the usefulness of crystal forms [62].

Ritonavir is a drug that has been used to treat HIV-1 infections since 1996. In 1998, a new metastable and unknown form posed major bioavailability problems. Afterwards, HT screening identified a total of five forms, the two well-known forms and three unknown ones [60].

A high-throughput co-crystal slurry screening study of indomethacin that used *in situ* Raman microscope and a multi-well plate not only provided information about co-crystal formation within one day, but also yielded data about the equilibrium of co-crystal formation and polymorphic transformation in just one screening [63].

5. CASE STUDIES OF POLYMORPHIC DRUGS

The following paragraphs report several examples of poorly soluble drugs for which polymorphic issues proved important. A summary is given in Table 2.

Table 2. Summarization of polymorphism of several drugs.

Drug Substance	Polymorphism Aspects	Bioavailability Issues
Chloramphenicol palmitate 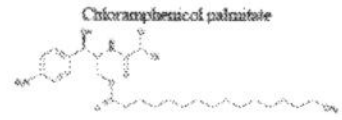	Chloramphenicol palmitate is a prodrug of chloramphenicol with antibiotic properties [64]. Chloramphenicol palmitate exist in three polymorphic forms: (A, B, C) [65,66], the stable form A (biologically inactive modification), the metastable form B (active modification) and unstable form C [67–69]. The three crystaline forms were also called α, β and γ. The α form is unstable at room temperature and gradually transforms to β on storage [70,71].	Form B (β) dissolves faster than Form A (α), and has a much higher solubility [72–74]. Low serum levels for the stable polymorph A were observed [75].
Oxytetracycline 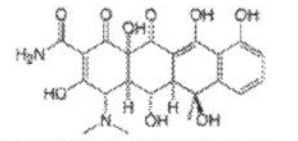	Oxytetracycline is a broad spectrum antibiotic. It exists in two different polymorphs [76].	Oxytetracycline showed differences in patients' blood levels [77] or differences in *in vitro* dissolution of tablets [78] because of differences in polymorphic forms.
Carbamazepine 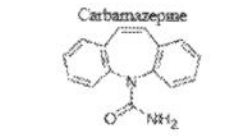	Carbamazepine is used in the treatment of epilepsy and trigeminal neuralgia. Different polymorphic forms were described [79–91]. Four anhydrous polymorphs were characterized: I, II, III, and IV, respectively identified as triclinic, trigonal, monoclinic, and monoclinic [77].	In spite different studies demonstrated similar pharmacokinetics in humans of anhydrous and dihydrate forms of carbamazepine [92] and no differences in bioavailability between a generic carbamazepine product and an innovator product [93], several clinical failures were reported concerning carbamazepine [94,95], in particular with generic carbamazepine tablets [96]. More recently, it was confirmed that the initial dissolution rate of carbamazepine was in the order of form III > form I > dihydrate, while the order of AUC values was form I > form III > dihydrate. This discrepancy may be attributed to the rapid transformation from form III to dihydrate in GI fluids [97]
Ritonavir 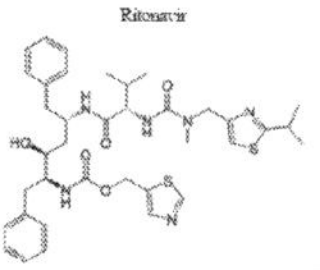	Ritonavir is an antiretroviral drug belonging to protease inhibitor class and used to treat HIV-1 infection. Ritonavir exhibits conformational polymorphism [98] and a total of five forms were described [60]. The forms I and II were more extensively characterized [98].	2 years after the launch of the first ritonavir product, several batches failed dissolution specifications because the presence of a different polymorphic form having ~50% lower intrinsic solubility of reference form [36].

5.1. Chloramphenicol Palmitate

A decades-old classic example of the importance of polymorphism to bioavailability is chloramphenicol palmitate, a prodrug of chloramphenicol with antibiotic properties, developed with the objective of obtaining a more pleasant flavored derivative [64]. Chloramphenicol palmitate exists in three polymorphic forms [65,66,70,71], the stable form A (biologically inactive modification), the metastable form B (active modification) and the unstable form C [67], which recently have been fully characterized thanks to advances in analytical methods [68,69]. Polymorph A is the thermodynamically stable one, but its absorption in humans is significantly lower than that of polymorph B [72], because Form B dissolves faster than Form A, and has much higher solubility [73]. This solubility difference probably results in the difference in ester hydrolysis rates, and thus in the difference in oral absorption, if one considers that chloramphenicol palmitate must be hydrolyzed by intestinal esterases before it can be absorbed [74]. These results were also proven by the low serum levels reached by the stable polymorph A, whereas the metastable polymorph yielded much higher serum levels when the same dose was administered [75].

5.2. Oxytetracycline

While for many years it has been known from various studies that patient blood levels of oxytetracycline differed according to the supplier of the oxytetracycline capsules, [77] or that *in vitro* dissolution performance of oxytetracycline tablets differed according to the various sources [78], only more recently have these differences been attributed to the presence of different polymorphs [76]. Tablets prepared from the form A polymorph dissolved significantly more slowly than tablets prepared from polymorph B: indeed, the tablets with form *A* polymorph exhibited about 55% dissolution at 30 min, while the tablets with form *B* polymorph exhibited almost complete (95%) dissolution at the same time. Further studies characterizing the physical and chemical properties of oxytetracycline polymorphs would be useful, as no recent works are available in the literature.

5.3. Carbamazepine

Highly different polymorphic forms of carbamazepine, a drug used in the treatment of epilepsy and trigeminal neuralgia, were discovered through classical crystallization methods and fully characterized from a physicochemical point of view [79,80,81,82,83,84,85,86,87,88,89]. More recently, a crystal engineering design strategy has facilitated supramolecular synthesis of 13 new crystalline phases of carbamazepine [90].

Even though different studies demonstrated that anhydrous and dihydrate forms of carbamazepine have similar pharmacokinetics in humans [92], and another indicated that there are no differences in bioavailability between a generic carbamazepine product and an innovator product [93], several clinical failures with carbamazepine were reported [94,95]. In particular, several problems were observed with Generic carbamazepine tablets, which were recalled due to clinical failures and dissolution changes [96]. It was suggested that discrepancies in clinical parameters and irreproducible clinical behavior within different batches and suppliers of the generic carbamazepine tablets were due to moisture uptake during storage. Actually, it is well known that anhydrous carbamazepine converts to the dihydrate within 1 h, when the anhydrous form is suspended in water [91]. More recently, it was confirmed that the initial dissolution rate of carbamazepine was in the order of form III > form I > dihydrate, while the order of AUC values was form I > form III > dihydrate. This discrepancy may be attributed to the rapid transformation from form III to dihydrate in GI fluids [97].

5.4. Ritonavir

Ritonavir, an antiretroviral drug of the protease inhibitor class used to treat HIV-1 infections, was found to have polymorphism that strongly impacts on solubility and dissolution rate. Originally, only one form was described, and was formulated as soft gel capsules containing an ethanol/water solution molecule. Two years after the launch of the product, several batches failed dissolution specifications. A new thermodynamically stable Form II was discovered, but this form precipitated out of solution, having ~50% lower intrinsic solubility than the reference form. This finally forced the manufacturer to recall the original formulation from the market [36] and reformulate it in an oily vehicle.

Using solid state spectroscopy and microscopy techniques including solid state NMR, Near Infrared Spectroscopy, powder X-ray Diffraction and Single crystal X-ray, ritonavir was found to exhibit conformational polymorphism with two unique crystal lattices that have significantly different solubility properties [98]. In addition, HT screening identified a total of five forms, the two well know forms and three unknown ones [60].

5.5. Atorvastatin Calcium

Atorvastatin calcium is an inhibitor of 3-hydroxy-3-methylglutaryl-coenzyme A (HMG-CoA) reductase, with strong ability to lower blood cholesterol. Atorvastatin, the most preferred molecule among statins, was developed and marketed by Pfizer under the trade name Lipitor® [121] and was the number one selling drug in the US until its patent expired in 2011. Atorvastatin is unstable and the hydroxyacid form (HF) is converted to a lactone form (LF), which is 15 times less soluble than the hydroxyacid form [103,104]. This instability of atorvastatin calcium leading to poor solubility (0.1 mg/mL) is the main cause for low bioavailability of the drug after oral administration: the absolute bioavailability of ATC is only 14% [105].

At least 60 polymorphic forms/solvates/hydrates have been patented [99,100,101] and several pharmaceutical companies are developing or have developed generic drug formulations based on different atorvastatin calcium polymorphs.

Due to the patent expiration, several companies produce the active pharmaceutical ingredient (API) of atorvastatin calcium, available on the market as stable crystalline polymorph I or amorphous form. It was not unusual to verify the presence of polymorphic impurities in the marketed atorvastatin calcium (API) with consequences on drug bioavailability and stability [102].

5.6. Axitinib

Axitinib is a tyrosine kinase inhibitor of endothelial growth factor that interrupts tumor angiogenesis and thus prevents the growth of cancer cells. Because of its strong molecular flexibility, 60 solvates, polymorphs of solvates, and five anhydrous forms have been discovered [106,107,108,109]. The commercial formulation under trade name Inlyta® contains the stable anhydrous form. Unusually, conventional crystallization methods did not lead to the discovery of this most stable polymorph; rather, it was obtained by the uncommon method of slurrying the solvates at high temperature. Understanding of the desolvation pathway was critical for obtaining the most stable polymorph of axitinib [107].

5.7. Phenylbutazone

Phenylbutazone is a potent anti-rheumatic drug that exists in different polymorphic [110,111,112] and solvated forms [113]. Different solubilities, dissolution rates and oral absorption were highlighted between two different polymorphic forms [114].

5.8. Rifaximin

Rifaximin is a synthetic derivative of rifamycin with very low gastrointestinal absorption, but that nonetheless displays a broad spectrum of antibacterial activity [115,116,117]. According to the European Pharmacopoeia, rifaximin shows crystal polymorphism [118] and several polymorphs (α, β, γ, δ, ε) have been described [119]. The most thermodynamically stable form, polymorph α, is the one used commercially. *In vitro* studies show different dissolution and solubility rates for these polymorphs, and *in vivo* investigations in dogs found different pharmacokinetic patterns, with δ and γ polymorphs displaying the highest systemic bioavailability [119]. Blandizzi *et al.*, [120] compared one generic rifaximin formulation with the branded product (the latter containing only polymorph-α) and found that most PK parameters such as highest concentration achieved in plasma (C_{max}), area under the concentration-time curve (AUC), and cumulative urinary excretion were significantly higher after administration of generic rifaximin. X-ray power diffraction analysis of the generic formulation showed the presence of both rifaximin-α and amorphous rifaximin, which could have contributed to the increased systemic bioavailability of the generic formulation.

6. REGULATORY CONSIDERATIONS

For approval of a new drug, the drug substance guideline of the US Food and Drug Administration (FDA) states that "appropriate" analytical procedures need to be used to detect polymorphs, hydrates and amorphous forms of the drug substance and also stresses the importance of controlling the crystal form of the drug substance during the various stages of product development [122].

Modern techniques such as ss-NMR and NIR can identify polymorphs in dosage forms (within limits), and should help improve mechanistic understanding of polymorphs in future studies [123]. Fast and easily applicable techniques such as DSC can determine the solubility of different polymorphs very rapidly and accurately [124]. The selection of crystal forms of improved solubility and bioavailability is possible when appropriate strategies are applied to guarantee the drug stability over the shelf life of the drug product. The evaluation of crystal transitions through appropriate analytical technologies serves to predict unwanted conversions during the drug product shelf life.

7. CONCLUSIONS

- The possibility of detecting drug polymorphism can be viewed in two opposite ways: as a risk of clinical failure when an undesired solid state conversion occurs, or as an advantage when more soluble polymorphs may be selected to overcome bioavailability problems. Thus, the pharmaceutical industry must carefully evaluate the presence of the phenomenon of the polymorphism for every drugs under development. In the past, when analytical techniques were not sophisticated enough to adequately detect polymorphism of drugs under development, several clinical failures emerged during the marketing phases, in some cases with serious repercussions for the pharmaceutical industry, such as the obligation to withdraw or reformulate the product. Now, the use of state-of-the-art technologies makes it possible to prevent this risk and to better and fully investigate the existence of different polymorphic forms of drugs in the industrial pipeline. In recent years, regulatory organisms such as the FDA and ICH have pressed the pharmaceutical industry to adopt methodologies and innovative analytical techniques that should provide better understanding of the polymorphism phenomenon for every drug under development, and enable Quality Control Departments to adequately evaluate the solid state of batches produced.

ACKNOWLEDGMENTS

The authors would like to thank Sheila Beatty for editing the English usage of the manuscript.

AUTHOR CONTRIBUTIONS

P.D.M. proposed the subject; P.D.M. and R.C. wrote the manuscript. Both authors read and approved the final manuscript.

CONFLICTS OF INTEREST

The authors declare no conflict of interest.

REFERENCES

1. Serajuddin, A.T.M. Salt formation to improve drug solubility. Adv. Drug Deliv. Rev. 2007, 59, 603–616.

2. Schultheiss, N.; Newman, A. Pharmaceutical cocrystals and their physicochemical properties. Cryst. Growth Des.2009, 9, 2950–2967.

3. Kawabata, Y.; Wada, K.; Nakatani, M.; Yamada, S.; Onouea, S. Formulation design for poorly water-soluble drugs based on biopharmaceutics classification system: Basic approaches and practical applications. Int. J. Pharm. 2011, 420, 1–10.

4. Khadka, P.; Ro, J.; Kim, H.; Kim, I.; Kim, J.T.; Kim, H.; Cho, J.M.; Yun, G.; Lee, J. Pharmaceutical particle technologies: An approach to improve drug solubility, dissolution and bioavailability. Asian J. Pharm. Sci. 2014, 9, 304–316.

5. Gao, L.; Zhang, D.; Chen, M. Drug nanocrystals for the formulation of poorly soluble drugs and its application as a potential drug delivery system. J. Nanopart. Res. 2008, 10, 845–862.

6. Chen, H.; Khemtong, C.; Yang, X.; Chang, X.; Gao, J. Nanonization strategies for poorly-soluble drugs. Drug Discov. Today 2011, 16, 354–360.

7. Saravana, K.K.; Prasanna, R.Y. Dissolution enhancement of poorly soluble drugs by using complexation technique. A review. J. Pharm. Sci. Res. 2013, 5, 120–124.

8. Challa, R.; Ahuja, A.; Ali, J.; Khar, R.K. Cyclodextrins in drug delivery: An updated review. AAPS Pharm. Sci. Technol.2005, 6, E329–E357.

9. Leuner, C.; Dressman, J. Improving drug solubility for oral delivery using solid dispersions. Eur. J. Pharm. Biopharm.2000, 50, 47–60.

10. Snider, D.A.; Addicks, W.; Owens, W. Polymorphism in generic drug product development. Adv. Drug Deliv. Rev.2004, 56, 391–395.

11. Huang, L.F.; Tong, W.Q. Impact of solid state properties on developability assessment of drug candidates. Adv. Drug Deliv. Rev. 2004, 56, 321–334.

12. Babu, N.J.; Nangia, A. Solubility advantage of amorphous drugs and pharmaceutical cocrystals. Cryst. Growth Des.2011, 11, 2662–2679.

13. Hancock, B.C.; Parks, M. What is the true solubility advantage for amorphous pharmaceuticals? Pharm. Res. 2000, 17, 397–404.

14. Murdande, S.B.; Pikal, M.J.; Shanker, R.M.; Bogner, R.H. Aqueous solubility of crystalline and amorphous drugs: challenges in measurement. Pharm. Dev. Technol. 2011, 16, 187–200.

15. Zhang, G.G.; Law, D.; Schmitt, E.A.; Qiu, Y. Phase transformation considerations during process development and manufacture of solid oral dosage forms. Adv. Drug Deliv. Rev. 2004, 56, 371–390.

16. Ku, M.S. Salt and polymorph selection strategy based on the biopharmaceutical classification system for early pharmaceutical Development. Am. Pharm. Rev. 2010, 20, 30.

17. Lachman, L.; Lieberman, H.; Kanig, J.L. The Theory and Practise of Industrial Pharmacy, 3rd ed.; Lea & Febiger: Philadelphia, PA, USA, 1986.

18. Merisko, E.; Liversidge, G.G. Nanocrystals: Resolving pharmaceutical formulation issues associated with poorly water-soluble compounds. In Particles; Marty, J.J., Ed.; Marcel Dekker: Orlando, FL, USA, 2002.

19. Takagi, T.; Ramachandran, C.; Bermejo, M.; Yamashita, S.; Yu, L.X.; Amidon, G.L. A provisional biopharmaceutical classification of the top 200 oral drug products in the United States, Great Britain, Spain, and Japan. Mol. Pharm. 2006,3, 631–643.

20. Ku, M.S.; Dulin, W. A biopharmaceutical classification-based Right-First-Time formulation approach to reduce human pharmacokinetic variability and project cycle time from First-In-Human to clinical Proof-Of-Concept. Pharm. Dev. Technol. 2012, 17, 285–302.

21. Dow Jones Newswires. GlaxoSmithKline on Track to Launch 11 Drugs by Dec. 2003; Dow Jones Newswires: New York, NY, USA, 2003.

22. Amidon, G.L.; Lennernäs, H.; Shah, V.P.; Crison, J.R. A theoretical basis for a biopharmaceutic drug classification: the correlation of in vitro drug product dissolution and in vivo bioavailability. Pharm. Res. 1995, 12, 413–420.

23. Williams, H.D.; Trevaskis, N.L.; Charman, S.A.; Shanker, R.M.; Charman, W.N.; Pouton, C.W.; Porter, C.J.H. Strategies to address low drug solubility in discovery and development. Pharmacol. Rev. 2013, 65, 315–499.

24. Food and Drug Administration. Guidance for Industry: Waiver of in Vivo Bioavailability and Bioequivalence Studies for Immediate-Release Solid Oral Dosage Forms Based on a Biopharmaceutics Classification System; Food and Drug Administation: Rockville, MD, USA, 2000.

25. Benet, L.Z. The role of BCS (Biopharmaceutics Classification System) and BDDCS (Biopharmaceutics Drug Disposition Classification System) in drug development. J. Pharm. Sci. 2013, 102, 34–42.

26. Savjani, K.T.; Gajjar, A.K.; Savjani, J.K. Drug Solubility: Importance and Enhancement Techniques. ISRN Pharm. 2012.

27. Vippagunta, S.R.; Brittain, H.G.; Grant, D.J.W. Crystalline solids. Adv. Drug Deliv. Rev. 2001, 48, 3–26.

28. Rodriguez-Spong, B.; Price, C.P.; Jayasankar, A.; Matzger, A.J.; Rodriguez-Hornedo, N. General principles of pharmaceutical solid polymorphism: a supramolecular perspective. Adv. Drug Deliv. Rev. 2004, 56, 241–274.

29. Kuhnert-Brandstätter, M. Thermomicroscopy in the Analysis of Pharmaceuticals; Pergamon Press: Oxford, UK, 1971.

30. Borka, L.; Haleblian, J.K. Crystal polymorphism of pharmaceuticals. Acta Pharm. Jugosl. 1990, 40, 71–94.

31. Borka, L. Review on crystal polymorphism of substances in the European Pharmacopoeia. Pharm. Acta Helv. 1991, 66, 6–22.

32. Giron, D. Thermal analysis and calorimetric methods in thecharacterization of polymorphs and solvates. Thermochim. Acta 1995, 248, 1–59.

33. Hilfiker, R.; Blatter, F.; von Raumer, M. Relevance of solid-state properties for pharmaceutical products polymorphism. In the Pharmaceutical Industry; Hilfiker, R., Ed.; WILEY-VCH Verlag GmbH & Co. KGaA: Weinheim, Germany, 2006.

34. Pudipeddi, M.; Serajuddin, A.T. Trends in solubility of polymorphs. J. Pharm. Sci. 2005, 94, 929–939.

35. Chemburkar, S.R.; Bauer, J.; Deming, K.; Spiwek, H.; Patel, K.; Morris, J.; Henry, R.; Spanton, S.; Dziki, W.; Porter, W.;et al. Dealing with the impact of ritonavir polymorphs on the late stages of bulk drug process development. Org. Process Res. Dev. 2000, 4, 413–417.

36. Dubbini, A.; Censi, R.; Martena, V.; Hoti, E.; Ricciutelli, M.; Malaj, L.; di Martino, P. Influence of pH and method of crystallization on the solid physical form of indomethacin. Int. J. Pharm. 2014, 473, 536–544.

37. Censi, R.; Rascioni, R.; di Martino, P. Changes in the solid state of anhydrous and hydrated forms of sodium naproxen under different grinding and environmental conditions: Evidence of the formation of new hydrated forms.Eur. J. Pharm. Biopharm. 2015, 92, 192–203.

38. Seddon, K.R. Pseudopolymorph: A Polemic. Cryst. Growth Des. 2004, 4.

39. Morris, K.R. Structural aspects of hydrates and solvates. In Polymorphism in Pharmaceutical Sciences, Drugs and the Pharmaceutical Sciences; Brittain, H., Ed.; Marcel Dekker: New York, NY, USA, 1999; Volume 95, pp. 125–181.

40. Khankari, R.J.; Grant, D.J.W. Pharmaceutical hydrates. Thermochim. Acta 1995, 248, 61–79.

41. Shefter, E.; Higuchi, T. Dissolution behavior of crystalline solvated and nonsolvated forms of some pharmaceuticals.J. Pharm. Sci. 1963, 52, 781–791.

42. Shan, N.; Zaworotko, M.J. Polymorphic Crystal Forms and Cocrystals in Drug Delivery (Crystal Engineering). Drug Dev. 2010.

43. Allen, P.V.; Rahn, P.D.; Sarapu, A.C.; Vanderwielen, A.J. Physical characterization of erythromycin: Anhydrate, monohydrate, and dihydrate crystalline solids. J. Pharm. Sci. 1978, 67, 1087–1093.

44. Blagden, N.; de Matas, M.; Gayan, P.T.; York, P. Crystal engineering of active pharmaceutical ingredients to improve solubility and dissolution rates. Adv. Drug Deliv. Rev. 2007, 59, 617–630.

45. Datta, S.; Grant, D.J.W. Crystal structures of drugs: Advances in determination, prediction and engineering. Nat. Rev. Drug Discov. 2004, 3, 42–57.

46. Di Martino, P.; Barthélémy, C.; Palmieri, G.F.; Martelli, S. Physical characterization of naproxen sodium hydrate and anhydrate forms. Eur. J. Pharm. Sci. 2001, 14, 293–300.

47. Di Martino, P.; Barthélémy, C.; Joiris, E.; Capsoni, D.; Masic, A.; Massarotti, V.; Gobetto, R.; Bini, M.; Martelli, S. A new tetrahydrated form of sodium naproxen. J. Pharm. Sci. 2007, 96, 156–167.

48. Malaj, L.; Censi, R.; di Martino, P. Mechanism for dehydration of three sodium naproxen hydrates. Cryst. Growth Des.2009, 9, 2128–2136.

49. Di Martino, P.; Malaj, L.; Censi, R.; Martelli, S. Physico-chemical and technological properties of sodium naproxen granules prepared in a high-shear mixer-granulator. J. Pharm. Sci. 2008, 97, 5263–5273.

50. Stahly, G.P. Diversity in Single- and Multiple-Component Crystals. The Search for and Prevalence of Polymorphs and Cocrystals. Cryst. Growth Des. 2007, 6, 1007–1026.

51. Pepinsky, R. Crystal engineering—A new concept in crystallography. Phys. Rev. 1955, 100, 971.

52. Schmidt, G.M.J. Photodimerization in the solid state. Pure Appl. Chem. 1971, 27, 647–678.

53. Desiraju, G.R. Chemistry beyond the molecule. Nature 2001, 412, 397–400.

54. Desiraju, G.R. Crystal engineering: A brief overview. J. Chem. Sci. 2010, 122, 667–675.

55. Biradha, K.; Su, C.Y.; Vittal, J.J. Recent developments in crystal engineering. Cryst. Growth Des. 2011, 11, 875–886.

56. Lipinski, C.A. Drug-like properties and the causes of poor solubility and poor permeability. J. Pharmacol. Toxicol. Met.2000, 44, 235–249.

57. Lipinski, C.A.; Lombardo, F.; Dominy, B.W.; Feeney, P.J. Experimental and computational approaches to estimate solubility and permeability

in drug discovery and development settings. Adv. Drug Del. Rev. 2001, 46, 3–26.

58. Caira, M.R. Crystalline Polymorphism of Organic Compounds. Top. Curr. Chem. 1998, 198, 163–208.

59. Price, S.L. The computational prediction of pharmaceutical crystal structures and polymorphism. Adv. Drug Deliv. Rev. 2004, 56, 301–319.

60. Morissette, S.L.; Soukasene, S.; Levinson, D.A.; Cima, M.J.; Almarsson, O. Elucidation of crystal form diversity of the HIV protease inhibitor ritonavir by high-throughput crystallization. Proc. Natl. Acad. Sci. USA 2003, 100, 2180–2184.

61. Peterson, M.L.; Morissette, S.L.; McNulty, C.; Goldsweig, A.; Shaw, P.; le Quesne, M.; Monagle, J.; Encina, N.; Marchionna, J.; Johnson, A.; et al. Iterative high-throughput polymorphism studies on acetaminophen and an experimentally derived structure for form III. J. Am. Chem. Soc. 2002, 124, 10958–10959.

62. Almarsson, O.; Hickey, M.B.; Peterson, M.L.; Morissette, S.L.; Soukasene, S.; McNulty, C.; Tawa, M.; MacPhee, J.M.; Remenar, J.F. High-Throughput surveys of crystal form diversity of highly polymorphic pharmaceutical compounds.Cryst. Growth Des. 2003, 3, 927–933.

63. Kojima, T.; Tsutsumi, S.; Yamamoto, K.; Ikeda, Y.; Moriwaki, T. High-throughput cocrystal slurry screening by use ofin situ Raman microscopy and multi-well plate. Int. J. Pharm. 2010, 399, 52–59.

64. Edgerton, W.H. Chloramphenicol Esters and Method for Obtaining Same. U.S. Patent 2,662,906, 15 December 1953.

65. Borka, L.; Backe-Hansen, K. IR spectroscopy of chloramphenicol palmitate. Polymorph alteration caused by the KBr disc technique. Acta Pharm. Suec. 1968, 5, 271–278.

66. Kanenewa, N.; Otsuka, M. Effect of grinding on the transformation of polymorphs of chloramphenicol palmitate.Chem. Pharm. Bull. 1985, 33, 1660–1668.

67. Burger, A. Neue untersuchungergebnisse von chloramphenicolpalmitat. Sci. Pharm. 1977, 45, 269–281.

68. Gamberini, M.C.; Baraldi, C.; Tinti, A.; Rustichelli, C.; Ferioli, V.; Gamberini, G. Solid state characterization of chloramphenicol

palmitate. Raman spectroscopy applied to pharmaceutical polymorphs. J. Mol. Struct. 2006, 785, 216–224.

69. Mishra, R.; Srivastava, A.; Sherma, A.; Tandon, P.; Baraldi, C.; Gamberini, M.C. Structural, electronic, thermodynamical and charge transfer properties of chloramphenicol palmitate using vibrational spectroscopy and DFT calculations. Spectrochim. Acta Part A Mol. Biomol. Spectr. 2013, 101, 335–342.

70. Eguchi, Y.; Iitaka, Y. The β-form of chloramphenicol palmitate. Acta Cryst. 1974, B30, 2781–2783.

71. Szulzewsky, K.; Kulpe, S.; Schulz, B.; Kunath, D. The structure of the b modification of chloramphenicol palmitate. A redetermination. Acta Cryst. 1981, B37, 1673–1676.

72. Aguiar, A.J.; Krc, J.; Kinkel, A.W.; Samyn, J.C. Effect of polymorphism on the absorption of chloramphenicol from chloramphenicol palmitate. J. Pharm. Sci. 1967, 56, 847–853.

73. Aguiar, A.J.; Zelmer, J.E. Dissolution behavior of polymorphs of chloramphenicol palmitate and mefanamic acid. J. Pharm. Sci. 1969, 58, 983–987.

74. Glazko, A.J.; Edgerton, W.H.; Dill, W.A.; Lenz, W.R. Chloromycetin palmitate—A synthetic ester of chloromycetin.Antibiot. Chemother. 1952, 2, 234–242.

75. Maeda, T.; Takenaka, H.; Yamahira, Y.; Noguchi, T. Use of rabbits for absorption studies on polymorphs of chloramphenicol palmitate. Chem. Pharm. Bull. 1980, 28, 431–436.

76. Liebenberg, W.; de Villiers, M.; Wurster, D.E.; Swanepoel, E.; Dekker, T.G.; Lotter, A.P. The effect of polymorphism on powder compaction and dissolution properties of chemically equivalent oxytetracycline hydrochloride powders.Drug Dev. Ind. Pharm. 1999, 25, 1027–1033.

77. Brice, G.W.; Hammer, H.F. Therapeutic nonequivalence of oxytetracycline capsules. J. Am. Med. Assoc. 1969, 208, 1189–1190.

78. Groves, M.J. Solution tests on generic brands of oxytetracycline tablets. Pharm. J. 1973, 210, 318–319.

79. Reboul, J.P.; Cristau, B.; Soyfer, J.C.; Astier, J.P. 5H-5-Dibenzyl[b,f]azepinecarboxamide (carbamazepine). Acta Crystallogr. Sect. B Struct. Commun. 1981, 37, 1844–1848.

80. Himes, V.L.; Mighell, A.D.; Decamp, W.H. Structure of carbamazepine-5H-dibenz[b,f]azepine-5-carboxamide. Acta Crystallogr. Sect. B Struct. Commun. 1981, 37, 2242–2245.

81. Chang, C.H.; Yang, D.S.C.; Yoo, C.S.; Wang, B.L.; Pletcher, J. The crystal structures of (S) and (R) baclofen and carbamazepine. Acta Crystallogr. 1981, A37.

82. Reck, G.; Dietz, G. The order-disorder structure of carbamazepine dihydrate: 5H-Dibenz[b,f] azepine-5-carboxamide dihydrate, C15H12N2O•••2 H2O. Cryst. Res. Technol. 1986, 21, 1463–1468.

83. Lowes, M.M.J.; Caira, M.R.; Lotter, A.P.; Vanderwatt, J.G. Physicochemical properties and X-ray structural studies of the trigonal polymorph of carbamazepine. J. Pharm. Sci. 1987, 76, 744–752.

84. Lisgarten, J.N.; Palmer, R.A.; Saldanha, J.W.J. Crystal and molecular structure of 5-carbamyl-5H-dibenzo[b,f]azepine.Crystallogr. Spectrosc. Res. 1989, 19, 641–649.

85. Ceolin, R.; Toscani, S.; Gardette, M.F.; Dzyabchenko, V.N.; Bachet, B. X-ray characterization of the triclinic polymorph of carbamazepine. J. Pharm. Sci. 1997, 86, 1062–1065.

86. Rustichelli, C.; Gamberini, G.; Ferioli, V.; Gamberini, M.C.; Ficarra, R.; Tommasini, S. Solid-state study of polymorphic drugs: carbamazepine. J. Pharm. Biomed. Anal. 2000, 23, 41–54.

87. Lang, M.D.; Kampf, J.W.; Matzger, A.J. Form IV of carbamazepine. J. Pharm. Sci. 2002, 91, 1186–1190.

88. Lang, M.D.; Grzesiak, A.L.; Matzger, A.J. The use of polymer heteronuclei for crystalline polymorph selection. J. Am. Chem. Soc. 2002, 124, 14834–14835.

89. Grzesiak, A.L.; Lang, M.; Kim, K.; Matzger, A.J. Comparison of the four anhydrous polymorphs of carbamazepine and the crystal structure of form I. J. Pharm. Sci. 2003, 92, 2260–2271.

90. Fleischman, S.G.; Kuduva, S.S.; McMahon, J.A.; Moulton, B.; Walsh, R.D.B.; Zaworotko, M.J.; Rodríguez-Hornedo, N. Crystal engineering of the composition of pharmaceutical phases: Multiple-component crystalline solids involving carbamazepine. Cryst. Growth Des. 2003, 3, 909–919.

91. Young, W.W.L.; Suryanarayanan, R. Kinetics of transition of anhydrous carbamazepine to carbamazepine dihydrate in aqueous suspensions. J. Pharm. Sci. 1991, 80, 496–500.

92. Kahela, P.; Aaltonen, R.; Lewing, E.; Anttila, M.; Kristoffersson, E. Pharmacokinetics and dissolution of two crystalline forms of carbamazepine. Int. J. Pharm. 1983, 14, 103–112.

93. Jumao-as, A.; Bella, I.; Craig, B.; Lowe, J.; Dasheiff, R.M. Comparison of steady-state blood levels of two carbamazepine formulations. Epilepsia 1989, 30, 67–70.

94. Koch, G.; Allan, J. Untoward effects of generic carbamazepine therapy. Arch. Neurol. 1987, 44, 578–579.

95. Sachdeo, R.; Chokroverty, S.; Beleldiuk, G. Risk of switching from brand-name to generic drugs in seizure disorder.Epilepsia 1987, 28, 581.

96. Meyer, M.C.; Straughn, A.B.; Jarvi, E.J.; Wood, G.C.; Pelsor, F.R.; Shah, V.P. The bioinequivalence of carbamazepine tablets with a history of clinical failures. Pharm. Res. 1992, 9, 1612–1616.

97. Kobayashi, Y.; Ito, S.; Itai, S.; Yamamoto, K. Physicochemical properties and bioavailability of carbamazepine polymorphs and dihydrate. Int. J. Pharm. 2000, 193, 137–146.

98. Bauer, J.; Spanton, S.; Henry, R.; Quick, J.; Dziki, W.; Porter, W.; Morris, J. Ritonavir: An extraordinary example of conformational polymorphism. Pharm. Res. 2001, 18, 859–866.

99. Mckenzie, A.T. Applicant: Warner-Lambert Company. Form III crystalline (R-(R*,R*)-2-(4-fluorophenyl)-beta, delta-dihyxory-5-(1-methyl-ethyl)-3-phenyl-4-phenylamino)carbonyl)-1H-pyreol-1-heptanoic acid hemi calcium salt (Atorvastatin). WO97/03958, 6 February 1997.

100. Jin, Y.S.; Ulrich, J. New crystalline solvates of atorvastatin calcium. Chem. Eng. Technol. 2010, 33, 839–844.

101. Chadha, R.; Kuhad, A.; Arora, P.; Kishor, S. Characterisation and evaluation of pharmaceutical solvates of atorvastatin calcium by thermoanalytical and spectroscopic studies. Chem. Cent. J. 2012, 6, 114–129.

102. Shete, G.; Puri, V.; Kumar, L.; Bansal, A.K. Solid state characterization of commercial crystalline and amorphous atorvastatin calcium samples. AAPS Pharm. Sci. Technol. 2010, 11, 598–609.

103. Kerc, J.; Salobir, M.; Bavec, S. Atorvastatin Calcium in a Pharmaceutical form Composition Thereof and Pharmaceutical Formulation Comprising Atorvastatin Calcium. U.S. Patent 7,030,151, 18 April 2006.

104. Kerc, J. Stable Pharmaceutical Formulation Comprising a HMGCoAreductase Inhibitor. U.S. Patent Application US 2009/0264497 A1, 22 October 2009.

105. Khan, F.N.; Dehghan, M.H.G. Enhanced bioavailability of atorvastatin calcium from stabilized gastric resident formulation. AAPS Pharm. Sci. Technol. 2011, 12, 1077–1086.

106. Chekal, B.P.; Campeta, A.M.; Abramov, Y.A.; Feeder, N.; Glynn, P.P.; McLaughlin, R.W.; Meenan, P.A.; Singer, R.A. The challenges of developing an API crystallization process for a complex polymorphic and highly solvating system. Part I. Org. Process. Res. Dev. 2009, 13, 1327–1337.

107. Campeta, A.M.; Chekal, B.P.; Abramov, Y.A.; Meenan, P.A.; Henson, M.J.; Shi, B.; Singer, R.A.; Horspool, K.R. Development of a targeted polymorph screening approach for a complex polymorphic and highly solvating API. J. Pharm. Sci. 2010, 99, 3874–3886.

108. Abramov, Y.A. QTAIM application in drug development: prediction of relative stability of drug polymorphs from experimental crystal structures. J. Phys. Chem. A 2011, 115, 12809–12817.

109. Vasileiadis, M.; Pantelides, C.C.; Adjiman, C.S. Prediction of the crystal structures of axitinib, a polymorphic pharmaceutical molecule. Chem. Eng. Sci. 2015, 121, 60–76.

110. Matsunaga, J.; Nambu, N.; Nagai, T. Physicochemical approach to biopharmaceutical phenomena. XXX. Polymorphism of phenylbutazone. Chem. Pharm. Bull. 1976, 24, 1169–1172.

111. Ibrahim, H.G.; Pisano, F.; Bruno, A. Polymorphism of phenylbutazone: Properties and comparisonal behaviour of crystals. J. Pharm. Sci. 1977, 66, 669–673.

112. Matsuda, Y.; Kawaguchi, S.; Kobayashi, H.; Nishijo, J. Polymorphism of phenylbutazone by spray dried methods. J. Pharm. Pharmacol. 1980, 32, 579–580.

113. Hosokawa, T.; Datta, S.; Sheth, A.R.; Grant, D.J.W. Relationships between crystal structures and thermodynamic properties of phenylbutazone solvates. Cryst. Eng. Commun. 2004, 6, 243–249.

114. Pandit, J.K.; Gupta, S.K.; Gode, K.D.; Mishra, B. Effect of crystal form on the oral absorption of phenylbutazone. Int. J. Pharm. 1984, 21, 129–132.

115. Marchi, E.; Montecchi, L.; Venturini, A.P.; Mascellani, G.; Brufani, M.; Cellai, L. 4-Deoxypyri-do[1),2):1,2]imidazo[5,4-c]rifamycin SV derivatives. A new series of semisynthetic rifamycins with high antibacterial activity and low gastroenteric absorption. J. Med. Chem. 1985, 28, 960–963.

116. Jiang, Z.D.; DuPont, H.L. Rifaximin: In vitro and in vivo antibacterial activity—Review. Chemotherapy 2005, 51, 67–72.

117. Adachi, J.A.; DuPont, H.L. Rifaximin: A novel nonabsorbed rifamycin for gastrointestinal disorders. Clin. Infect. Dis.2006, 42, 541–547.

118. European Pharmacopoeia. Rifaximin (Revised Monograph). 2011; Suppl 7.1:2362. Available online: http://www.edqm.eu/en/european-pharmacopoeia-8th-edition-1563.html (accessed on 15 September 2014).

119. Viscomi, G.C.; Campana, M.; Barbanti, M.; Grepioni, F.; Polito, M.; Confortini, D.; Rosini, G.; Righi, P.; Cannata, V.; Braga, D. Crystal forms of rifaximin and their effect on pharmaceutical properties. Cryst. Eng. Commun. 2008, 10, 1074–1081.

120. Blandizzi, C.; Viscomi, G.C.; Scarpignato, C. Impact of crystal polymorphism on the systemic bioavailability of rifaximin, an antibiotic acting locally in the gastrointestinal tract, in healthy volunteers. Drug Des. Dev. Ther. 2015, 9, 1–11.

121. Lipitor. Package Insert, Pfizer Ireland Pharmaceuticals, Dublin, Ireland. Parke Davis; Division of Pfizer Inc.: New York, NY, USA, 2009.

122. Byrn, S.; Pfeiffer, R.; Ganey, M.; Hoiberg, C.; Poochikian, G. Pharmaceutical solids: A strategic approach to regulatory consideration. Pharm. Res. 1995, 12, 945–954.

123. Singhal, D.; Curatolo, W. Drug polymorphism and dosage form design: A practical perspective. Adv. Drug Deliv. Rev.2004, 23, 335–347.

124. Park, K.; Evans, J.M.B.; Myerson, A.S. Determination of solubility of polymorphs using Differential Scanning Calorimetry. J. Cryst. Growth Des. 2003, 3, 991–995.

Chapter 9

Influence of Solid Drug Delivery System Formulation on Poorly Water-Soluble Drug Dissolution and Permeability

Marko Krstić [1], Miljana Popović [1], Vladimir Dobričić [2] and Svetlana Ibrić [1,*]

[1]*Department of Pharmaceutical Technology and Cosmetology, Faculty of Pharmacy, University of Belgrade, Vojvode Stepe 450, P.O. Box 146, Belgrade 11221, Serbia*

[2]*Department of Pharmaceutical Chemistry, Faculty of Pharmacy, University of Belgrade, Vojvode Stepe 450, P.O. Box 146, Belgrade 11221, Serbia*

ABSTRACT

The majority of drugs have a low dissolution rate, which is a limiting step for their absorption. In this manuscript, solid dispersions (SD), solid self-microemulsifying drug delivery systems (S-SMEDDS) and solid self-nanoemulsifying drug delivery systems (S-SNEDDS) were evaluated as potential formulation strategies to increase the dissolution rate of carbamazepine. Influence of increased dissolution rate on permeability of carbamazepine was evaluated using PAMPA test. In S-SMEDDS and S-SNEDDS formulations, the ratio of liquid SMEDDS/SNEDDS and solid carrier (Neusilin® UFL2) was varied, and carbamazepine content was constant. In SD formulations, the ratio of carbamazepine and Neusilin® UFL2, was varied. Formulations that showed the best dissolution rate of carbamazepine (SD_1:6, SMEDDS_1:1, SNEDDS_1:6) were mutually compared, characterization of these formulations was performed by DSC, PXRD and FT-IR analyses, and a PAMPA test was done. All formulations have shown a significant increase in dissolution rate compared to pure carbamazepine and immediate-release carbamazepine tablets. Formulation S-SMEDDS_1:1 showed the fastest release rate and permeability of carbamazepine. DSC, PXRD and FT-IR analyses confirmed that in S-SMEDDS and S-SNEDDS carbamazepine remained in polymorph form III, and that it was converted to an amorphous state in SD formulations. All formulations showed increased permeability of carbamazepine, compared to pure carbamazepine.

Keywords: carbamazepine; dissolution rate; PAMPA test; solid dispersions; S-SMEDDS; S-SNEDDS; characterization

1. INTRODUCTION

Oral administration of drugs is considered to be the most natural, uncomplicated, convenient and safe method. Since nearly one third of drugs are poorly water soluble, oral bioavailability of those drugs could be an issue [1]. Carbamazepine is an antiepileptic agent with low solubility (0.15 ± 0.07 mg/mL, 25 °C, [2]) and high permeability, which makes it a great candidate for this study. Also, carbamazepine is prescribed in high doses, so this also represents a challenge for researchers. According to Biopharmaceutical Classification System (BCS) that classifies drugs into four categories according to their permeability and solubility properties, carbamazepine is classified as a class II drug. As dissolution of poorly water-soluble drugs (e.g., carbamazepine) is the rate-limiting step for absorption, improvement of dissolution rate should increase drug absorption and bioavailability [3,4]. Other than that, carbamazepine is susceptible to polymorphic transition, so it is a great candidate in the evaluation of the influence of excipients on polymorphic form.

Various techniques have been introduced in order to enhance the dissolution rate and solubility of drugs, such as physical (particle size reduction, drug dispersion in carriers, complexation and solubilization by surfactants), chemical and other methods [1,5]. Various methods for enhancing the solubility of carbamazepine were evaluated: dispersion of carbamazepine in carriers [3,6,7,8], complexation [9] and solubilization by surfactants [2,10,11,12].

Solid dispersions are molecular mixtures of poorly water-soluble drugs in hydrophilic carriers, where the drug release profile is driven by polymer properties. Management of the drug release profile using solid dispersions is achieved by manipulation of the carrier and solid dispersion particle properties. Parameters, such as carrier molecular weight and composition, drug crystallinity and particle porosity and wettability, when successfully controlled, can produce improvements in bioavailability [13].

Self-nanoemulsifying drug delivery systems (SNEEDS) are isotropic mixtures of oil, surfactant (HLB > 12) and co-surfactant. These systems are spontaneously emulsified when exposed to GIT fluids and form oil-in-water nanoemulsions. The droplet size of oil in these nanoemulsions is 100–250 nm [14,15].

Self-microemulsifying drug delivery systems (SMEDDS) are isotropic mixtures of oil, surfactant (HLB > 12) and co-surfactant. In comparison with SNEDDS, the content of hydrophilic surfactants and cosurfactants is increased and lipid content is reduced. The risk of drug precipitation in SMEDDS is higher due to lower lipid content [16]. Under the influence of GIT movements they form a fine oil-in-water emulsion. After the dispersion in aqueous media, these systems form homogeneous, transparent, isotropic and thermodynamically stable microemulsions [17]. The droplet size of oil in these microemulsions is less than 100 nm [15].

However, self-emulsifying formulations are normally prepared as liquids that have some disadvantages, for example, high production costs, low stability and portability, low drug loading and different dosage forms. Irreversible drugs/excipients precipitation may also be problematic. More importantly, the large quantity of surfactants in the formulations can induce gastrointestinal irritation. In order to avoid these problems, solid self-emulsifying drug delivery systems (SSEDDS) have been investigated as alternative formulations. Solid self-emulsifying drug delivery systems combine the advantages of liquid SEEDS (*i.e.*, enhanced solubility and bioavailability) with those of solid dosage forms (e.g., low production cost, convenience of process control, high stability and reproducibility, better patient compliance) [18].

Dissolution and membrane permeability of drugs through the mucosa are the key parameters in the gastrointestinal absorption. [4]. While a significant increase of apparent solubility may be achieved by some solubility-enabling formulations, the impact of these formulations on permeability of a lipophilic drug is often overlooked. Use of surfactants as a strategy to increase the apparent aqueous solubility of lipophilic drugs may lead to increased, decreased or unchanged membrane permeability. Micellar solubilization of drugs allows extraordinary increases of solubility, but also results in a decreased free fraction of drug available for intestinal membrane permeation. Consequently, permeability decreases with the increase of surfactant concentration above the critical micelle concentration (CMC). A direct tradeoff exists between the apparent solubility and permeability irrespectively of the free fraction, which must be taken into account when developing solubility-improving formulations in order to maximize the overall oral absorption [19].

Several *in vitro* methods for the intestinal absorption assessment have been developed. These methods include Caco-2 and other "intestinal-like" cells (MDCK, TC-7, HT29-MTX, 2/4/A1, T-84, LLC-PK1, A549...), Everted sac and Ussing chamber, *in-situ* rat intestinal perfusion, artificial membrane and Follicle-Associated Epithelium Model (FAE) [20,21].

The artificial membrane is used as an alternative model for GI membrane in the assessment of drug absorption potential. A majority of drugs are absorbed primarily or partially through passive transport. Therefore, the rate of permeation through a simple artificial membrane, which mimics passive transcellular transport, provides a good indication of a drug's absorption potential [22].

The most used *in vitro* methods for prediction of permeability are cell-based (especially Caco-2 cells) and PAMPA (parallel artificial membrane permeability assay). PAMPA is considered a quick and simple test that measures passive permeability in the absence of efflux systems or transporters [21]. Since carbamazepine is a substance that is absorbed by passive transcellular transport, PAMPA test is a convenient method for evaluation of its permeability.

The aim of this study was to investigate the possibility of increasing the dissolution rate of carbamazepine by formulation of different types of solubility-enabling formulations. Using carbamazepine as a model substance, three different types of formulations were prepared: solid dispersions, self-microemulsifying drug delivery systems (SMEDDS) and self-nanoemulsifying drug delivery systems (SNEDDS). Additionally, the aim was to evaluate the influence of increased solubility on permeability of carbamazepine, using PAMPA test.

2. RESULTS AND DISCUSSION

2.1. Droplet Size Analysis

The droplet size of SMEDDS and SNEDDS without carbamazepine was analyzed. It can be concluded that upon high water dilution, SMEDDS is capable of forming oil-in-water microemulsions, because the droplet size is less than 100 nm. Polydispersity is the ratio of standard deviation to mean droplet size and indicates the uniformity of droplet size within the formulation. It is a general rule that the higher the polydispersity, the lower the uniformity of the droplet size in the formulation [23]. SMEDDS had monomodal droplet size distribution with a PDI value of 0.104 ± 0.005 and droplet size of (17.66 ± 0.17 nm), which indicates that a stable microemulsion was formed. SNEDDS had a monomodal droplet size distribution with a PDI value of 0.184 ± 0.012 and a droplet size of 156.3 ± 0.28 nm, which indicates that a stable nanoemulsion was formed (Figure 1).

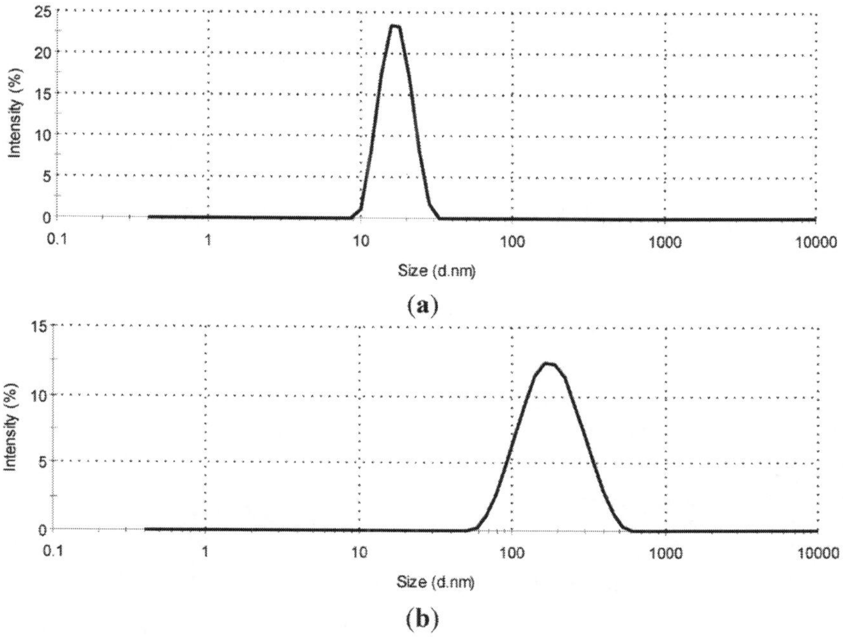

Figure 1. The droplet size distribution by intensity for SMEDDS (self-microemulsifying drug delivery systems) (**a**) and SNEDDS (self-nanoemulsifying drug delivery systems) (**b**) emulsions after dilution with water phase up to 90% *w/w*.

2.2. Fourier-Transform Infrared Spectroscopy (FT-IR)

FT-IR analysis of selected formulations and pure carbamazepine were performed in order to characterize the physical state of carbamazepine and to identify potential interactions between carbamazepine and Neusilin® UFL2. The FT-IR spectra of carbamazepine showed a characteristic peak at 3462 cm^{-1} (–NH valence vibration), 1676 cm^{-1} (–CO–R vibration), 1598 cm^{-1} (–C=C– and –C=O vibration) and 1384 cm^{-1} (–NH deformation) [24]. Spectrum of SD_1:6 shows significant diminution of characteristic peaks in fingerprint region, which indicates presence of interactions between carbamazepine and Neusilin® UFL2. Peak of NH vibration is present at 3462 cm^{-1}, but it is blunt and diminished. Other three characteristic peaks are also present, but not very prominent. FT-IR spectra of S-SMEDDS_1:1

and S-SNEDDS_1:6 showed all characteristic peaks for carbamazepine, indicating that no interactions occurred between carbamazepine and solid carrier, so it can be concluded that in these formulations carbamazepine stayed in polymorph form III. The line intensity is slightly weaker, which is expected, due to the smaller share of carbamazepine in formulations. FT-IR spectra of evaluated formulations are shown in Figure 2.

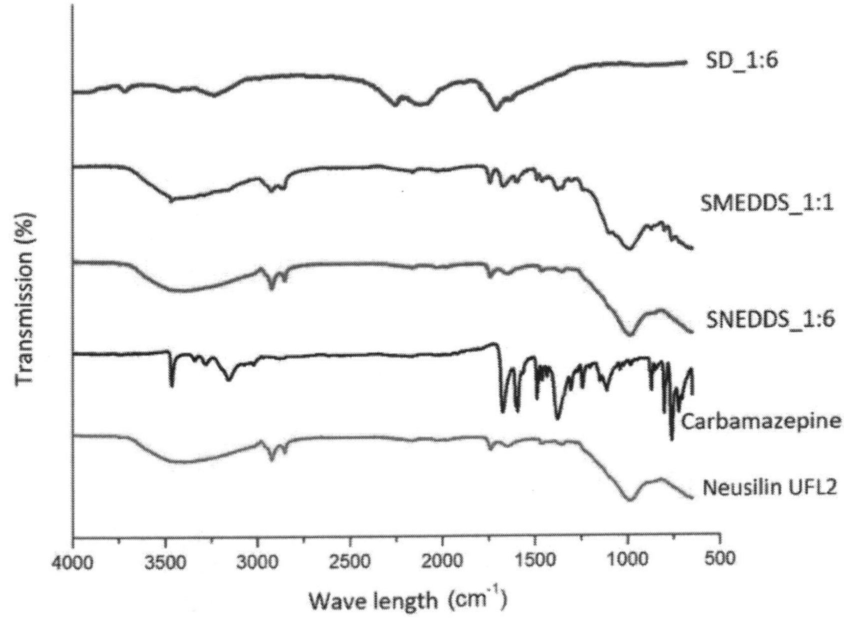

Figure 2. FT-IR spectra of evaluated formulations and pure carbamazepine.

2.3. Differential Scanning Calorimetry (DSC)

DSC curves of pure carbamazepine and tested formulations are shown in Figure 3. During heating, the sample of pure carbamazepine melts, which corresponds to the first endothermic peak at 175.08 °C, contributing to melting of polymorph form III, after which the exothermic peak is observed at 176.63 °C, which comes from the recrystallization of polymorph form III to polymorph form I. Another endothermic peak is observed at 190.95 °C, which comes from melting the polymorph form I of carbamazepine. These results are in accordance with our previous results and results obtained from Grzesiak *et al.* [25]. In SMEDDS_1:1 and SNEDDS_1:6 formulations, characteristic peaks of melting of carbamazepine polymorph form III were observed at the temperature range of 150–175.8 °C, indicating that in these formulations carbamazepine probably has not changed its polymorph form. In SD_1:6 formulation, this peak is not noticed, which indicates that in this formulation carbamazepine was probably converted into an amorphous form, which will be confirmed later by PXRD.

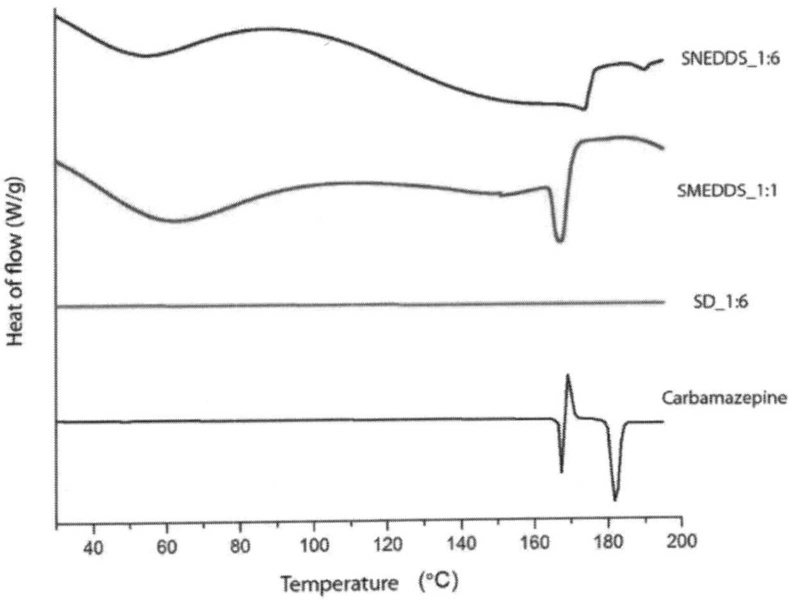

Figure 3. DSC curves of evaluated formulations and pure carbamazepine.

2.4. Powder X-ray Diffraction Analysis (PXRD)

Figure 4 illustrates the PXRD patterns of pure CBZ and the optimal formulation. The PXRD pattern of CBZ exhibits characteristic high-intensity diffraction peaks at 13.02°, 15.22°, 15.78°, 19.40°, 24.92°, 27.50° and 31.86 2θ, which is in accordance with diffractograms previously reported for crystal form III [25,26]. This characteristic peaks are observed in S-SMEDDS_1:1 and S-SNEDDS_1:6, which confirms assumptions given by DSC and FT-IR analyses that carbamazepine remains in polymorph form III. In the diffractogram of SD_1:6, no peaks are observed, which indicates that carbamazepine was converted to amorphous form.

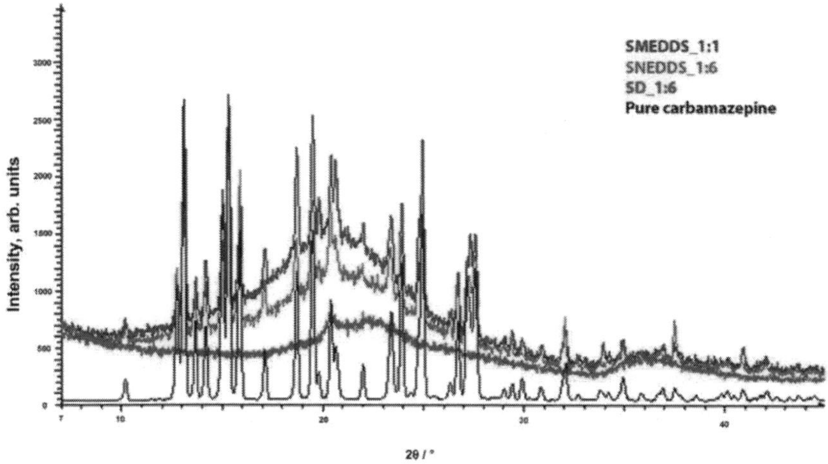

Figure 4. PXRD patterns of evaluated formulations and pure carbamazepine.

2.5. In Vitro Drug Release Profile

All formulations of solid dispersions, except SD_1:1, had faster release rates than pure carbamazepine and commercial immediate-release carbamazepine tablets. The fastest release rate was from SD_1:6, followed by SD_1:4 and SD_1:2. The increase of Neusilin® UFL2 content in formulations resulted in the increase of release rate of carbamazepine from formulations. Improvement of dissolution rate from solid dispersions can be explained by several mechanisms: decreased crystallinity, particle size reduction, improved wettability and porosity of particles in formulations [13].

All S-SMEDDS formulations showed a significant increase of dissolution rate in comparison with pure carbamazepine and commercial immediate-release carbamazepine tablets (Figure 5). The extent of carbamazepine release from carbamazepine-loaded S-SMEDDS can be explained by a high specific area of adsorbent which adsorbs liquid SMEDDS inside the pores, limiting drug exposure to the surface [27,28]. The presence of SMEDDS most likely enables solubilization of carbamazepine by creating microemulsions. In the first 30 min, the highest dissolution rate was reached from SMEDDS_1:1 (90.63%), followed by SMEDDS_1:2, SMEDDS_1:4 and SMEDDS_1:3 (the release rates from each of these formulations were over 80%). From 45 min onward, the dissolution rate of carbamazepine was slightly higher from the SMEDDS_1:3, compared to other three formulations. A high specific surface area of Neusilin® UFL2 provides better wetting of powder, which increases the release rate of carbamazepine. [29]. Due to a higher ratio of carrier

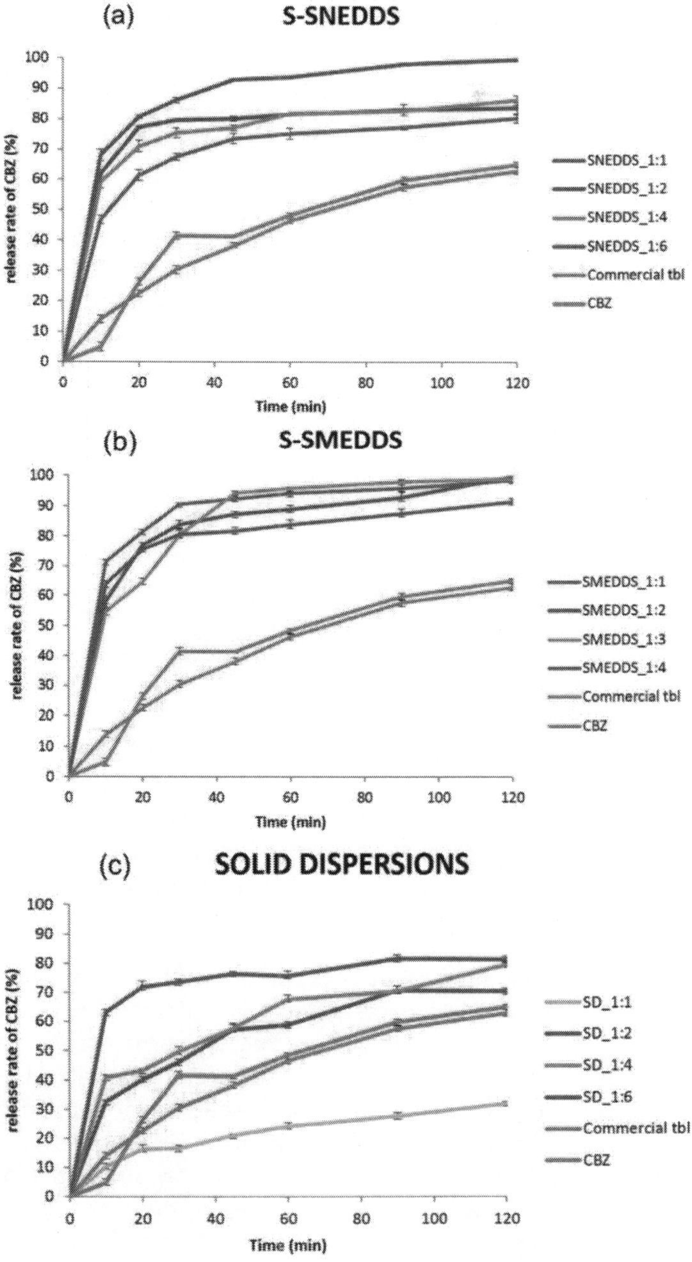

Figure 5. *In vitro* dissolution profiles ofcarbamazepine and commercial tablets in comparison with formulations prepared in our laboratory: S-SNEDDS (**a**); S-SMEDDS (**b**); and solid dispersions (**c**).

(SMEDDS_1:3 and SMEDDS_1:4), liquid SMEDDS probably becomes entrapped in the long and narrow pores of the carrier, which could cause slower initial release of carbamazepine. Another assumption is that due to the smaller ratio of liquid SMEDDS in formulation, a smaller amount of carbamazepine is solubilized, which can lead to initial slower release.

CBZ dissolution profiles from SMEDDS, calculating difference (f1) and similarity factors (f2), it can be concluded that there is no statistically significant difference between these formulations, since f1 values are between 0 and 15 and f2 values greater than 50 for each comparison.

All S-SNEDDS formulations showed faster release rates in comparison with pure carbamazepine and commercial immediate-release carbamazepine tablets. SNEDDS_1:6 showed the fastest release rate of all SNEEDS formulations (68.17% in the first 10 min, 80.82% after 20 min and 93.01% after 45 min). This could be attributed to a high ratio of Neusilin® UFL2, an adsorbent with a high specific surface (300 m^2/g). Droplets of SNEEDS with carbamazepine are entrapped in the pores of the adsorbent, which limits drug exposure to the surface of Neusilin® UFL2. A high level of exposure to the surface may result in precipitation of the drug, which decreases the extent of drug release [29]. SNEDDS_1:2 and SNEDDS_1:4 showed a similar drug release profile: 62.01% and 58.92% at 10 min, 77.28% and 71.04% at 20 min, 79.71% and 75.52% at 30 min, respectively and almost identical release rates after 60 min. The lowest release rate of all SNEEDS observed for SNEDDS_1:1 formulation were: 46.83% at 10 min, 61.38% at 20 min, 67.57% at 30 min. From 45 min onward, the dissolution rate was slightly lower in comparison with SNEDDS_1:4. The lowest dissolution rate of SNEDDS_1:1 may be attributed to low adsorbent content in the formulation, compared to other three formulations.

According to the results obtained from dissolution test, all formulations had faster release rates of carbamazepine in comparison with pure carbamazepine and commercial immediate-release carbamazepine tablets. This could be due to the highly specific surface area of Neusilin® UFL2 (300 m^2/g), as well as carbamazepine–carrier interactions. In comparison with solid dispersions, S-SMEDDS and S-SNEDDS had faster release rates of carbamazepine (50%–70% in the first 10 min), probably due to the influence of two factors: highly specific surface of Neusilin® UFL2 and presence of SMEDDS/SNEDDS. Difference factors (f1) and similarity factors (f2) were calculated in order to compare formulations that had the fastest release rates in comparison with pure carbamazepine and commercial immediate-release carbamazepine tablets. These parameters are presented in Table 1. Two compared formulations are equivalent if f1 values are between 0 and 15 and f2 values greater than 50 (50–100) [24]. According to the results presented in Table 1, all four are significantly different from pure carbamazepine and commercial immediate-release tablets.

Table 1. Difference factors (f1) and similarity factors (f2) of formulation and pure carbamazepine and commercial tablets.

		Pure Carbamazepine	Commercial Tablets
SMEDDS_1:1	f1	62.17	56.50
	f2	18.49	15.95
SNEDDS_1:6	f1	55.33	49.05
	f2	24.40	22.17
SD_1:6	f1	54.86	44.02
	f2	24.41	25.62

2.6. Parallel Artificial Membrane Permeability Assay (PAMPA)

The results of PAMPA test are given in Table 2. From each formulation type, those that showed the fastest release rate were chosen (SD_1:6, SMEDDS_1:1, SNEDDS_1:6), as well as pure carbamazepine, and then a PAMPA test was performed. The apparent permeability of pure carbamazepine was in accordance with previously reported results for the same drug in a similar PAMPA system [22]. Results of PAMPA test showed that all tested formulations have a higher permeability rate than pure carbamazepine because of the existence of interactions between carbamazepine and carrier.

Table 2. PAMPA test: values of apparent permeability coefficients (Papp) of tested formulations and pure carbamazepine.

Formulations	Papp × 10^{-6} (cm/s)
Carbamazepine	11.77 ± 0.34
SD_1:6	16.12 ± 2.27
SNEDDS_1:6	19.12 ± 0.87
SMEDDS_1:1	21.42 ± 1.67

Increase of permeability in SD formulations may be due to particle size reduction and amorphization of the drug. Also, the increase in permeability may be a result of supersaturation from amorphous solid dispersion formulation. As opposed to other solubilization methods, the use of amorphous solid dispersions does not alter the equilibrium solubility of the drug; rather, it enables an unstable supersaturated solution to be attained. Solubilization via surfactants and cosolvents may decrease the apparent membrane permeability. This is because the apparent membrane/aqueous phase partition coefficient, which is a primary determinant of the intestinal membrane permeability, decreases with increasing drug solubility, thereby reducing the thermodynamic driving force for membrane permeation [30]. On the other hand, apparent solubility increased via supersaturation does not result in a decrease in a partition coefficient. This is because the apparent partition coefficient is dictated by the equilibrium solubility of the drug. Since supersaturation has kinetic/nonequilibrium solubility, it will not affect the apparent partition coefficient. Hence, apparent membrane permeability is unaffected by supersaturation while flux (the product of apparent solubility × apparent permeability) increases dramatically. Small structures of amorphous drug/carrier "nanoaggregates" formed in solution play a critical role in maintaining supersaturation by crystallization inhibition [30].

The use of surfactants for solubility increments may also change the permeability in various ways. Surfactants may increase membrane permeability for drugs with inherently low-permeability and high aqueous solubility (*i.e.*, BCS class III compounds) through the disruption of membrane integrity to increase paracellular transport. However, for lipophilic drugs with inherently high transcellular membrane permeability (e.g., BCS class II), surfactants can decrease the free fraction of the drug which results in decreased permeability. In using surfactants for the solubilization of lipophilic drugs, it is well recognized that above the critical micelle concentration (CMC), drugs may be incorporated into surfactant micelles. Additionally, it is this micellization process which drives the extraordinary increase in apparent aqueous solubility afforded by surfactants. However, the micellar solubilization also results in decreased free fraction of drug. This incorporation of the drug in micelles significantly decreases the amount of free drug available for permeation through the intestinal membrane. Decreased free fraction is directly translated to lower concentration gradient and hence thermodynamic driving force for membrane permeation [19]. It is reported that SEEDS formulations enhance passive permeability of highly lipophilic drugs, mainly through modulation of membrane fluidity [31]. Among tested formulations, the increase in permeability is especially expressed in formulations SMEDDS_1:1 and SNEDDS_1:6, probably due to interactions within surfactants and artificial membrane, along with carbamazepine–carrier interactions. S-SMEDDS showed the highest permeability rate, which was expected, due to the small particle size of all evaluated formulations.

3. EXPERIMENTAL SECTION

3.1. Materials

Cremophor® EL (BASF, Ludwigshafen, Germany), was used as a surfactant in the preparation of SMEDDS. Macrogol 400 was used as a cosurfactant in the preparation of SMEDDS.

Labrasol® (Gattefosse GmbH, Bad Krozingen, Germany), was used as a surfactant in the preparation of SNEDDS. Polysorbate 80 (Tween® 80, Sigma-Aldrich Chemie GmbH, Steinheim, Germany), was used as a surfactant in the preparation of the SNEDDS. Transcutol® HP (Gattefosse GmbH), was used as a cosurfactant in the preparation of the SNEDDS. Mygliol® 812 (Sasol GmbH, Hamburg, Germany), was used as oil in preparation of the SMEDDS and SNEDDS. Neusilin® UFL2 (Fuji Chemical Industry Co., Ltd., Toyama, Japan) was used as a solid carrier for all formulations. CBZ (Ph. Eur. 8.0) was used as a model of poorly soluble active ingredient. Galepsin® (Galenika AD, Belgrade, Serbia) was used as a commercial immediate-release formulation (200 mg). Dodecan (Sigma-Aldrich Chemie GmbH) and egg lecithin (Lipoid GmbH, Ludwigshafen, Germany) were used in the PAMPA test. The water used in all experiments was double-distilled. Absolute ethanol was Ph. Eur. 8.0 grade. All other reagents that were employed for HPLC analysis were of analytical grade.

In this study, several formulation strategies were evaluated as a possible solution to enhance the solubility of carbamazepine: solid dispersions, solid self-microemulsifying drug delivery systems (S-SMEDDS) and solid self-nanoemulsifying drug delivery systems (S-SNEDDS).

3.2. Methods

3.2.1. Solid Dispersions

Solid dispersions were prepared by the solvent casting method. Carbamazepine was dissolved in a minimal amount of absolute ethanol and this solution was mixed with Neusilin® UFL2 in a mortar with a pestle. The solvent was then removed by heating at 25 °C in a dry-heater for 24 h. Four different formulations were made in which carbamazepine: Neusilin® UFL2 ratio was varied: 1:1, 1:2, 1:4, 1:6 (marked as SD_1:1, SD_1:2, SD_1:4 and SD_1:6, respectively).

3.2.2. Solid Self-Microemulsifying Drug Delivery Systems (S-SMEDDS)

In these formulations, the surfactant phase was a mixture of Cremophor® EL as the surfactant and Macrogol 400 as the co-surfactant (in a 3:1 ratio). Mygliol® 812 was used as an oil phase. Surfactant phase: the oil phase ratio was 9:1. The ratio of components within SMEDDS was determined in previous screening studies and droplet size and PDI confirmed by photon correlation spectroscopy. SMEDDS was then adsorbed on a mixture of Neusilin® UFL2 as a solid carrier and carbamazepine, previously prepared by thoroughly mixing these two components in a mortar with a pestle. Four formulations were prepared in which the SMEDDS: Neusilin® UFL2 ratio was varied: 1:1, 1:2, 1:3, 1:4 (marked as SMEDDS_1:1, SMEDDS_1:2, SMEDDS_1:3 and SMEDDS_1:4, respectively). Carbamazepine content was constant in all formulations (20%).

3.2.3. Solid Self-Nanoemulsifying Drug Delivery Systems (S-SNEDDS)

The SNEDDS formulation contains 17.29% Mygliol® 812 as oil phase, 46.28% Labrasol®/Polisorbate 80 (1:1) as the surfactant, and 36.43% Transcutol® HP as the co-surfactant. The ratio of the components within SNEDDS was determined in previous screening studies and droplet size and PDI confirmed by photon correlation spectroscopy. SNEDDS was then adsorbed on a mixture of Neusilin® UFL2 as a solid carrier and carbamazepine. Four formulations were made, in which the SNEDDS: Neusilin® UFL2 ratio was varied: 1:1, 1:2, 1:4, 1:6 (marked as SNEDDS_1_1, SNEDDS_1_2, SNEDDS_1_3, SNEDDS_1_4, respectively). Carbamazepine content was constant in all formulations (20%).

3.3. Photon Correlation Spectroscopy (PCS)

The droplet size of drug-unloaded SMEDDS and SNEDDS was evaluated using photon correlation spectroscopy (PCS) after appropriate dilution with water. PCS measures droplet size by analyzing random intensity fluctuations in light scattering due to Brownian motion of the particles. The intensity correlation function provides information on the translational diffusion coefficient of the scattering particles and hence the hydrodynamics radius according to Stokes-Einstein equation. The droplet size was determined using the apparatus Nano ZS90 (Malvern Instruments, Malvern, UK) equipped with a

He–Ne laser at 633 nm at 20 ± 0.2 °C. The size measurements were carried out at a fixed angle of 90°. Droplet size measurements were performed in the samples diluted with the water phase up to 90% *w/w*. Computer software performs statistical analysis of data and calculates the average droplet size (Z-Ave) and polydispersity index (PDI) from intensity distribution. The results are the mean and standard deviation (S.D.) of three consecutive measurements for each sample [24,32].

3.4. In Vitro Drug Release Studies

Dissolution profiles of SMEDDS, SNEDDS and solid dispersions as well as pure carbamazepine and commercial immediate-release carbamazepine tablets were determined using a rotating paddle apparatus (Erweka DT70, Erweka GmbH, Hausenstamm, Germany). The dissolution conditions were: 37 °C, 900 mL of water (medium), 50 rpm. Samples of 5 mL were withdrawn from the medium at fixed time intervals (10, 20, 30, 45, 60, 90 and 120 min). All samples were filtered and assayed for carbamazepine. Carbamazepine concentration was determined spectrophotometrically at 287 nm (Evolution 300, Thermo Fisher Scientific, Hemel Hempstead, England, UK). The weight of every sample was such that every sample contains 200 mg of carbamazepine.

Release profiles of carbamazepine from prepared formulations were compared with the dissolution profiles of pure carbamazepine and commercial immediate-release carbamazepine tablets by calculating the difference factor (f1) and similarity factor (f2) [24].

3.5. Characterization of Selected Formulations

From each type of formulation, the one that showed the fastest release rate of carbamazepine was selected, and then PAMPA test was performed, in order to evaluate the permeability of carbamazepine. Also, characterization of these formulations was performed, using DSC, PXRD and FT-IR spectroscopy.

3.5.1. Fourier-Transform Infrared Spectroscopy (FT-IR)

FT-IR spectra in the region of 600–4000 cm^{-1} for pure carbamazepine and some formulations that are suspected to have an inadequate increase of dissolution rate, were obtained using a Shimadzu IR-Prestige-21 FT-IR spectrometer coupled with a horizontal Golden Gate MKII single-reflection ATR system (Specac, Kent, UK) and equipped with a Zn Se lens.

3.5.2. Differential Scanning Calorimetry (DSC)

DSC analysis were carried out on a computer-interfaced differential scanning calorimeter (DSC Q2000, TA Instruments). DSC was used to determine the presence of polymorphic form of crystalline carbamazepine in samples and also the degree of crystallinity for the encapsulated drug. The instrument was calibrated for temperature and energy using indium standards. The samples were accurately weighed (1–2 mg) and heated from 20–200 °C at a rate of 10 °C/min, under a nitrogen purge gas flow of 50 mL/min. The presence of an endothermic and/or exothermic peak was used as a marker of crystalline form of carbamazepine.

3.5.3. Powder X-ray Diffraction Analysis (PXRD)

Powder X-ray diffraction patterns were measured in order to evaluate crystalline/amorphous character of prepared samples. Measurements were performed using a Bruker AXS D8 Advance powder diffractometer, equipped with copper cathode (λ = 0.15418 nm, 40 kV, 40 mA). Patterns were obtained with step width of 0.02° and a detector in 2θ between 4° and 40° at ambient temperature. Samples, ground into powders with an agate mortar and pestle, were measured on a low background quartz plate in an aluminum holder.

3.6. PAMPA Test

The PAMPA test was used to predict passive human gastrointestinal absorption of pure carbamazepine and formulations of each formulation type that had the fastest release rate.

In this test, a hydrophilic polyvinylidene difluoride (PVDF) 96-well filtration plate (Millipore, Bedford, MA, USA) was used as the carrier of the artificial membrane and as the receiving plate. The filter material of each well in the filtration (receiving) plate was coated with 5 µL of egg lecithin solution in dodecane (1%, *w/v*). The receiving plate was placed onto the donor plate (Millipore, Bedford, MA, USA), which had previously been filled with 300 µL of donor solutions (100–200 µM solutions of carbamazepine and its formulations in phosphate buffer pH 5.5). Subsequently, 300 µL of the phosphate buffer solution pH 5.5 was added to each well of the receiving plate. The system was covered with a plastic lid to prevent evaporation and was incubated for two hours at room temperature. After incubation, the concentrations of carbamazepine in the corresponding wells of the receiving plate as well as in the starting solutions were determined by use of previously described HPLC method [18]. The apparent permeability coefficients (Papp) were calculated using the following equations:

$$\%T = 100 \cdot \frac{A_R \cdot V_R}{A_{D0} \cdot V_D} \quad (1)$$

$$P_{app} = \frac{V_D \cdot V_R}{(V_D + V_R) \cdot S \cdot t} \ln\left[\frac{100 \cdot V_D}{100 \cdot V_D - \%T \cdot (V_D + V_R)}\right] \quad (2)$$

V_D and V_R are the volumes of the donor and receiving solutions, respectively (mL); A_{D0} and A_R are the HPLC peak areas of the initial and receiving solutions, respectively; S is the surface area of the artificial membrane (0.28 cm^2, according to the manufacturer); t is the incubation time (s) and $\%T$ is extent of transport across the membrane [22].

4. CONCLUSIONS

An increase in carbamazepine dissolution was achieved in all formulations in comparison with pure carbamazepine and commercial immediate-release carbamazepine tablets. S-SMEDDS and S-SNEDDS formulations had higher release rates of carbamazepine solid dispersions. The highest dissolution rate and permeability of carbamazepine was accomplished from SMEDDS_1:1. These formulations weighed approximately 500 mg, and contain 20% carbamazepine, which is far more acceptable than other formulations found in the literature. The impact of solubility enhancement on permeability of carbamazepine was assessed by use of the PAMPA test. It was also demonstrated that the *in vitro* permeability rate of carbamazepine was increased in tested formulations. The results of this work suggest that the proper selection of the ratio of components in solubility-enhancing formulations may be crucial for achieving appropriate bioavailability of poorly soluble drugs. Also, the impact of solubility increase on permeability of the drug must be considered when preparing such formulations, in order to achieve an optimal solubility–permeability balance, and maximize the overall absorption of these drugs.

ACKNOWLEDGMENTS

This work was supported by Project TR34007, funded by Ministry of Education, Science and Technological Development, Republic of Serbia.

AUTHOR CONTRIBUTIONS

S.I. and M.K. designed the research and discussed the results, V.D. performed the PAMPA assay, M.P. prepared formulations and performed the dissolution test, M.K. participated in all phases of experimental work. M.K. wrote the paper. All authors read and approved the final manuscript.

CONFLICTS OF INTEREST

The authors declare no conflict of interest.

REFERENCES

1. Neduri, K.; Bontha, V.K.; Vemula, S.K. Different techniques to enhance the dissolution rate of lovastatin: Formulation and evaluation. Asian J. Pharm. Clin. Res. 2013, 6, 56–60.

2. Nesamony, J.; Karla, A.; Majrad, M.S.; Boddu, S.H.S.; Jung, R.; Williams, F.E.; Schnapp, A.M.; Nauli, S.M.; Kalinoski, A.L. Development and characterization of nanostructured mists with potential for actively targeting poorly water-soluble compounds into the lungs. Pharm. Res. 2013, 30, 2625–2639.

3. Wang, Z.; Chen, B.; Quan, G.; Li, F.; Wu, Q.; Dian, L.; Dong, Y.; Li, G.; Wu, C. Increasing the oral bioavailability of poorly water-soluble carbamazepine using immediate-release pellets supported on SBA-15 mesoporous silica. Int. J. Nanomed. 2012, 7, 5807–5818.

4. Löbenberg, R.; Amidon, G.L. Modern bioavailability, bioequivalence and biopharmaceutics classifcation system. New scientifc approaches to international regulatory standards. Eur. J. Pharm. Biopharm. 2000, 50, 3–12.

5. Mohanachandran, P.S.; Sindhumol, P.G.; Kiran, T.S. Enhancement of solubility and dissolution rate: An overview.Pharm. Glob. IJCP 2010, 4, 1–10.

6. Raghavendra Rao, N.G.; Upendra, K. Development of carbamazepine fast dissolving tablets: Effect of functionality of hydrophilic carriers on solid dispersion technique. Asian J. Pharm. Clin. Res. 2010, 3, 114–117.

7. Rane, Y.; Mashru, R.; Sankalia, M.; Sankalia, J. Effect of hydrophilic swellable polymers on dissolution enhancement of carbamazepine solid dispersions studied using response surface methodology. AAPS PharmSciTech 2007, 8, E1–E11.

8. Hirasawa, N.; Okamoto, H.; Danjo, K. Lactose as a low molecular weight carrier of solid dispersions for carbamazepine and ethenzamide. Chem. Pharm. Bull. 1999, 47, 417–420.

9. Mirza, M.A.; Agarwal, S.P.; Iqbal, Z. Effect of fulvic acid on oral delivery of carbamazepine. Sci. Adv. Mater. 2011, 3, 1–10.

10. Nisha, G.S.; Vaishali, P.; Geeta, R.; Prabhakar, P.; Harish, N.M.; Marina, K.; Charayulu, R.N. Formulation and evaluation of self microemulsifying drug delivery system of carbamazepine. Int. J. Res. Pharm. Sci. 2011, 2, 162–169.

11. Kokare, C.R.; Kumbhar, S.A.; Patil, A. Formulation and evaluation of self-emulsifying drug delivery system of carbamazepine. Ind. J. Pharm. Educ. Res. 2013, 47, 172–177.

12. Nan, Z.; Lijun, G.; Tao, W.; Dongqin, Q. Evaluation of carbamazepine (CBZ) supersaturable self-microemulsifying (S-SMEDDS) formulation in vitro and in vivo. Iran. J. Pharm. Res. 2012, 11, 257–264.

13. Vasconcelos, T.; Sarmento, B.; Costa, P. Solid dispersions as strategy to improve oral bioavailability of poor water soluble drugs. Drug Discov. Today 2007, 12, 1068–1075.

14. Basalious, E.B.; Shawky, N.; Badr-Eldin, S.M. SNEDDS containing bioenchancers for improvement of dissolution and oral absorption of lacidipine. I: Development and optimization. Int. J. Pharm. 2010, 391, 203–211.

15. Pouton, C.W. Formulation of poorly water-soluble drugs for oral administration: physicochemical and physiological issues and the lipid formulation classification system. Eur. J. Pharm. Sci. 2006, 29, 278–287.

16. Katteboina, S.; Chandrasekhar, P.; Balaji, S. Approaches for the development of solid self-emulsifying drug delivery systems and dosage forms. Asian J. Pharm. Sci. 2009, 4, 240–253.

17. Ingle, L.M.; Wankhade, V.P.; Udasi, T.A.; Tapar, K.K. New approaches for development and characterization of SMEDDS. Int. J. Pharm. Pharm. Sci. Res. 2013, 3, 7–14.

18. Krstic, M.; Razic, S.; Djekic, L.; Dobricic, V.; Momcilovic, M.; Vasiljevic, D.; Ibric, S. Application of a mixture experimental design in the optimization of the formulation of solid self-emulsifying drug delivery systems containing carbamazepine. Lat. Am. J. Pharm. 2015, 34, 885–894.

19. Dahan, A.; Miller, J.M. The solubility-permeability interplay and its implications in formulation design and development for poorly soluble drugs. AAPS J. 2012, 14, 244–251.

20. Gibaud, S.; Attivi, D. Microemulsions for oral administration and their therapeutic applications. Expert Opin. Drug Deliv. 2012, 9, 937–951.

21. Shahbazi, M.; Santos, H.A. Improving oral absorption via drug-loaded nanocarriers: Absorption mechanisms, intestinal models and rational fabrication. Curr. Drug Metab. 2013, 14, 28–56.

22. Zhu, C.; Jiang, L.; Chen, T.M.; Hwang, K.K. A comparative study of artificial membrane permeability assay for high throughput profiling of drug absorption potential. Eur. J. Med. Chem. 2002, 37, 399–407.

23. Shakeel, F.; Baboota, S.; Ahuja, A.; Ali, J.; Aqil, M.; Shafiq, S. Nanoemulsions as vehicles for transdermal delivery of aceclofenac. AAPS PharmSciTech 2007, 8, 191–199.

24. Suresh, S.; Shivakumar, H.N.; Kumar, K. Effect of β cyclodextrin complexation on the solubility and dissolution rate of carbamazepine from tablets. Indian J. Pharm. Sci. 2006, 68, 301–307.

25. Grzesiak, A.L.; Lang, M.; Kim, K.; Matzeger, A.J. Comparison of the Four Anhydrous Polymorphs of Carbamazepine and the Crystal Structure of Form. Int. J. Pharm. Sci. 2003, 92, 2260–2271.

26. Rustichelli, C.; Gamberini, G.; Ferioli, V.; Gamberini, M.C.; Ficarra, R.; Tommasini, S. Solid-state study of polymorphic drugs: Carbamazepine. J. Pharm. Biomed. Anal. 2000, 23, 41–54.

27. Emami, J. In vitro-in vivo correlation: From theory to applications. J. Pharm. Pharm. Sci. 2006, 9, 169–189.

28. Milović, M.; Djuriš, J.; Djekić, L.; Vasiljević, D.; Ibrić, S. Characterization and evaluation of solid self-microemulsifying drug delivery systems

with porous carriers as systems for improved carbamazepine release. Int. J. Pharm. 2012, 436, 58–65.

29. Agarwal, V.; Siddiqui, A.; Ali, A.; Nazzal, S. Dissolution and powder flow characterization of solid self-emulsified drug delivery system (SEEDS). Int. J. Pharm. 2009, 366, 44–52.

30. Miller, J.M.; Beig, A.; Carr, R.A.; Spence, J.K.; Dahan, A. A win-win solution in oral delivery of lipophilic drugs: Supersaturation via amorphous solid dispersions increases apparent solubility without sacrifice of intestinal membrane permeability. Mol. Pharm. 2012, 9, 2009–2016.

31. Alqahtani, S.; Alayoubi, A.; Nazzal, S.; Sylvester, P.W.; Kaddoumi, A. Enhanced solubility and oral bioavailability of γ-tocotrienol using a self-emulsifying drug delivery system (SEEDS). Lipids 2014, 49, 819–829.

32. Djekic, L.; Primorac, M.; Jockovic, J. Phase behavior, microstructure and ibuprofen solubilization capacity of pseudo-ternary nonionic microemulsions. J. Mol. Liq. 2011, 160, 81–87.

Chapter 10

Enhanced Supersaturation and Oral Absorption of Sirolimus Using an Amorphous Solid Dispersion Based on Eudragit® E

Youngseok Cho [1], Eun-Sol Ha [2], In-Hwan Baek [3], Min-Soo Kim [2,*], Cheong-Weon Cho [1,*] and Sung-Joo Hwang [4,*]

[1]*College of Pharmacy, Chungnam National University, Daejeon 305-764, Korea*

[2]*College of Pharmacy, Pusan National University, Busan 609-735, Korea*

[3]*College of Pharmacy, Kyungsung University, Busan 608-736, Korea*

[4]*College of Pharmacy and Yonsei Institute of Pharmaceutical Sciences, Yonsei University, Incheon 406-840, Korea*

ABSTRACT

The present study aimed to investigate the effect of Eudragit® E/HCl (E-SD) on the degradation of sirolimus in simulated gastric fluid (pH 1.2) and to develop a new oral formulation of sirolimus using E-SD solid dispersions to enhance oral bioavailability. Sirolimus-loaded solid dispersions were fabricated by a spray drying process. A kinetic solubility test demonstrated that the sirolimus/E-SD/TPGS (1/8/1) solid dispersion had a maximum solubility of 196.7 μg/mL within 0.5 h that gradually decreased to 173.4 μg/mL after 12 h. According to the dissolution study, the most suitable formulation was the sirolimus/E-SD/TPGS (1/8/1) solid dispersion in simulated gastric fluid (pH 1.2), owing to enhanced stability and degree of supersaturation of E-SD and TPGS. Furthermore, pharmacokinetic studies in rats indicated that compared to the physical mixture and sirolimus/HPMC/TPGS (1/8/1) solid dispersion, the sirolimus/E-SD/TPGS (1/8/1) solid dispersion significantly improved oral absorption of sirolimus. E-SD significantly inhibited the degradation of sirolimus in a dose-dependent manner. E-SD also significantly inhibited the precipitation of sirolimus compared to hydroxypropylmethyl cellulose (HPMC). Therefore, the results from the present study suggest that the sirolimus-loaded E-SD/TPGS solid dispersion has great potential in clinical applications.

Keywords: sirolimus; supersaturation; bioavailability; Eudragit® E; solid dispersion

1. INTRODUCTION

Sirolimus (also known as rapamycin) is a macrocyclic lactone produced by *Streptomyces hygroscopicus* that inhibits interleukin (IL)-2 and other cytokine receptor-dependent signal transduction mechanisms [1]. Sirolimus is an immunosuppressive agent indicated for the prophylaxis of organ rejection in patients aged ≥ 13 years receiving renal transplants [2]. Unfortunately, sirolimus belongs to the biopharmaceutics classification system (BCS) class II drug category because of its low solubility and high permeability [3]. To improve the biological performance of sirolimus, various formulations, such as inclusion complexes with cyclodextrins, liposomal formulations, nanocrystals, and solid dispersion, have been developed [4,5,6,7,8]. Our group recently reported that of the 71 combination formulations we evaluated, the most efficient ternary solid dispersion for the enhanced bioavailability of sirolimus was hydroxypropylmethyl cellulose (HPMC)/d-α-tocopheryl polyethylene glycol 1000 succinate (TPGS) [9]. Recently, Petruševska *et al.* reported that HPMC effectively created a desired microenviroment that both maintained a supersaturated solution and prevented sirolimus precipitation [10]. In addition, it was observed that *in vitro* supersaturated dissolution data strongly correlated with *in vivo*pharmacokinetic parameters [9,10]. In fact, the oral bioavailability of sirolimus can be improved by enhancing *in vitro*supersaturation via an amorphous solid dispersion.

Eudragit® E is a cationic polymer based on dimethylaminoethyl, butyl and methyl methacrylate. The Eudragit® E solid dispersion is a well-established formulation system that enhances the bioavailability of poorly water-soluble active pharmaceutical ingredients (APIs) [11,12,13]. An amorphous Eudragit® E solid dispersion containing poorly water-soluble APIs can be manufactured via the principle of solvent evaporation, melting, and/or solvent-mediated melting using spray drying, or the hot-melt extrusion process [14,15]. Unfortunately, Eudragit® E dissolves in aqueous solutions below pH 5.0; thus, drug dissolution from the Eudragit® E solid dispersion decreases at a pH above 6.0 [16]. Recently, Yoshida *et al.* developed the Eudragit® E/HCl (E-SD) system, which was prepared by partial neutralization of Eudragit® E using hydrochloric acid for the enhancement of supersaturation and oral absorption of poorly water-soluble APIs [17,18]. They demonstrated that E-SD forms a micelle-like structure in different conditions, similar to a polymeric surfactant [19].

The present study aimed to investigate the effect of E-SD on the degradation of sirolimus in simulated gastric fluid (pH 1.2), as well as to develop a new oral formulation of sirolimus using the E-SD solid dispersion to enhance oral bioavailability. Sirolimus-loaded solid dispersions were fabricated via the spray drying process. *In vitro* and *in vivo*characterization of the solid dispersions was conducted to focus on the effect of the E-SD solid dispersion on supersaturation and the oral absorption of sirolimus.

2. RESULTS AND DISCUSSION

Sirolimus is known to be very unstable in electrolyte solutions, especially acidic ones, and is degraded via ring opening/fragmentation [20,21]. Previously, we reported that a rapid degradation of sirolimus was observed in simulated gastric fluid (pH 1.2) compared to intestinal fluid (pH 6.8), acetate buffer (pH 4.0), and distilled water [22]. Thus, the effect of E-SD on the degradation of sirolimus was further investigated in simulated gastric fluid (pH 1.2). As shown in Figure 1, the degradation of sirolimus occurred rapidly and the percent remaining was below 2% at 30 min. HPMC had no influence on degradation, while E-SD demonstrated a significant dose-dependent inhibitory effect. As expected, the pseudo-first-order degradation constant of sirolimus decreased with increasing concentrations of E-SD (Table 1). In particular, the kinetic rate of degradation of 5 mg/mL E-SD was about 10-fold lower than that of the drug alone. TPGS, a nonionic polymeric surfactant, dramatically inhibited the degradation of sirolimus by forming micelles [23]. Recently, the critical micelle concentration of E-SD was reported to be 100 µg/mL [19]. HPMC, a hydrophilic polymer, does not form micelle-like structures, while E-SD, similar to TPGS, forms micelle-like structures. Thus, the enhanced stability of sirolimus with increasing amounts of E-SD could be attributed to the formation of micelle-like structures.

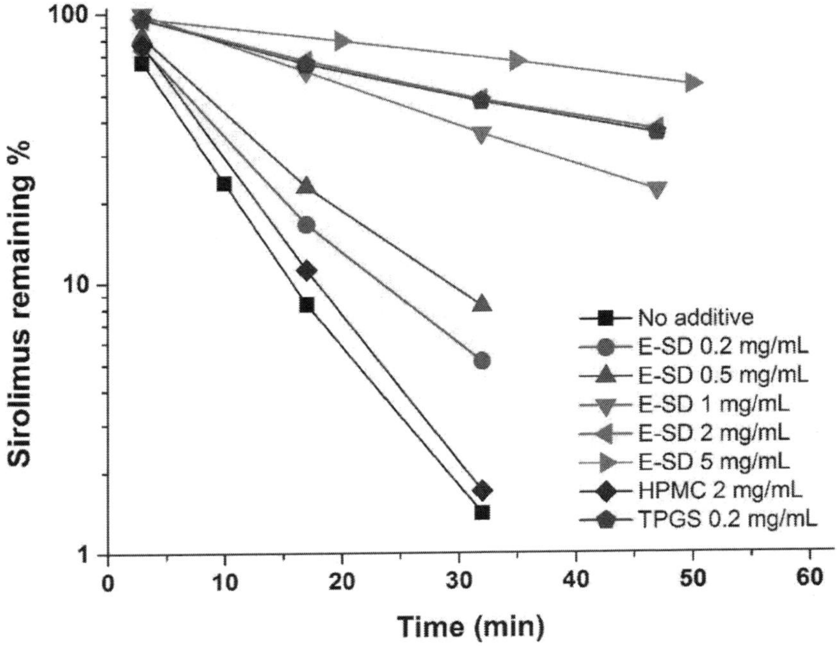

Figure 1. Effect of E-SD on the degradation of sirolimus in simulated gastric fluid (pH 1.2).

To investigate the effect of the sirolimus/E-SD ratio on the improvement of sirolimus super-saturation, E-SD solid dispersions were fabricated via the spray drying method. For comparison, sirolimus-loaded HPMC and HPMC/TPGS solid dispersions were also prepared.

Table 1. Effect of E-SD on the degradation of sirolimus in simulated gastric fluid (pH 1.2).

Composition	k_{obs} (min^{-1})	$t_{1/2}$ (min)
No additives	0.1322 ± 0.0011	5.24 ± 0.04
E-SD 0.2 mg/mL	0.0925 ± 0.0013	7.49 ± 0.10
E-SD 0.5 mg/mL	0.0794 ± 0.0006	8.73 ± 0.07
E-SD 1 mg/mL	0.0345 ± 0.0005	20.07 ± 0.29
E-SD 2 mg/mL	0.0219 ± 0.0004	31.65 ± 0.58
E-SD 5 mg/mL	0.0125 ± 0.0005	55.64 ± 2.02
HPMC 2 mg/mL	0.1319 ± 0.0020	5.25 ± 0.08
TPGS 0.2 mg/mL	0.0225 ± 0.0009	30.83 ± 1.16

The pseudo-first-order rate constants (k_{obs}) and half-life ($t_{1/2}$) were calculated according to the equations $\ln[C_t] = \ln[C_0] - k_{obs} \cdot t$ and $t_{1/2} = 0.693/k_{obs}$, respectively. $[C_0]$ is the initial concentration of sirolimus and $[C_t]$ is the percentage remaining at time t, which allowed for the calculation of k_{obs} as the slopes of the lines obtained by linear regression analysis. Data are expressed as the mean ± standard deviation ($n = 3$).

As shown in Figure 2 and Table 2, all solid dispersions had irregular-shaped particles with similar mean particle size (6–8 μm). There were no significant size differences between the solid dispersions ($p > 0.05$). The encapsulation efficiency in each solid dispersion was almost equal to that of theoretical values, as shown by HPLC analysis, indicating that sirolimus was not degraded during the spray drying process.

As depicted in Figure 3, the crystalline state of sirolimus within the solid dispersion was determined by powder X-ray diffractometer (PXRD, Bruker AXS GmbH, Karlsruhe, Germany). Raw sirolimus had the characteristic diffraction patterns at 2θ values of 7.2°, 9.9°, 14.5°, 16.2°, 20.0°, and 20.4°. However, the characteristic diffraction patterns of crystalline sirolimus were not observed in the PXRD patterns of all sirolimus-loaded solid dispersions, respectively. This indicates that the crystallinity of sirolimus significantly decreased, and that sirolimus exists in an amorphous form within solid dispersions.

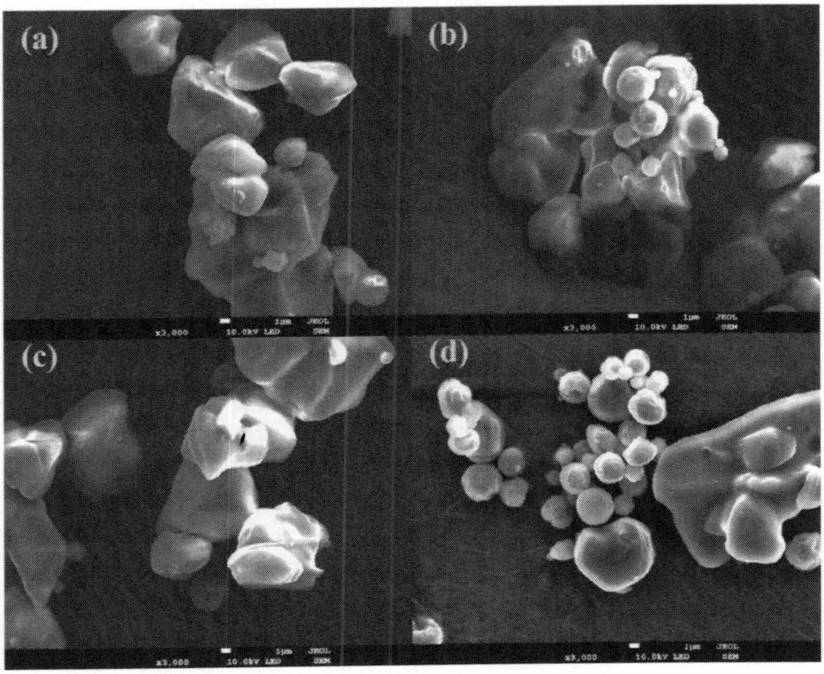

Figure 2. Scanning electron micrographs of sirolimus-loaded solid dispersions. (**a**) Sirolimus/E-SD solid dispersion (1/9); (**b**) Sirolimus/E-SD/TPGS solid dispersion (1/8/1); (**c**) Sirolimus/E-SD/HPMC solid dispersion (1/8/1); (**d**) Sirolimus/HPMC/TPGS solid dispersion (1/8/1). Total magnification of 3000× with a scale bar corresponding to 1 μm.

Table 2. Formulation, encapsulation efficiency, and particle size of sirolimus-loaded solid dispersions.

Formulation (Weight)	Encapsulation Efficiency (%)	Mean Particle Size (μm)
Sirolimus/E-SD = 1/1	97.9 ± 2.1	6.72 ± 1.82 (1.89) #
Sirolimus/E-SD = 1/2	98.1 ± 1.8	7.43 ± 1.99 (1.90)
Sirolimus/E-SD = 1/9	96.7 ± 2.5	7.21 ± 1.79 (1.95)
Sirolimus/E-SD/TPGS = 1/8/1	98.2 ± 1.5	7.56 ± 1.95 (2.10)
Sirolimus/E-SD/HPMC = 1/8/1	96.9 ± 1.8	7.89 ± 2.03 (2.02)
Sirolimus/HPMC = 1/9	99.6 ± 1.2	6.92 ± 1.88 (1.92)
Sirolimus/HPMC/TPGS = 1/8/1	98.9 ± 1.5	6.43 ± 1.71 (2.15)

The encapsulation efficiency (%) = weight of loaded drug/weight of the feeding drug × 100. The particle size of the respective gelatin microparticle-containing self-microemulsifying formulation in the solid state was measured using a HELOS laser diffraction analyzer. # SPAN = $(d_{90\%} - d_{10\%})/d_{50\%}$, where $d_{10\%}$, $d_{50\%}$, and $d_{90\%}$ are the diameter sizes and the given percentage value is the percentage of the particles smaller than that size.

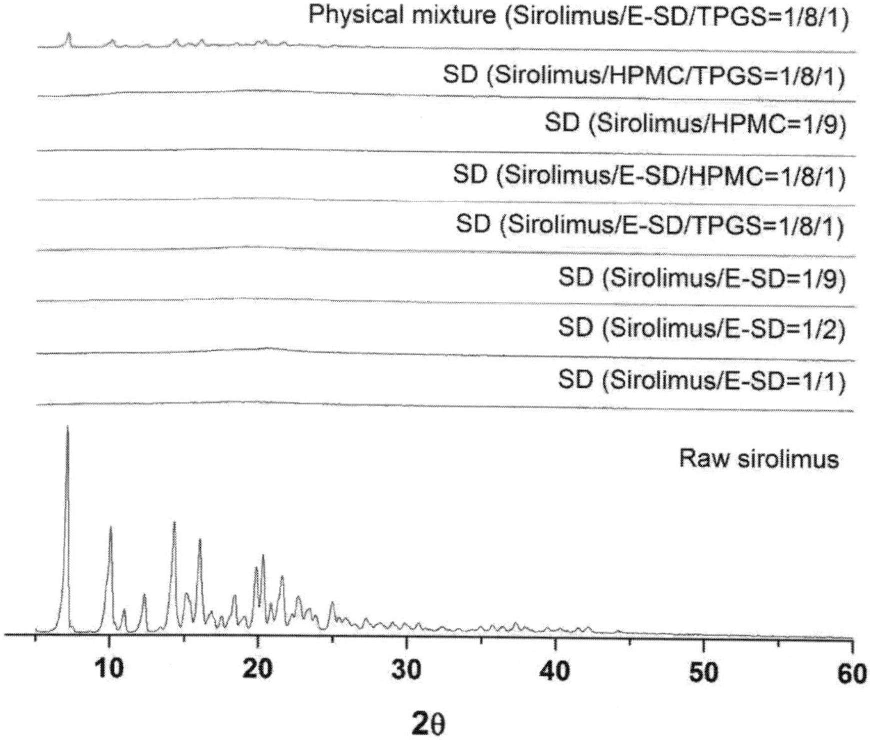

Figure 3. Powder X-ray diffraction patterns of sirolimus-loaded solid dispersions.

Kinetic solubility tests were performed in distilled water to investigate degree of supersaturation of sirolimus-loaded solid dispersions. As shown in Figure 4, the maximum solubility (supersaturation concentration) of sirolimus in the solid dispersions was 49.8–196.7 µg/mL, and the concentration at 12 h was 35.1–173.4 µg/mL. In particular, the maximum solubility of sirolimus dramatically increased with increasing amounts of E-SD in the solid dispersions. For example, the sirolimus/E-SD (1/9) solid dispersion showed a maximum solubility of 170 µg/mL within 0.5 h, whereas the sirolimus/HPMC (1/9) solid dispersion had a solubility of 49.8 µg/mL at 2 h, which decreased to 35.1 µg/mL after 12 h. Compared with HPMC, E-SD demonstrated higher supersaturation and prolonged superstation of sirolimus. These results suggest that E-SD might inhibit drug precipitation from the supersaturated state. Therefore, the effect of E-SD and HPMC on the recrystallization of a supersaturated sirolimus solution was investigated by the addition of a sirolimus solution in DMSO into water or simulated intestinal fluid (pH 6.8). As shown in Figure 5, sirolimus gradually precipitated in both water and simulated intestinal fluid (pH 8) without E-SD and HPMC; sirolimus is practically insoluble in water (2.6 µg/mL) and contains functional groups that are ionizable in the pH range 1–10 [24]. However, precipitation was significantly inhibited by E-SD as E-SD maintained the concentration of sirolimus above 160 µg/mL for at least 12 h. In addition, E-SD exhibited a superior inhibitory effect when compared to HPMC. The inhibition of precipitation by E-SD might be attributed to the inhibition of nucleation and/or crystal growth in a highly supersaturated state by blocking the active surface and providing steric stabilization by acting like a surfactant [25,26,27,28]. It may also be owing to the specific interactions between sirolimus and E-SD, such as hydrophobic interactions or hydrogen bonding [29,30,31]. Recently, Hiashi *et al*reported that hydrophobic and hydrophilic interactions between mefenamic acid and Eudragit EPO molecules play a key role in the formation of a stable supersaturated solution by using high-resolution magic-angle spinning NMR measurements [32]. However, this hypothesis requires further study.

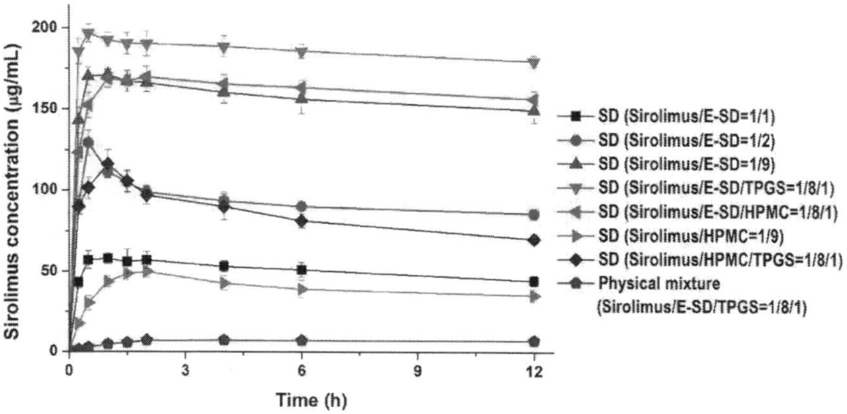

Figure 4. Kinetic solubility profiles of sirolimus-loaded solid dispersions. Data are expressed as the mean ± standard deviation (*n* = 3).

The addition of TPGS to sirolimus/E-SD solid dispersions further increased supersaturation. The sirolimus/E-SD/TPGS (1/8/1) solid dispersion demonstrated a maximum solubility of 196.7 µg/mL within 0.5 h that gradually decreased to 173.4 µg/mL after 12 h. The solubility of the sirolimus/E-SD/TPGS (1/8/1) solid dispersion was higher than that of the sirolimus/HPMC/TPGS (1/8/1) solid dispersion at all timepoints. Furthermore, the sirolimus/E-SD/TPGS (1/8/1) solid dispersion showed a maximum supersaturation that was approximately 28-fold greater than the concentration of its physical mixture.

Dissolution profiles were obtained for sirolimus-loaded solid dispersions in simulated gastric fluid (pH 1.2) containing 0.05% w/v sodium lauryl sulfate (SLS), and simulated intestinal fluid (pH 6.8) containing 0.05% w/v SLS as dissolution medium. As shown in Figure 6, rapid dissolution profiles were obtained for the solid dispersions in simulated intestinal fluid (pH 6.8), while characteristic convex curves were observed for dispersions in the simulated gastric fluid (pH 1.2) owing to the degradation of sirolimus. The dissolution from solid dispersions was dramatically higher than that from the physical mixture. Of the tested samples, the most suitable formulation was the sirolimus/E-SD/TPGS (1/8/1) solid dispersion in simulated gastric fluid (pH 1.2) owing to the enhanced stability and supersaturation by E-SD and TPGS. In simulated intestinal fluid (pH 6.8), approximately 90% of the drug dissolved from the sirolimus/E-SD/TPGS (1/8/1) solid dispersion, while only 10% of the drug dissolved from the physical mixture at 30 min. The dissolution profile of the sirolimus/HPMC/TPGS (1/8/1) solid dispersion was similar to the sirolimus/E-SD/TPGS (1/8/1) solid dispersion in simulated intestinal fluid (pH 6.8).

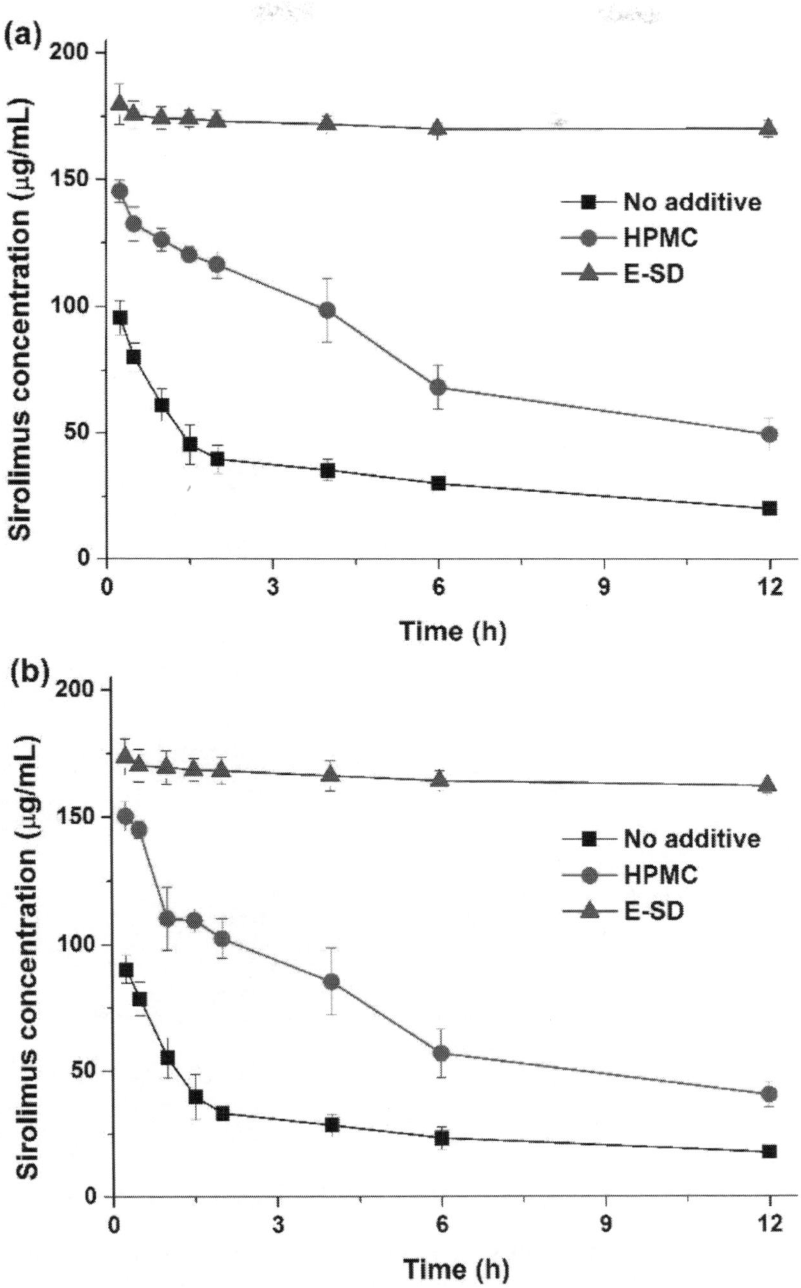

Figure 5. Effect of E-SD and HPMC on the recrystallization of sirolimus in two different media. Water (**a**) intestinal fluid (pH 6.8) (**b**). Data are expressed as the mean ± standard deviation (n = 3).

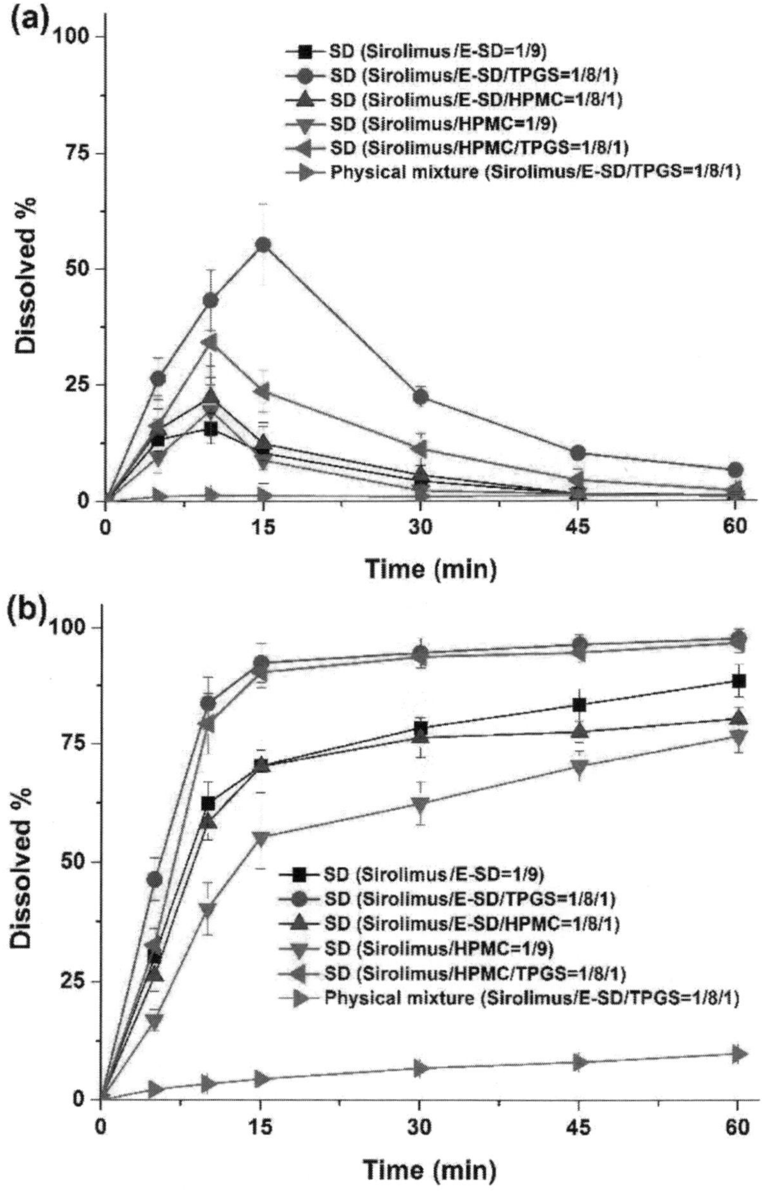

Figure 6. Dissolution profiles of sirolimus-loaded solid dispersions in media with different pHs. Simulated gastric fluid (pH 1.2) containing 0.05% w/v SLS (**a**); intestinal fluid (pH 6.8) containing 0.05% w/v SLS (**b**). Data are expressed as the mean ± standard deviation ($n = 4$).

The enhanced supersaturation and dissolution property of sirolimus in an amorphous solid dispersion based on E-SD could be directly related to the enhanced oral bioavailability of sirolimus. Thus, we investigated the oral bioavailability of sirolimus-loaded E-SD solid dispersions in SD rats. Figure 7 shows the blood concentration-time profile of sirolimus after oral administration of both solid dispersions and the physical mixture. Pharmacokinetic parameters ($AUC_{0\rightarrow12\ h}$, C_{max}, and T_{max}) are presented in Table 3. As shown in Figure 7, the blood concentration of the sirolimus/E-SD/TPGS (1/8/1) solid dispersion with the rapid drug absorption rate was dramatically higher than that of the physical mixture at all timepoints. The $AUC_{0\rightarrow12\ h}$, C_{max}, and T_{max} were 459.5 ± 93.6 ng·h/mL, 97.4 ± 19.2 ng/mL, and 1.2 ± 0.3 h, respectively. The oral absorption of the E-SD/TPGS (8/1) solid dispersion was significantly higher than that of the physical mixture and HPMC/TPGS (8/1) solid dispersion, with approximately 7.2- and 1.4-fold increases in $AUC_{0\rightarrow12\ h}$, respectively (Table 3). The bioavailability of the sirolimus/E-SD/TPGS (1/8/1) solid dispersion was markedly higher than that of the physical mixture and sirolimus/HPMC/TPGS (1/8/1) solid dispersion, due to higher supersaturation and supersaturated concentration over an increased period of time, as well as enhanced stability in simulated gastric fluid (pH 1.2) [33]. These results suggest that the supersaturation and oral absorption of sirolimus were significantly increased by the amorphous E-SD/TPGS solid dispersion.

Table 3. Pharmacokinetic parameters of sirolimus-loaded solid dispersions in rats.

Formulation	$AUC_{0\rightarrow12\ h}$ (ng·h/mL)	C_{max} (ng/mL)	T_{max} (h)
Physical mixture (sirolimus/E-SD/TPGS = 1/8/1)	64.0 ± 11.8	9.6 ± 2.6	3.8 ± 1.1
Solid dispersion (sirolimus/E-SD/TPGS = 1/8/1)	459.5 ± 93.6 [a,b]	97.4 ± 19.2 [a,b]	1.2 ± 0.3
Solid dispersion (sirolimus/HPMC/TPGS = 1/8/1)	325.5 ± 46.4 [a]	61.0 ± 14.2 [a]	1.6 ± 0.5

[a] $p < 0.05$ *vs.* physical mixture; [b] $p < 0.05$ *vs.* sirolimus/HPMC/TPGS. Data are expressed as the mean ± standard deviation ($n = 5$).

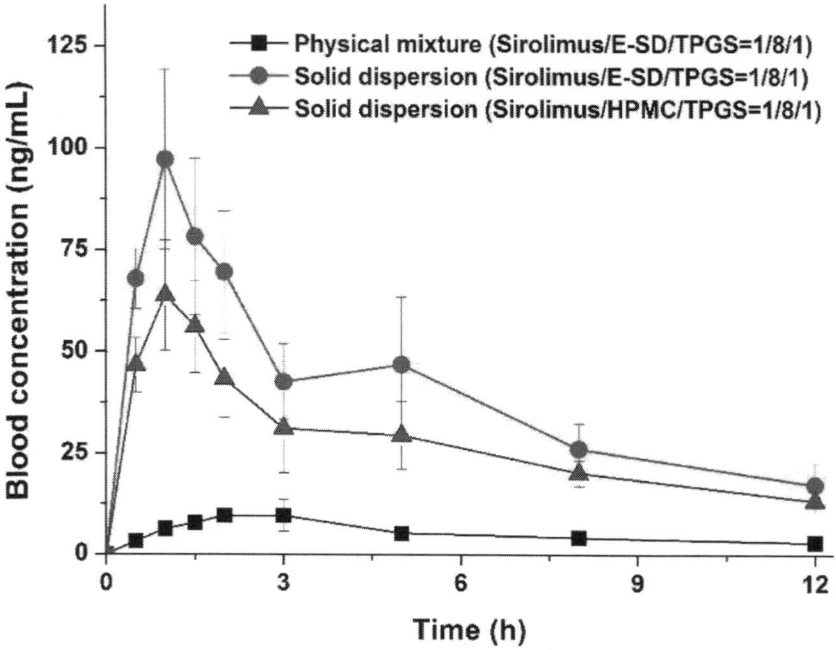

Figure 7. Blood concentration-time profile of sirolimus in rats after oral administration of sirolimus-loaded solid dispersions and the physical mixture. Data are expressed as the mean ± standard deviation (n = 5).

3. EXPERIMENTAL SECTION

3.1. Materials

Sirolimus (purity 99.4%) and everolimus (an internal standard for LC-MC analysis, purity 95.9%) were purchased from Beijing Everbright Science & Trading Co. (Beijing, China) and Sigma-Aldrich (St. Louis, MO, USA), respectively. Eudragit®E (poly(butylmethacrylateco-(2-dimethylaminoethyl)methacrylate-co-methyl methacrylate) was provided by BASF Co. Ltd. (Ludwigshafen, Germany). Hydroxypropyl-methyl cellulose (HPMC 2910) and D-α-tocopheryl polyethylene glycol 1000 succinate (TPGS) were purchased from Shin-Etsu Chemical Co Ltd (Tokyo, Japan) and Eastman Co. (Kingsport, TN, USA), respectively. Acetonitrile, dimethyl sulfoxide (DMSO) and methanol were of high-performance liquid chromatography (HPLC) grade. All other chemicals were of reagent grade.

3.2. Preparation of Eudragit® E/HCl (E-SD) Using Spray Drying Method

Eudragit® E/HCl (E-SD) was prepared by a spray drying process as previously reported [17,18]. Eudragit® E (100 g) was dispersed into a mixture solution consisting of 10% (w/v) HCl (73.5 g) and 526.5 g of water; the components were mixed by mechanical stirring until a clear solution was obtained. This solution was spray dried using a B-191 mini spray dryer (Buchi, Flawil, Switzerland) under the following conditions: inlet temperature, 130–145 °C; outlet temperature, 75–85 °C; feed rate, 2–10 mL/min; and atomization air pressure, 5 kPa. Eudragit® E/HCl (E-SD) was obtained as a powder.

3.3. Effect of E-SD, HPMC, and TPGS on the Degradation of Sirolimus in Solution

The effect of E-SD, HPMC, and TPGS on the degradation of sirolimus in solution was investigated. Sirolimus was dissolved in DMSO at a concentration of 1 mg/mL. A 100-μL aliquot of the solution was added to 50 mL simulated gastric fluid (pH 1.2, without pepsin) containing different concentrations of E-SD, HPMC, and TPGS at 37 °C in the dark. Samples were removed at various time intervals, and the concentration of sirolimus was determined by HPLC. HPLC analysis of all samples was performed with a Waters HPLC system (Waters, Milford, MA, USA) equipped with a Waters 600 controller pump, and Waters 486 tunable absorbance detector. The mobile phase was methanol/water (84:16, v/v) at a flow rate of 1.0 mL/min. All chromatographic analysis was performed on a C18 analytical column (Luna C18 (2), 5 μm, 4.6 mm × 250 mm, Phenomenex, Torrance, CA, USA) at 60 °C. Detection was performed at a wavelength of 278 nm.

3.4. Effect of E-SD and HPMC on the Recrystallization of Sirolimus in a Supersaturated Solution

To investigate the effect of E-SD and HPMC on the recrystallization of a supersaturated sirolimus solution, sirolimus was dissolved in DMSO at a concentration of 30 mg/mL. A 2-mL aliquot of the sirolimus solution was added to 300 mL of water or simulated intestinal fluid (pH 6.8) with/without 600 mg of an additive (E-SD or HPMC), and agitated with an USP rotating paddle apparatus (Electrolab, Mumbai, India) at 37 °C and 50 rpm. Samples (2 mL) were removed at various time intervals and filtered using a 0.45 μm glass fiber syringe filter. The filtered samples were diluted with methanol, and the concentration of sirolimus was determined by HPLC.

3.5. Preparation of Sirolimus-Loaded Solid Dispersions

Sirolimus-loaded solid dispersions were fabricated via the spray drying method using a Buchi B-191 nozzle type mini spray dryer. Sirolimus was dissolved in a mixture of ethanol and methylene chloride (55:45, w/w) containing E-SD, HPMC, and/or TPGS at a concentration of 3% w/w of solute. The detailed composition of sirolimus-loaded solid dispersions are described in Table 2. The resulting solution was delivered to the inner line of the nozzle at a flow rate of 3–6 mL/min using a peristaltic pump, and solidified under the following conditions: inlet temperature, 65–80 °C; outlet temperature, 45–55 °C; and atomization air pressure, 5 kPa. For comparison, the physical mixture was also prepared by simply mixing sirolimus with E-SD, HPMC, and/or TPGS using a spatula in a glass vial.

3.6. Characterization of Sirolimus-Loaded Solid Dispersions

The sirolimus content in solid dispersions was determined by HPLC analysis. About 20 mg of each solid dispersion was dissolved in 50 mL of ethanol and methylene chloride (55:45, w/w). The morphology of sirolimus-loaded solid dispersions was observed by scanning electron microscopy (SEM; JSM-7100f, Jeol Ltd., Tokyo, Japan). The particle size and size distribution of sirolimus-loaded solid dispersions were characterized using a HELOS laser diffraction spectrometer (SYMPATEC Ltd., Clausthal-Zellerfeld, Germany). Powder X-ray diffraction patterns of sirolimus-loaded solid dispersions were determined using a D8 Advance X-ray diffraction system (Bruker AXS GmbH, Karlsruhe, Germany) with Ni-filtered Cu-Kα radiation. The samples were analyzed between 5° and 60°. Kinetic solubility profiles of the sirolimus-loaded solid dispersions were obtained in a water-jacketed vessel linked to a temperature-controlled water bath held at 37 °C. The solid dispersions (containing 40 mg sirolimus) were combined with 200 mL water and agitated constantly at 100 rpm. At pre-determined intervals, samples from various time points were filtered using a 0.45-µm glass fiber syringe filter followed by dilution with methanol, and the amount of drug dissolved in each sample was determined by HPLC. Dissolution profiles of sirolimus from solid dispersions were determined using an USP rotating paddle apparatus (Electrolab) at 37 °C and 100 rpm in 500 mL simulated gastric fluid (pH 1.2, without pepsin) containing 0.05% w/v SLS, and simulated intestinal fluid (pH 6.8) containing 0.05% w/v SLS as dissolution medium. After samples containing 2 mg sirolimus were placed in the dissolution medium, 3 mL samples were collected for analysis at pre-determined intervals and replaced with 3 mL of fresh dissolution medium after each collection. Samples from various time points were filtered using a 0.45-µm glass fiber syringe filter followed by dilution with methanol. The amount of drug dissolved in each sample was determined by HPLC.

3.7. In Vivo Study in Rats

The animal study protocol was in compliance with the institutional guidelines for the care and use of laboratory animals, and was approved by the ethics committee of Chungnam National University (Daejeon, Korea). Fifteen male Sprague-Dawley (SD) rats (250 ± 10 g, Orient Bio Inc., Seongnam, Korea) were divided into three treatment groups of five rats each. Prior to the study, the rats fasted for 18 hours; each group received either the physical mixture or solid dispersions at sirolimus doses of 5 mg/kg (dose/rat weight) for each formulation by oral gavage. The physical mixture and sirolimus-loaded solid dispersions were dispersed in 1 mL water immediately prior to oral dosing. Approximately 0.45 mL blood samples were collected in heparinized tubes from the retro-orbital plexus of rats at 0.5, 1, 1.5, 2, 3, 5, 8, and 12 h after dosing. Blood concentrations of sirolimus were determined by liquid chromatography with mass spectrometry (LC-MS), according to our previously reported method [21]. Pharmacokinetic analysis of the data was carried out with WinNonlin Standard Edition software, version 5.3 (Pharsight Corp., St. Louis, MO, USA, 2009). The area under the curve ($AUC_{0\rightarrow12\ h}$) was calculated according to the trapezoidal method. The peak blood concentration (C_{max}) and the time to reach C_{max}(T_{max}) of sirolimus in the plasma were taken directly from the data. Statistical analysis for the dissolution data and pharmacokinetic parameters was performed using a one-way analysis of variance (ANOVA) test followed by the Student-Newman-Keuls (SNK) and least-squares difference (LSD) tests with SPSS 21.0 software (IBM SPSS Statistics, Armonk, NY, USA, 2013).

4. CONCLUSIONS

In this study, E-SD significantly inhibited the degradation of sirolimus in a dose-dependent manner. In addition, E-SD exhibited a superior inhibitory effect on the precipitation of sirolimus compared to HPMC. Sirolimus-loaded solid dispersions were fabricated by the spray drying process. In both the kinetic solubility and dissolution test, the most suitable formulation was the sirolimus/E-SD/TPGS (1/8/1) solid dispersion owing to the enhanced stability and degree of supersaturation of E-SD and TPGS. Furthermore, pharmacokinetic studies in rats indicated that compared to the physical mixture and sirolimus/HPMC/TPGS (1/8/1) solid dispersion, the sirolimus/E-SD/TPGS (1/8/1) solid dispersion significantly improved the oral absorption of sirolimus. Therefore, the results from our study suggest that the sirolimus-loaded E-SD/TPGS solid dispersion has great potential in clinical applications.

ACKNOWLEDGMENTS

This work was supported by the National Research Foundation of Korea (NRF) grant funded by the Korea government (MSIP) (No.2009-0083538).

AUTHOR CONTRIBUTIONS

Conceived and designed the experiments: M.-S.K., C.-W.C., and S.-J.H.; Performed the experiments: Y.C., E.-S.H., and I.-H.B.; Analyzed the data: Y.C., E.-S.H., I.-H.B., and M.-S.K.; Wrote the paper: Y.C.; All authors read and approved the final manuscript.

CONFLICTS OF INTEREST

The authors declare no conflict of interest.

REFERENCES

1. Kahan, B.D. Sirolimus: A comprehensive review. Expert. Opin. Pharmacother. 2001, 2, 1903–1917.
2. Toso, C.; Merani, S.; Bigam, D.L.; Shapiro, A.M.J.; Kneteman, N.M. Sirolimus-based immunosuppression is associated with increased survival after liver transplantation for hepatocellular carcinoma. Hepatology 2010, 51, 1237–1243.
3. Vasquez, E.M. Sirolimus: A new agent for prevention of renal allograft rejection. Am. J. Health -Syst. Pharm. 2000, 57, 437–448.
4. Buech, G.; Bertelmann, E.; Pleyer, U.; Siebenbrodt, I.; Borchert, H.H. Formulation of sirolimus eye drops and corneal permeation studies. J. Ocul. Pharmacol. Ther. 2007, 23, 292–303.
5. Alemdar, A.Y.; Sadi, D.; McAlister, V.C.; Mendez, I. Liposomal formulations of tacrolimus and rapamycin increase graft survival and fiber outgrowth of dopaminergic grafts. Cell Transplant. 2004, 13, 263–271.

6. Preetham, A.C.; Satish, C.S. Formulation of a poorly water-soluble drug sirolimus in solid dispersions to improve dissolution. J. Disper. Sci. Technol. 2011, 32, 778–783.

7. Junghanns, J.U.; Muller, R.H. Nanocrystal technology, drug delivery and clinical applications. Int. J. Nanomed. 2008, 3, 295–310.

8. Hu, X.; Lin, C.; Chen, D.; Zhang, J.; Liu, Z.; Wu, W.; Song, H. Sirolimus solid self-microemulsifying pellets: Formulation development, characterization and bioavailability evaluation. Int. J. Pharm. 2012, 438, 123–133.

9. Kim, M.S.; Kim, J.S.; Cho, W.; Cha, K.H.; Park, H.J.; Park, J.; Hwang, S.J. Supersaturatable formulations for the enhanced oral absorption of sirolimus. Int. J. Pharm. 2013, 445, 108–116.

10. Petruešvska, M.; Homar, M.; Petek, B.; Resman, A.; Kocjan, D.; Urleb, U.; Peternel, L. Hydroxypropyl methylcellulose mediated precipitation inhibition of sirolimus: From a screening campaign to a proof-of-concept human study. Mol. Pharm. 2013, 10, 2299–2310.

11. Tang, J.; Xu, N.; Ji, H.; Liu, H.; Wang, Z.; Wu, L. Eudragit Nanoparticles Containing Genistein: Formulation, Development, and Bioavability Assessment. Int. J. Nanomed. 2011, 6, 2429–2435.

12. Yoo, S.D.; Lee, S.H.; Kang, E.; Jun, H.; Jung, J.Y.; Park, J.W.; Lee, K.H. Bioavailability of itraconazole in rats and rabbits after administration of tablets containing solid dispersion particles. Drug Dev. Ind. Pharm. 2000, 26, 27–34.

13. Baek, I.H.; Kim, M.S. Improved supersaturation and oral absorption of dutasteride by amorphous solid dispersions.Chem. Pharm. Bull. 2012, 60, 1468–1473.

14. Chokshi, R.J.; Sandhu, H.K.; Iyer, R.M.; Shah, N.H.; Malick, A.W.; Zia, H. Characterization of physico-mechanical properties of indomethacin and polymers to assess their suitability for hot-melt extrusion processs as a means to manufacture solid dispersion/solution. J. Pharm. Sci. 2005, 94, 2463–2474.

15. Janssens, S.; Zeure, A.D.; Paudel, A.; Humbeeck, J.V.; Rombaut, P.; Mooter, G.V. Influence of preparation methods on solid state supersaturation of amorphous solid dispersions: A case study with itraconazole and Eudragit E100.Pharm. Res. 2010, 27, 775–785.

16. Dai, J.; Nagai, T.; Wang, X.; Zhang, T.; Meng, N.; Zhang, Q. pH-sensitive nanoparticles for improving the oral bioavailability of cyclosporine A. Int. J. Pharm. 2004, 280, 229–240.

17. Yoshida, T.; Kurimoto, I.; Yoshihara, K.; Umejima, H.; Ito, N.; Watanabe, S.; Sako, K.; Kikuchi, A. Aminoalkyl methacrylate copolymers for improving the solubility of tacrolimus. I: Evaluation of solid dispersion formulations.Int. J. Pharm. 2012, 428, 18–24.

18. Yoshida, T.; Kurimoto, I.; Yoshihara, K.; Umejima, H.; Ito, N.; Watanabe, S.; Sako, K.; Kikuchi, A. Effect of aminoalkyl methacrylate copolymer E/HCl on in vivo absorption of poorly water-soluble drug. Drug Dev. Ind. Pharm. 2013, 39, 1698–1705.

19. Yoshida, T.; Kurimoto, I.; Umejima, H.; Watanabe, S.; Sako, K.; Kikuchi, A. Effects of dissolved state of aminoalkyl methacrylate copolymer E/HCl on solubility enhancement effect for poorly water-soluble drugs. Colloid Polym. Sci.2013, 291, 1191–1199.

20. Rouf, M.A.; Bilensoy, E.; Vural., I.; Hincal, A.A. Determination of stability of rapamycin following exposure to different conditions. Eur. J. Pharm. Sci. 2007, 32, S46–S46.

21. Luengo, J.I.; Konialian, A.L.; Holt, D.A. Studies on the chemistry of rapamycin-novel transformations under lewis-acid catalysis. Tetrahedron Lett. 1993, 34, 991–994.

22. Kim, M.S.; Kim, J.S.; Park, H.J.; Cho, W.K.; Cha, K.H.; Hwang, S.J. Enhanced bioavailability of sirolimus via preparation of solid dispersion nanoparticles using a supercritical antisolvent process. Int. J. Nanomed. 2011, 6, 2997–3009.

23. Kim, M.S.; Kim, J.S.; Cho, W.K.; Hwang, S.J. Enhanced solubility and oral absorption of sirolimus using D-α-tocopheryl polyethylene glycol succinate micelles. Artif. Cells Nanomed. Biotechnol. 2013, 41, 85–91.

24. Simamora, P.; Alvarez, J.M.; Yalkowsky, S.H. Solubilization of rapamycin. Int. J. Pharm. 2001, 213, 25–29.

25. Ha, E.S.; Choo, G.H.; Baek, I.H.; Kim, M.S. Formulation, characterization, and in vivo evaluation of celecoxib-PVP solid dispersion nanoparticles using supercritical antisolvent process. Molecules 2014, 19, 20325–20339.

26. Takano, R.; Takata, N.; Saito, R.; Furumoto, K.; Higo, S.; Hayashi, Y.; Machida, M.; Aso, Y.; Yamashita, S. Quantitative analysis of the effect

of supersaturation on in vivo drug absorption. Mol. Pharm. 2010, 7, 1431–1440.

27. Kim, M.S. Influence of hydrophilic additives on the supersaturation and bioavailability of dutasteride-loaded hydroxypropyl-β-cyclodextrin nanostructures. Int. J. Nanomed. 2013, 8, 2029–2039.

28. Aggarwal, A.K.; Jindal, P. Interaction of terbinafine hydrochloride with nicotinamide in solution and solid state. J. Pharm. Investig. 2014, 44, 23–32.

29. Ha, E.S.; Baek, I.H.; Cho, W.; Hwang, S.J.; Kim, M.S. Preparation and evaluation of solid dispersion of atorvastatin calcium with Soluplus® by spray drying technique. Chem. Pharm. Bull. 2014, 62, 545–551.

30. Kojima, T.; Higashi, K.; Suzuki, T.; Tomono, K.; Moribe, K.; Yamamoto, K. Stabilization of a supersaturated solution of mefenamic acid from a solid dispersion with Eudragit EPO. Pharm. Res. 2012, 29, 2777–2791.

31. Park, M.H.; Baek, J.S.; Lee, C.A.; Kim, D.C.; Cho, C.W. The effect of Eudragit type on BSA-loaded PLGA nanoparticles. J. Pharm. Investig. 2014, 44, 339–349.

32. Higashi, K.; Yamamoto, K.; Pandey, M.K.; Mroue, K.H.; Moribe, K.; Yamamoto, K.; Ramamoorthy, A. insights into atomic-level interaction between mefenamic acid and Eudragit EPO in a supersaturated solution by high-resolution magic-angle spinning NMR Spectroscopy. Mol. Pharm. 2014, 11, 351–357.

33. Cho, W.; Kim, M.S.; Kim, J.S.; Park, J.; Park, H.J.; Cha, K.H.; Park, J.S.; Hwang, S.J. Optimized formulation of solid self-microemulsifying sirolimus delivery systems. Int. J. Nanomed. 2013, 8, 1673–1682.

Chapter 11

Highly Stable Tetra-Phenolato Titanium(IV) Agent Formulated into Nanoparticles Demonstrates Anti-Tumoral Activity and Selectivity

Sigalit Meker [1], Ori Braitbard [2], Katrin Margulis-Goshen [1], Shlomo Magdassi [1], Jacob Hochman [2] and Edit Y. Tshuva [1,*]

[1]The Institute of Chemistry, The Hebrew University of Jerusalem, Jerusalem 91904, Israel

[2]Department of Cell and Developmental Biology, Alexander Silberman Institute of Life Science, The Hebrew University of Jerusalem, Jerusalem 91904, Israel

ABSTRACT

Titanium(IV) complexes exhibit high potential as anti-tumor agents, particularly due to their low intrinsic toxicity and cytotoxicity toward cisplatin resistant cells. Nevertheless, Ti(IV) complexes generally undergo rapid hydrolysis that previously hampered their utilization as anticancer drugs. We recently overcame this difficulty by developing a highly stable Ti(IV) complex that is based on tetra-phenolato, hexadentate ligand, formulated into organic nanoparticles. Herein we investigated the activity of this complex *in vitro* and *in vivo*. Although inactive when tested directly due to poor solubility, when formulated, this complex displayed (a) high cytotoxicity toward cisplatin resistant human ovarian cells, A2780-cp, with resistance factor of 1.1; (b) additive behavior in combination with cisplatin toward ovarian and colon cancer cells; (c) selectivity toward cancer cells as implied by its mild activity toward non-cancerous, fibroblast lung cells, MRC-5; (d) high stability and durability as manifested by the ability to maintain cytotoxicity, even following one week of incubation in 100% aquatic medium solution; and (e) *in vivo* efficacy toward solid tumors of human colon cancer cells, HT-29, in nude mice without any clinical signs of toxicity. These features support the formulated phenolato Ti(IV) complex being an effective and selective anti-tumoral agent.

Keywords: titanium(IV); phenolato ligands; cisplatin; cytotoxicity; anti-tumor; cisplatin resistant cells

1. INTRODUCTION

Inorganic titanium(IV) complexes exhibit high potential as non-toxic, effective anti-tumor agents [1,2,3,4,5,6,7,8,9,10,11,12]. They have been investigated as such for several decades due to their high activity toward various cancer cells and low intrinsic toxicity of the titanium metal [2,3,4,5,7,8,13,14], which represent significant advantages over the clinically used cisplatin [1,11,15]. The latter, as well as its derivatives such as carboplatin, suffer from not only severe side effects but also from inherent or acquired resistance of some cell types (Scheme 1) [16,17,18].

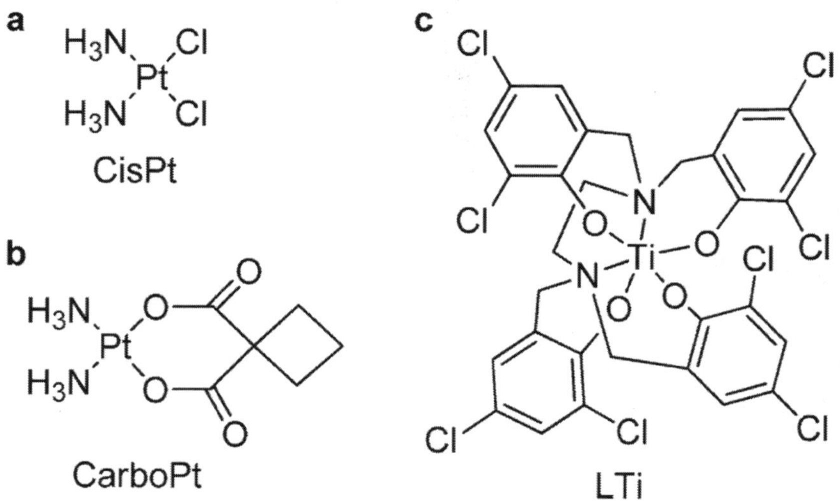

Scheme 1. (**a**) Cisplatin; (**b**) Carboplatin; and (**c**) LTi, a cytotoxic Ti(IV) complex that bears tetra-phenolato ligand.

The specific Ti(IV) complexes that entered phase I and II clinical trials are budotitane ($(bzac)_2Ti(OEt)_2$) and titanocene dichloride (Cp_2TiCl_2); these complexes had demonstrated high activity toward cisplatin resistant cancer cells and low toxicity in pre-clinical trials [3,4,13,19,20,21]. Nevertheless, the complexes failed the trials mainly due to the rapid hydrolysis of the complexes to form inactive, unidentified, polynuclear species [19,20,21,22,23]. Since then, the rapid hydrolysis of Ti(IV) complexes has been an obstacle in utilizing them as potential anti-tumor drugs. Nevertheless, we introduced a third family of cytotoxic Ti(IV) agents that is based on phenolato ligands [24,25,26,27,28,29,30]. Complexes of this type demonstrate high *in vitro* activity toward various cancer cell-lines, *in vivo* efficacy, and markedly improved hydrolytic stability compared with budotitane and titanocene dichloride [31,32]. Furthermore, this family of complexes exhibited strong correlation between their properties and ligand structure [27,28,33,34,35,36]; in particular, phenolato groups, compared with aliphatic ligand derivatives, enhanced the complex cytotoxicity as well as its hydrolytic stability [26]. Under this notion, a Ti(IV) complex LTi that bears tetra-phenolato ligand was synthesized and characterized (Scheme 1) [37,38]. This complex was stable for weeks in 10% D_2O solutions, but had low solubility and biological accessibility, which impaired its cytotoxicity [38]. Specifically, this complex demonstrated activity toward some murine cell-lines but was inactive toward the more resistant human colon cancer cells, HT-29 [37]. As accessibility is a crucial factor in determining the compound true reactivity [39,40,41], the complex was formulated into organic nanoparticles, by a rapid solvent removal process from volatile oil-in-water microemulsions, which incorporated the complexes in their nanodroplets [37,38,42,43]. As nanoparticles, this complex also demonstrated high cytotoxicity toward HT-29 cells, complete solubility in water and high hydrolytic stability, thus overcoming the difficulties encountered with the previously known Ti(IV) complexes.

Herein, we further examine the potential of this promising compound in its dispersible form for anticancer therapy. Profound *in vitro* studies toward cancerous and non-cancerous cells, as well as those toward cisplatin-sensitive and -resistant cells—alone and in combination with platinum drugs, and evaluation following pre-incubation in medium—all point to high stability, selectivity, and general efficacy of the reported compound. Additionally, *in vivo* measurements further support that the compound has high potential as an anticancer drug.

2. RESULTS AND DISCUSSION

LH_4 and the complex LTi were synthesized according to published procedures [38,44]. LTi was then formulated into nanoparticles of unified size via rapid solvent evaporation from microemulsions, as previously described [38].

2.1. Cytotoxicity toward Cisplatin Resistant Cells

Resistance to cisplatin leads in some cases to cross-resistance to additional drugs or treatments, therefore overriding cisplatin resistance is extremely valuable for the clinical utility of newly developed drugs. Thus, the cytotoxicity of LTi was measured toward human ovarian cisplatin-sensitive, A2780, and its equivalent cisplatin-resistant, A2780-cp, cancer cell lines. A2780-cp cells, in addition to cisplatin, have cross resistance toward alkylation agent Melphalan, intercalating DNA agent Doxorubicin, and irradiation, due to an increased ability to repair DNA damage as well as cytogenetic abnormalities that include altered DNA damaging recognition and inefficiency of the mismatch repair (MMR) pathway [45,46,47]. The activity of the LTi compared with cisplatin and carboplatin was measured via the methylthiazolyldiphenyl-tetrazolium bromide (MTT) assay, as previously described (Figure 1, Table 1) [48].

LTi demonstrated high anti-proliferative activity toward both the sensitive and the resistant cell-lines, with resistance factor of 1.1. The small resistance factor, *i.e.*, similar activity toward both sensitive and resistant cells, implies the ability of LTi to override cisplatin resistance. Although cisplatin demonstrated slightly higher activity than LTi toward the sensitive line A2780, its activity decreased 11-folds toward the resistant line, A2780-cp [47]. The clinically approved carboplatin was less active than LTi toward both cell lines and demonstrated a resistance factor of 10 [47], similarly to the parent Pt drug, thus supporting the notion that both operate by related mechanisms [49]. These results overall imply differences in the mechanism of action of the titanium complex compared with that of the platinum compounds, where the Ti(IV) complex is capable of overriding the abnormalities related to the DNA repair pathways in the A2780-cp cells [47]. Nevertheless, it cannot be determined unequivocally at this point whether DNA is indeed the target for LTi as it is for the platinum drugs.

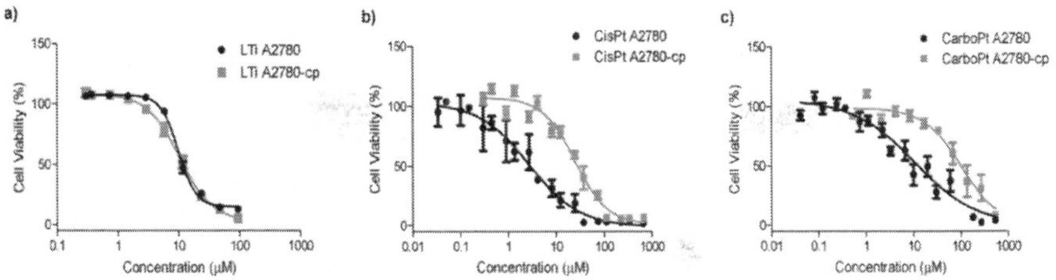

Figure 1. Dependence of cell viability on administered concentration of: (**a**) LTi, (**b**) cisplatin, and (**c**) carboplatin toward human ovarian sensitive A2780, and resistant to cisplatin A2780-cp cancer cell lines.

Table 1. Relative IC_{50} (μM) values toward human ovarian cisplatin-sensitive A2780, and -resistant A2780-cp cancer cell lines and resistance factors.

Complex	IC_{50} (μM)		Resistance Factor [a]
	A2780	A2780-cp	
LTi	9 ± 1	10 ± 4	1.1
Cisplatin	3 ± 2	33 ± 2	11
Carboplatin	21 ± 14	210 ± 90	10

[a] Resistance factor is the ratio between IC_{50} value toward the resistant cell-line, A2780-cp, and IC_{50} value toward the sensitive cell line, A2780. IC_{50} values presented are the mean ± SD of at least three independent experiments conducted on different days, each with at least three replicates.

2.2. In Vitro Combination with Platinum Drugs

To further examine the differences in the mechanisms of action of LTi and the platinum compounds, their activity as a combination was measured toward HT-29, A2780, and A2780-cp cell-lines. The compounds were administered to the cells as a DMSO/medium solution at a 1:1 ratio. The combination was incubated with the cells for three days and cell viability was measured by the MTT assay. The IC_{50} of the combination was compared with those of each compound when administered alone, and the isobolographic plots are presented in Figure 2 [50,51]. Isobolographic analysis is a method to identify the nature of the interactions of two potential drugs, whether they operate synergistically (benefit one another), additively (work independently), or antagonistically (interfere with one another). The IC_{50} values of the drugs when administrated alone are plotted as axial points of the axes, connected by the additive isobole and its accompanying error range. The results of the experiments are presented as the concentration of both compounds at the calculated IC_{50} point of the combination.

The combination of LTi and cisplatin, at a ratio of 1:1, demonstrated mostly additive behavior toward all lines examined. These results are in good agreement with the notion of different mechanisms of action for cisplatin and the titanium complex. Nevertheless, the activity of 1:1 combination of carboplatin and LTi demonstrated antagonistic behavior toward both of the ovarian, A2780 and A2780-cp cell-lines, with borderline additive effect toward HT-29 cells (Supplementary Figure S1). This does not necessarily result from related mechanisms of the two drugs, but may reflect the effect of indirect parameters such as solubility and cellular penetration. Interestingly, in a previous study on combination of diamino bis-phenolato Ti(IV) complexes and cisplatin [52], one Ti(IV) complex had exhibited highly synergistic effect with cisplatin toward HT-29 cells, whereas a different Ti(IV) derivative showed a clear antagonistic effect. These results overall suggest that the particular derivative selected for treatment is of importance, as ligand substitution may affect cellular penetration, solubility, and interactions with a combined drug. Nevertheless, the results altogether support the utility potential of LTi, both alone and in combinations, particularly with cisplatin, the more effective of the two platinum drugs tested.

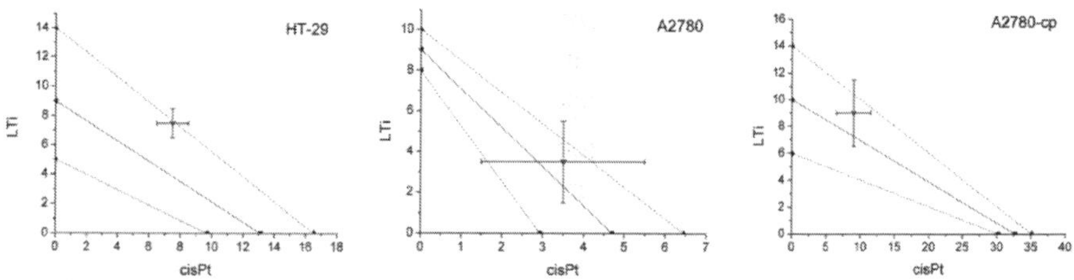

Figure 2. Isobolographic analysis of the cytotoxicity of the combination of LTi and cisplatin, at 1:1 ratio, toward human colon HT-29, ovarian cisplatin-sensitive A2780, and -resistant A2780-cp cell lines. The IC_{50} values of the compounds alone are the axial points, connected to provide the additive line with its error range. Results within the additive range indicate additive behavior, those below indicate synergism, and those above indicate antagonism.

2.3. Activity Following Incubation in Cell Growth Media

The Ti(IV) complex LTi has previously shown significantly higher hydrolytic stability compared with other phenolato Ti(IV) complexes. This complex demonstrated resistance to hydrolysis for weeks in comparative measurements that included exposure to 10% D_2O solutions [38]. To evaluate its hydrolytic stability in conditions that more closely mimic the biological environment, the complex as nanoparticles was exposed to complete RPMI-1640 cell growth media, including 10% fetal bovine serum and 1% antibiotics, for varying periods up to one week (168 h), at 37 °C, and 5% CO_2. Consequently, the complex was incubated for additional three days with HT-29 human colon cells and its effect on cell viability was measured by the crystal violet assay [53] (Figure 3).

LTi demonstrated high anti-proliferative activity toward HT-29 cells, maintaining its activity even following one week of exposure to complete medium, which substantially exceeds the standard period of incubation with cells in *in vitro* studies. This result is noteworthy considering that previously known titanium compounds demonstrated substantial activity decrease following 5–24 h in medium [31,36]. This observation is consistent with the observed high hydrolytic stability of the complex in aquatic solutions. This implies a long shelf life and durability of this complex.

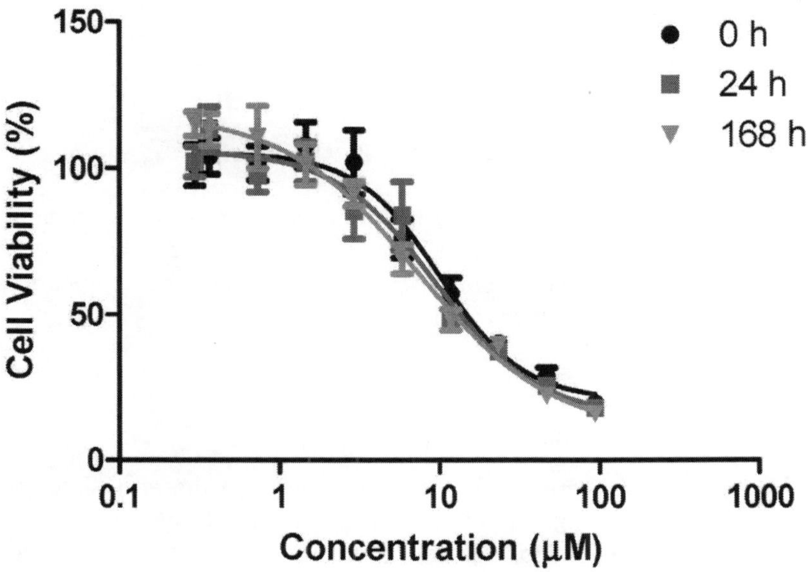

Figure 3. Dependence of human colon HT-29 cell viability on administered concentration of LTi following pre-incubation in complete cell growth medium for varying periods (h) prior to administration to the cells. IC_{50} values (µM), as the mean ± SD of at least three independent experiments conducted on different days, each with at least three replicates: 0 h: 9 ± 4; 24 h: 9 ± 4; 168 h: 8 ± 2.

2.4. Selectivity to Cancer Cells

The selectivity of the anti-proliferating activity of LTi was investigated by measuring its effect on cell viability of MRC-5, non-cancerous, human lung fibroblast cells (Figure 4) [54]. LTi affected moderately the MRC-5 cell viability, thus implying that the anti-proliferative activity of LTi can be selective toward cancer cells.

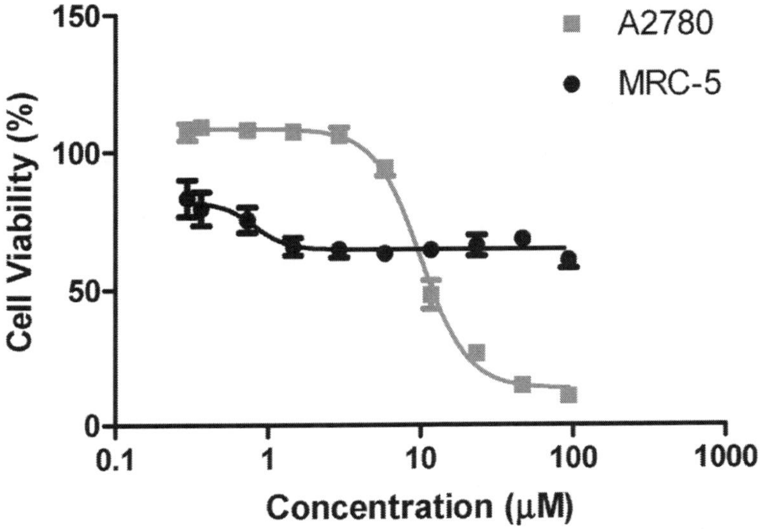

Figure 4. Anti-proliferative activity of LTi toward non-cancerous human lung MRC-5 cells, and human ovarian cancer A2780 cells demonstrated for comparison. IC_{50}values could not be calculated because the compound did not reach 50% inhibition of cell viability.

2.5. In Vivo Anti-Tumor Activity

In vivo experiments were conducted to demonstrate the anti-tumorigenic effect of LTi. First, a group of five mice inoculated with human colon adenocarcinoma cells and injected with LTi demonstrated significant decrease in tumor growth relative to control untreated mice (Figure 5). At the end of the experiment, the tumors were taken out for evaluation; whereas the average weight of the tumors developed in the control group was 0.57 ± 0.16 g, that of the treated mice was 0.13 ± 0.18 g, where two of the five treated mice did not develop any tumor. These results were repetitive in a second corroboration experiment, which was conducted with a larger group of 10 mice, out of which three did not develop tumors (Figure 5). These findings clearly demonstrate an anti-tumorigenic effect of LTi

(Figure 5). Significantly, no clinical signs of toxicity were observed in the mice treated with LTi, thus emphasizing the potential efficiency and safety of this anti-tumoral agent.

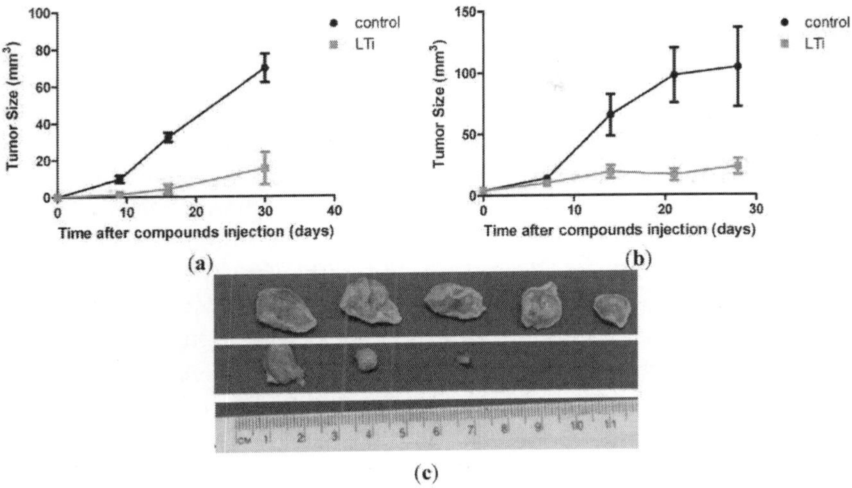

Figure 5. LTi reduces the development of solid tumors from colon cancer cells: immune deficient (nude) mice (**a**) N = 5; (**b**) N = 10 were inoculated S.C. with 5 × 10^6HT-29, human colon adenocarcinoma cells and subjected to IP injection of LTi (100 µg/0.2 mL), every other day for three weeks. Control mice were injected with PBS. (**c**) Solid tumors at the termination of the experiment. Upper panel: control mice; Lower panel: treated mice (two-fifths were devoid of tumors).

3. EXPERIMENTAL SECTION

3.1. Ligand and Complex Synthesis

Ligand LH_4 and the complex LTi were synthesized according to published procedures [38,44]. LTi was formulated into nanoparticles by evaporation of microemulsions to form nanopowder of LTi, as previously described [38,42].

3.2. Materials

Cisplatin crystalline, carboplatin, methylthiazolyldiphenyl-tetrazolium bromide ≥97.5% suitable for cell culture, and crystal violet 1% aqueous solution, were purchased from Sigma-Aldrich Chemical Company Inc (Sigma-Aldrich, St. Louis, MO, USA).

3.3. In Vitro Cytotoxicity Assays

Cytotoxicity was measured on HT-29, human colon cells obtained from ATCC Inc (Manassas, VA, USA); A2780, A2780-cis, cisplatin-sensitive and -resistant human ovarian cells; and MRC-5 non-cancerous human fibroblast lung cells, all obtained from ECACC Inc (Salisbury, UK). Cytotoxicity was measured using the methylthiazolyldiphenyl-tetrazolium bromide (MTT) assay [48] or by crystal violet assay [53]. Cells (0.6×10^6) in medium containing 1% penicillin/streptomycin antibiotics; 1% L-glutamine; 10% fetal bovine serum (FBS); and 88% medium RPMI-1640, all purchased from Biological Industries Inc., were seeded into a 96-well plate and allowed to attach for a day. The cells were consequently treated with the reagent tested at 10 different concentrations. Solutions of cisplatin and carboplatin were prepared by dissolving 1–8 mg in 100–200 μL of DMSO, with further diluting to obtain the 10 different concentrations. Organic solutions were then further diluted with medium to give 10% DMSO solutions, followed by applying 10 μL of each final solution to wells already containing 200 μL of the above solution of cells in the medium; giving final concentrations of the complexes up to 1000 mg/L, in 0.5% DMSO solutions. LTi was diluted similarly into RPMI-1640 medium without any organic solvent. Combination measurements were conducted similarly by preparing stock solutions of LTi and the platinum compound in 1:1 ratio and diluted it to 10 different concentrations. Control wells were treated with similar amounts of DMSO or RPMI-1640.

MTT assay: After a standard of 3 days incubation at 37 °C in 5% CO_2 atmosphere, MTT (0.1 mg in 20 μL) was added to the cells; cells were then incubated for additional 3 h with MTT. The MTT solution was removed, and the cells were dissolved in 200 μL isopropanol. The absorbance at 550 nm was measured by a Bio-Tek EL-800 microplate reader spectrophotometer.

Crystal violet assay: After a standard of 3 days incubation at 37 °C in 5% CO_2 atmosphere, the cells medium was removed and the cells were fixated by incubating them in 200 μL formaldehyde 37%, at room temperature, for 5 min. The formaldehyde was removed and 100 μL of 1% crystal violet aqueous solution was added. Following 20 min of incubation of the cells with crystal violet at room temperature, the crystal violet solution was removed and the cells were washed twice with water. The cells were dissolved in 200 μL of methanol and the absorbance at 570 nm was measured by a Bio-Tek EL-800 microplate reader spectrophotometer.

Each measurement was repeated at least 3 × 3 times, namely, three repeats per plate, all repeated at least 3 times on different days (9 repeats altogether). Relative IC_{50} values were determined by a non-linear regression of a variable slope (four parameters) model. IC_{50} values presented in text as the mean ± SD of at least three independent experiments.

3.4. Activity Following Incubation in Cell Growth Media

LTi activity following pre-incubation in complete cell growth media was accomplished by diluting LTi dispersions in cell growth media containing 1% penicillin/streptomycin antibiotics, 1% L-glutamine, 10% fetal bovine serum (FBS), and 88% medium RPMI-1640, as described above. The dispersions were incubated at 37 °C in 5% CO_2 atmosphere for 0, 24, and 168 h before applying them to cells that were seeded as detailed above. Cells were incubated with the dispersions for 3 additional days and cell viability was measured by crystal violet assay, as described.

3.5. In Vivo Anti-Tumor Activity

In vivo analysis was conducted as followed: Nude Balb/C mice (6–8 weeks old females) were obtained from Harlan (Israel), held in a specific-pathogen-free (SPF) facility (AAALAC accreditation #1285) and treated in accordance with the national institutes of health (NIH) guidelines and approval by the institutional committee for ethics in animal experimentation. Mice were injected S.C. with 5×10^6 HT-29, human colon adenocarcinoma cells. Starting 24 h after the challenge, mice were subjected to IP injections of LTi, every other day for 3 weeks. Mice were followed up to 28 days post inoculation.

4. CONCLUSIONS

LTi formulated into nanoparticles is a remarkably effective anti-tumoral agent. Not only is it active *in vitro* toward a number of cells, but it is also active toward cisplatin-resistant cells, and shows *in vivo* efficacy. The high efficacy is accompanied by low toxicity as expected of the titanium metal, as manifested by the inactivity toward non-cancerous cells as well as the absence of clinical signs of toxicity in the treated animals. Moreover, the superiority of LTi particularly lays in the high hydrolytic stability of the compound, especially as no change in activity was obtained after up to a week in complete aquatic medium, implying a long shelf life of this complex, which is essential for therapeutic applications. The possibility of additive effects when combined with known drugs adds to the high potential of this complex for use in a clinical setting.

It is interesting to recall LTi was inactive when administered directly as is, due to insufficient solubility; thus, the nano-formulation enabled the compound to demonstrate its true behavior [38]. It is thus obvious that solubility and general biological accessibility are crucial factors in the development of any potential drugs, where nano-formulations presents an effective way to overcome such limitations.

Our preliminary studies suggest a different mechanistic pathway for LTi than that of cisplatin, as may be deduced by the high activity of LTi on cisplatin-resistant cells and by the additive influence of the drugs in *in vitro* combination analyses that generally indicate independent operations. Nevertheless, full mechanistic determination is still lacking. Thus, due to the incredibly high potential of LTi as displayed by the results presented herein, intensive profound mechanistic analyses as well as development of related derivatives are underway in our laboratory.

ACKNOWLEDGMENTS

We thank the office of the Chief Scientist for funding under the Kamin program..

AUTHOR CONTRIBUTIONS

E.Y.T. and J.H. conceived the experiments; S.Me. designed the experiments, prepared the compounds, measured the cytotoxicity toward the cell-lines including stability measures and combinations activity, and analyzed the data; O.B. and J.H. conducted the *in vivo* testing; K.M.-G. and S.Ma. contributed the microemulsions for the nano-formulation; S.Me., O.B., and E.Y.T. wrote the manuscript, with the helpful comments of J.H., and S.Ma.; and E.Y.T. supervised the study.

CONFLICTS OF INTEREST

The authors declare no conflict of interest.

REFERENCES

1. Ott, I.; Gust, R. Non platinum metal complexes as anti-cancer drugs. Arch. Pharm. 2007, 340, 117–126.
2. Tshuva, E.Y.; Ashenhurst, J.A. Cytotoxic titanium(IV) complexes: Renaissance. Eur. J. Inorg. Chem. 2009, 2203–2218.
3. Koepf-Maier, P.; Koepf, H. Non-platinum group metal antitumor agents. history, current status, and perspectives.Chem. Rev. 1987, 87, 1137–1152.
4. Köpf-Maier, P.; Köpf, H. Transition and main-group metal cyclopentadienyl complexes: Preclinical studies on a series of antitumor agents of different structural type. Bioinorg. Chem. 1988, 70, 103–185.
5. Melendez, E. Titanium complexes in cancer treatment. Crit. Rev. Oncol. Hematol. 2002, 42, 309–315.
6. Kostova, I. Titanium and vanadium complexes as anticancer agents. Anticancer Agents Med. Chem. 2009, 9, 827–842.
7. Caruso, F.; Rossi, M. Antitumor titanium compounds. Mini-Rev. Med. Chem. 2004, 4, 49–60.
8. Caruso, F.; Rossi, M.; Pettinari, C. Anticancer titanium agents. Expert Opin. Ther. Pat. 2001, 11, 969–979.
9. Buettner, K.M.; Valentine, A.M. Bioinorganic chemistry of titanium. Chem. Rev. 2012, 112, 1863–1881.
10. Zhang, C.X.; Lippard, S.J. New metal complexes as potential therapeutics. Curr. Opin. Chem. Biol. 2003, 7, 481–489.
11. Bruijnincx, P.C.; Sadler, P.J. New trends for metal complexes with anticancer activity. Curr. Opin. Chem. Biol. 2008, 12, 197–206.
12. Jakupec, M.A.; Galanski, M.; Arion, V.B.; Hartinger, C.G.; Keppler, B.K. Antitumour metal compounds: More than theme and variations. Dalton Trans. 2008, 183–194.
13. Keppler, B.K.; Friesen, C.; Moritz, H.G.; Vongerichten, H.; Vogel, E. Tumor-inhibiting bis(beta-diketonato) metal-complexes—Budotitane, cis-diethoxybis(1-phenylbutane-1,3-dionato)titanium (IV)—The 1st transition-metal complex after platinum to qualify for clinical-trials. Struct. Bond. 1991, 78, 97–127.

14. Desoize, B. Metals and metal compounds in cancer treatment. Anticancer Res. 2004, 24, 1529–1544.

15. Muhammad, N.; Guo, Z. Metal-based anticancer chemotherapeutic agents. Curr. Opin. Chem. Biol. 2014, 19, 144–153.

16. Kelland, L. The resurgence of platinum-based cancer chemotherapy. Nat. Rev. Cancer 2007, 7, 573–584.

17. Arnesano, F.; Natile, G. Mechanistic insight into the cellular uptake and processing of cisplatin 30 years after its approval by FDA. Coord. Chem. Rev. 2009, 253, 2070–2081.

18. Muggia, F. Platinum compounds 30 years after the introduction of cisplatin: Implications for the treatment of ovarian cancer. Gynecol. Oncol. 2009, 112, 275–281.

19. Kröger, N.; Kleeberg, U.R.; Mross, K.; Edler, L.; Hossfeld, D.K. Phase II clinical trial of titanocene dichloride in patients with metastatic breast cancer. Oncol. Res. Treat. 2000, 23, 60–62.

20. Lümmen, G.; Sperling, H.; Luboldt, H.; Otto, T.; Rübben, H. Phase II trial of titanocene dichloride in advanced renal-cell carcinoma. Cancer Chemother. Pharmacol. 1998, 42, 415–417.

21. Schilling, T.; Keppler, K.B.; Heim, M.E.; Niebch, G.; Dietzfelbinger, H.; Rastetter, J.; Hanauske, A.R. Clinical phase I and pharmacokinetic trial of the new titanium complex budotitane. Invest. New Drugs 1996, 13, 327–332.

22. Caruso, F.; Massa, L.; Gindulyte, A.; Pettinari, C.; Marchetti, F.; Pettinari, R.; Ricciutelli, M.; Costamagna, J.; Canales, Juan C.; Tanski, J.; et al. (4-acyl-5-pyrazolonato)titanium derivatives: Oligomerization, hydrolysis, voltammetry, and dft study. Eur. J. Inorg. Chem. 2003, 2003, 3221–3232.

23. Toney, J.H.; Marks, T.J. Hydrolysis chemistry of the metallocene dichlorides M(.Eta.5-C5H5)2Cl2, M = titanium, vanadium, or zirconium. Aqueous kinetics, equilibria, and mechanistic implications for a new class of antitumor agents. J. Am. Chem. Soc. 1985, 107, 947–953.

24. Tshuva, E.Y.; Peri, D. Modern cytotoxic titanium(IV) complexes; insights on the enigmatic involvement of hydrolysis.Coord. Chem. Rev. 2009, 253, 2098–2115.

25. Peri, D.; Meker, S.; Shavit, M.; Tshuva, E.Y. Synthesis, characterization, cytotoxicity, and hydrolytic behavior of C-2- and C-1-symmetrical Ti-IV

complexes of tetradentate diamine bis(phenolato) ligands: A new class of antitumor agents. Chem. Eur. J. 2009, 15, 2403–2415.

26. Shavit, M.; Peri, D.; Manna, C.M.; Alexander, J.S.; Tshuva, E.Y. Active cytotoxic reagents based on non-metallocene non-diketonato well-defined C-2-symmetrical titanium complexes of tetradentate bis(phenolato) ligands. J. Am. Chem. Soc. 2007, 129, 12098–12099.

27. Peri, D.; Meker, S.; Manna, C.M.; Tshuva, E.Y. Different ortho and para electronic effects on hydrolysis and cytotoxicity of diamino bis(phenolato) "Salan" Ti(IV) complexes. Inorg. Chem. 2011, 50, 1030–1038.

28. Meker, S.; Manna, C.M.; Peri, D.; Tshuva, E.Y. Major impact of N-methylation on cytotoxicity and hydrolysis of salan Ti(IV) complexes: Sterics and electronics are intertwined. Dalton Trans. 2011, 40, 9802–9809.

29. Peri, D.; Manna, C.M.; Shavit, M.; Tshuva, E.Y. Ti-IV complexes of branched diamine bis(phenolato) ligands: Hydrolysis and cytotoxicity. Eur. J. Inorg. Chem. 2011, 2011, 4896–4900.

30. Tzubery, A.; Tshuva, E.Y. Trans titanium(IV) complexes of salen ligands exhibit high antitumor activity. Inorg. Chem.2011, 50, 7946–7948.

31. Manna, C.M.; Braitbard, O.; Weiss, E.; Hochman, J.; Tshuva, E.Y. Cytotoxic salan-titanium(IV) complexes: High activity toward a range of sensitive and drug-resistant cell lines, and mechanistic insights. ChemMedChem 2012, 7, 703–708.

32. Immel, T.A.; Grutzke, M.; Spate, A.K.; Groth, U.; Ohlschlager, P.; Huhn, T. Synthesis and x-ray structure analysis of a heptacoordinate titanium(IV)-bis-chelate with enhanced in vivo antitumor efficacy. Chem. Commun. 2012, 48, 5790–5792.

33. Glasner, H.; Tshuva, E.Y. A marked synergistic effect in antitumor activity of salan titanium(IV) complexes bearing two differently substituted aromatic rings. J. Am. Chem. Soc. 2011, 133, 16812–16814.

34. Manna, C.M.; Armony, G.; Tshuva, E.Y. New insights on the active species and mechanism of cytotoxicity of salan-Ti(IV) complexes: A stereochemical study. Inorg. Chem. 2011, 50, 10284–10291.

35. Immel, T.A.; Groth, U.; Huhn, T. Cytotoxic titanium salan complexes: Surprising interaction of salan and alkoxy ligands. Chem. Eur. J. 2010, 16, 2775–2789.

36. Tzubery, A.; Tshuva, E.Y. Cytotoxicity and hydrolysis of trans-Ti(IV) complexes of salen ligands: Structure-activity relationship studies. Inorg. Chem. 2012, 51, 1796–1804.

37. Meker, S.; Margulis-Goshen, K.; Weiss, E.; Braitbard, O.; Hochman, J.; Magdassi, S.; Tshuva, E.Y. Anti-proliferative activity of nano-formulated phenolato titanium(IV) complexes against cancer cells. ChemMedChem 2014, 9, 1294–1298.

38. Meker, S.; Margulis-Goshen, K.; Weiss, E.; Magdassi, S.; Tshuva, E.Y. High antitumor activity of highly resistant salan-titanium(IV) complexes in nanoparticles: An identified active species. Angew. Chem. Int. Ed. 2012, 51, 10515–10517.

39. Severin, G.W.; Nielsen, C.H.; Jensen, A.I.; Fonslet, J.; Kjær, A.; Zhuravlev, F. Bringing radiotracing to titanium-based antineoplastics: Solid phase radiosynthesis, pet and ex vivo evaluation of antitumor agent [45Ti](salan)Ti(dipic). J. Med. Chem. 2015, 58, 7591–7595.

40. Glasner, H.; Meker, S.; Tshuva, E.Y. Cationic phenolato titanium(IV) complexes of enhanced solubility as active and biologically accessible anti-tumor compounds. J. Organomet. Chem. 2015, 788, 33–35.

41. Miller, M.; Tshuva, E.Y. Cytotoxic titanium(IV) complexes of chiral diaminobis(phenolato) ligands: Better combination of activity and stability by the bipyrrolidine moiety. Eur. J. Inorg. Chem. 2014, 2014, 1485–1491.

42. Margulis-Goshen, K.; Magdassi, S. Formation of simvastatin nanoparticles from microemulsion. Nanomedicine 2009, 5, 274–281.

43. Margulis-Goshen, K.; Magdassi, S. Organic nanoparticles from microemulsions: Formation and applications. Curr. Opin. Colloid Interface Sci. 2012, 17, 290–296.

44. Higham, C.S.; Dowling, D.P.; Shaw, J.L.; Cetin, A.; Ziegler, C.J.; Farrell, J.R. Multidentate aminophenol ligands prepared with mannich condensations. Tetrahedron Lett. 2006, 47, 4419–4423.

45. Behrens, B.C.; Hamilton, T.C.; Masuda, H.; Grotzinger, K.R.; Whangpeng, J.; Louie, K.G.; Knutsen, T.; McKoy, W.M.; Young, R.C.; Ozols, R.F. Characterization of a cis-diamminedichloroplatinum(II)-resistant human ovarian-cancer cell-line and its use in evaluation of platinum analogs. Cancer Res. 1987, 47, 414–418.

46. Masuda, H.; Ozols, R.F.; Lai, G.M.; Fojo, A.; Rothenberg, M.; Hamilton, T.C. Increased DNA-repair as a mechanism of acquired-resistance to

cis-diamminedichloroplatinum(II) in human ovarian-cancer cell-lines. Cancer Res. 1988, 48, 5713–5716.

47. Aird, R.E.; Cummings, J.; Ritchie, A.A.; Muir, M.; Morris, R.E.; Chen, H.; Sadler, P.J.; Jodrell, D.I. In vitro and in vivoactivity and cross resistance profiles of novel ruthenium(II) organometallic arene complexes in human ovarian cancer. Brit. J. Cancer 2002, 86, 1652–1657.

48. Ganot, N.; Meker, S.; Reytman, L.; Tzubery, A.; Tshuva, E.Y. Anticancer metal complexes: Synthesis and cytotoxicity evaluation by the mtt assay. J. Vis. Exp. 2013, 81.

49. Rabik, C.A.; Dolan, M.E. Molecular mechanisms of resistance and toxicity associated with platinating agents. Cancer Treat. Rev. 2007, 33, 9–23.

50. Chou, T.-C. Theoretical basis, experimental design, and computerized simulation of synergism and antagonism in drug combination studies. Pharmacol. Rev. 2006, 58, 621–681.

51. Tallarida, R.J.; Porreca, F.; Cowan, A. Statistical-analysis of drug-drug and site-site interactions with isobolograms.Life Sci. 1989, 45, 947–961.

52. Ganot, N.; Redko, B.; Gellerman, G.; Tshuva, E.Y. Anti-proliferative activity of the combination of salan Ti(IV) complexes with other organic and inorganic anticancer drugs against HT-29 and NCI-H1229 cells: Synergism with cisplatin. Rsc. Adv. 2015, 5, 7874–7879.

53. Flick, D.A.; Gifford, G.E. Comparison of invitro-cell cyto-toxic assays for tumor necrosis factor. J. Immunol. Methods1984, 68, 167–175.

54. Zhang, X.D.; Nguyen, T.; Thomas, W.D.; Sanders, J.E.; Hersey, P. Mechanisms of resistance of normal cells to trail induced apoptosis vary between different cell types. FEBS Lett. 2000, 482, 193–199.

Index